The Patellofemoral Joint

Alberto Gobbi
João Espregueira-Mendes
Norimasa Nakamura
Editors

The Patellofemoral Joint

State of the Art in Evaluation and Management

Springer

Editors
Alberto Gobbi
Orthopaedic Arthroscopic Surgery
International (O.A.S.I.) Bioresearch
Foundation, Gobbi N.P.O.
Milan
Italy

Norimasa Nakamura
Department of Orthopaedics
Osaka Health Science University
Osaka
Japan

Department of Rehabilitation Science
Osaka University
Center for Advanced Medical
Engineering and Informatics,
Osaka
Japan

João Espregueira-Mendes
Clínica do Dragão – Espregueira-
Mendes Sports Centre – FIFA
Medical Centre of Excellence
Porto
Portugal

ICVS/3B's Research Group – Biomaterials
Biodegradables and Biomimetics
University of Minho
Headquarters of the European
Institute of Excellence on Tissue
Engineering and Regenerative Medicine
Guimarães
Portugal

Department of Orthopaedics
Minho University
Braga
Portugal

ISBN 978-3-642-54964-9 ISBN 978-3-642-54965-6 (eBook)
DOI 10.1007/978-3-642-54965-6
Springer Heidelberg New York Dordrecht London

Library of Congress Control Number: 2014942019

Printed on acid-free paper

Springer is part of Springer Science+Business Media (www.springer.com)

"Dedicated to my father Augusto and my son Nicolò, who together with my Family, taught me that life is a constant journey. Dedicated to all my Friends, who believed this dream might become true."

Preface

It is our great pleasure to introduce the new book *The Patellofemoral Joint: State of the Art in Evaluation and Management.*

A medical education, where basic science together with medical innovations and state of the art in surgical techniques are indispensable, is an important challenge. ISAKOS wants to provide an educational umbrella in which all would collaborate and benefit. This is intended to leverage the education skills in arthroscopy, knee surgery and orthopedic sports medicine around the world.

Patellofemoral joint pathologies represent common but difficult-to-treat entities, due to the difficulty in elucidating the etiology of anterior knee pain as well as in restoring back to normal patellar tracking and stability. If untreated, repetitive trauma due to altered joint surface contact pressures can result in significant loss of articular cartilage, progressive degenerative changes of the patellofemoral joint and development of osteoarthritis. Currently, there has been remarkable progress in anatomy, biomechanics and biology related to patellofemoral joint treatment. Improved rehabilitation strategies are now available together with novel conservative or surgical procedures, with the aim to address biological problems utilizing biological solutions.

ISAKOS has given us a special opportunity to invite international orthopedic surgeons and researchers all over the world to provide their specific insights into patellofemoral problems. Originally conceived as a small booklet at the ISAKOS Congress in Toronto in June 2013, this work quickly turned into a major project, involving more than 35 authors worldwide. We invited orthopedic surgeons, physiotherapists, and researchers from all over the world to provide their specific research works related to patellofemoral problems. As a result, we have organized a comprehensive review on a global overview of the physiology, pathology, diagnosis and treatment options. We are confident this special issue will not only cover all the essential issues to be learned by young doctors and researchers, but also will manage to offer the most advanced suggestions specifically in the new treatment options and diagnosis.

Our hope is that this book will be also valuable for all clinicians and researchers interested in the patellofemoral joint and its disorders and will represent an outstanding reference in the future for the treatment of this unique joint structure. Our effort in this book was to give no less.

Milan, Italy Alberto Gobbi
Porto, Portugal João Espregueira-Mendes
Osaka, Japan Norimasa Nakamura

Contents

1 Introduction

Alberto Gobbi, João Espregueira-Mendes, and Norimasa Nakamura

Medical education represents a challenge worldwide. ISAKOS intends to leverage education and plays a role in the field of orthopedic sports medicine around the world, providing equal opportunities among its members. Uneven realities described and emphasized by fellows and residents, arriving from all around the world, along with their extraordinary learning skills and strong motivation, made us realize that ISAKOS had the responsibility to provide an educational umbrella in which all agents could collaborate and profit. This assumed an important role and it is today an admirable ongoing reality. Therefore, in one unparalleled determination and effort supported by many, we can bring you into high-performing educational sets, no matter what zip code you live in. We will join you or you will be joining us in this priceless educational mission. Consequently, this book shows the most advanced techniques with ultimate technologies under the guidance of globally renowned experts.

In this book dedicated to patellofemoral joint, the reader will be able to get acquainted with the state of the art in this subject. This knowledge conveys a comprehensive and friendly resource of education. It is a secure value and an important reflex of authors' commitment to ISAKOS educational mission. The intelligibility, interest, and actuality of the reading you are about to begin consubstantiate a catch-up that, we believe, will contribute to provide the best health care to our patients. Science and skills brought to you by this group of authors arise from an intense willing of helping on behalf of better quality of life. ISAKOS strives to give no less.

A. Gobbi (✉)
Orthopaedic Arthroscopic Surgery International (O.A.S.I.) Bioresearch Foundation, Gobbi N.P.O., Milan, Italy
e-mail: gobbi@cartilagedoctor.it

J. Espregueira-Mendes
Clínica do Dragão – Espregueira-Mendes Sports Centre – FIFA Medical Centre of Excellence, F.C. Porto Stadium, 4350-415 Porto, Portugal

ICVS/3B's Research Group – Biomaterials, Biodegradables and Biomimetics, University of Minho, Headquarters of the European Institute of Excellence on Tissue Engineering and Regenerative Medicine, Caldas das Taipas, Guimarães, Portugal

Department of Orthopaedics, Minho University, Braga, Portugal
e-mail: jem@espregueira.com

N. Nakamura
Department of Orthopaedic Surgery, Osaka University Graduate School of Medicine, 2-2 Yamadaoka, Suita, Osaka 565-0871, Japan
e-mail: norimasa.nakamura@ohsu.ac.jp

A. Gobbi et al. (eds.), *The Patellofemoral Joint*,
DOI 10.1007/978-3-642-54965-6_1, © ISAKOS 2014

Morphogenesis of the Patellofemoral Joint

2

Pedro Guillen-Garcia
and José Francisco Rodríguez-Vázquez

Contents

The development of the patella and the femoropatellar joint is intimately related to the development of the tendon of quadriceps and the knee joint which are integrated morphologically and functionally, providing the articular component for the extension mechanism.

After 6 weeks of development, a band of mesenchymal tissue is displayed between the femur and tibia which were in a chondrification process that corresponded to the articular homogeneous interzone [3, 12]. The anlage of the patella appeared as a slight mesenchymal condensation located between the femoral condyles and the anlage of the tendon of the quadriceps muscle [3, 10]. The latter is differentiated from the patellar area forming a continuous condensed mesenchymal band, which is extended from the quadriceps muscle to the anlage of the tibia (Fig. 2.1).

The differentiation of the anlage of the patella is much more evident after 7 weeks of development (Fig. 2.2). It appears dorsally, in contact with the quadriceps tendon and separated from the inferior end of the femur by an area of loose mesenchymal tissue. The chondrification continues at the inferior end of the femur and at the superior end of the tibia, thus resulting in the origination of the femoral condyles and the superior surface of the tibia (Fig. 2.3). The femorotibial interzone is formed by two eccentric bands of mesenchymal tissue which cover the condyles of the femur and the superior surface of the tibia and a medial band that is loose in comparison with the eccentric bands, which corresponded to the three-layered interzone (Figs. 2.2 and 2.3).

P. Guillen-Garcia (✉)
Trauma Unit, Clinica Cemtro, Madrid, Spain

Unidad de Investigation Biomedica, Madrid, Spain

Universidad Catolica San Antonio
de Murcia, Murcia, Spain
e-mail: pedroguillen612@gmail.com

J.F. Rodríguez-Vázquez
Clinica Cemtro, Madrid, Spain

A. Gobbi et al. (eds.), *The Patellofemoral Joint*,
DOI 10.1007/978-3-642-54965-6_2, © ISAKOS 2014

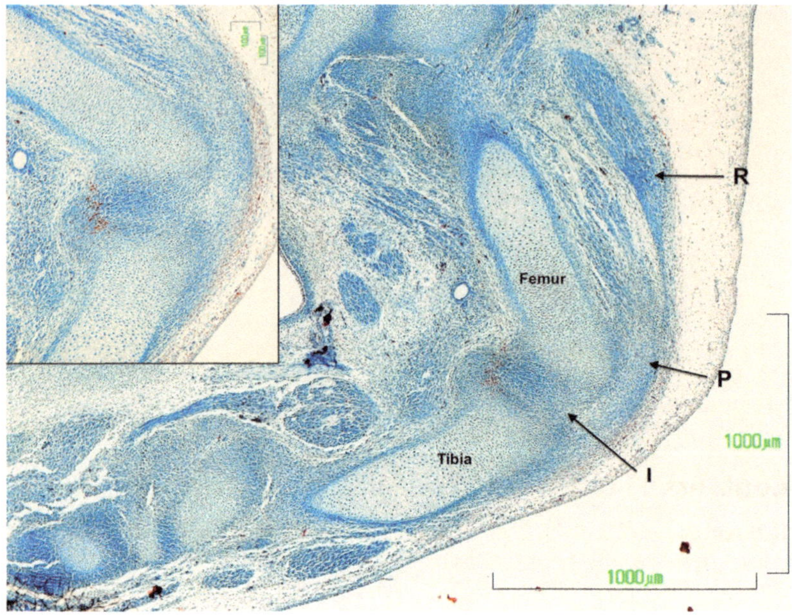

Fig. 2.1 Human fetus at 6 weeks of development (16 mm GL, 18 Carnegie stage). Sagittal section of the left lower limb. Formation of the interzone between the femur and the tibia (*I*) and appearance of the anlage of the patella (*P*). Rectus femoris muscle (*R*)

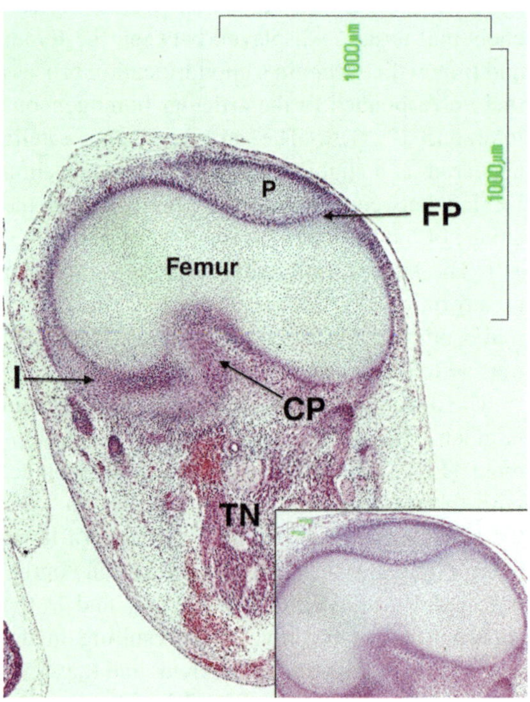

Fig. 2.2 Human fetus at 7 weeks of development (20 mm GL, 20 Carnegie stage). Transverse section of the femoropatellar joint. Differentiation of the patella (*P*) and formation of the interzone of the femoropatellar joint (*FP*). Cruciate posterior ligament (*CP*). Femorotibial interzone (*I*). Tibial nerve (*TN*)

The patellar anlage increases, configuring a harmonious and uniform formation [7]. The more advanced chondrification of both the femur and the anlage of the patella clearly defines them, identifying a fine mesenchymal band which forms the femoropatellar interzone (Fig. 2.2). The mesenchyme condenses cranially and caudally at the patella making both the tendon of quadriceps muscle and the patellar tendon evident (Fig. 2.3). At this stage, the formation of the cruciate ligaments has also begun, the posterior cruciate ligament becoming identifiable earlier than its counterpart [10] (Fig. 2.2).

Two anlagen appear at the end of the 7 weeks: the suprapatellar bursa which was observed as a small cavity, dorsally to the tendon of quadriceps and cranially to the patella, and the first sign of the cavitation of the femoropatellar joint [4, 9, 3]. The fibers of the quadriceps tendon begin to appear visible at its attachment, on the patella base (Fig. 2.4).

After 8 weeks of development, the cavitation of the femoropatellar interzone is evident, while the patella is in a clear phase of chondrogenesis (Fig. 2.5). At the femoral and tibial condyles, a dense band of connective tissue is formed, which marks the first sign of organization of the articular cartilage. The densification to initiate the formation

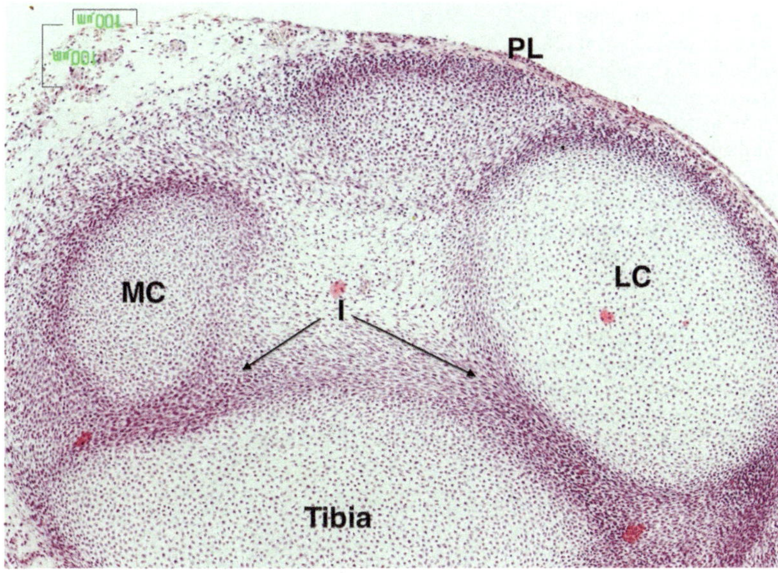

Fig. 2.3 Human fetus at 6 weeks of development (20 mm GL, 20 Carnegie stage). Transverse section of the patellar tendon (*PL*). Femorotibial interzone (*I*). Lateral condyle (*LC*). Medial condyle (*MC*)

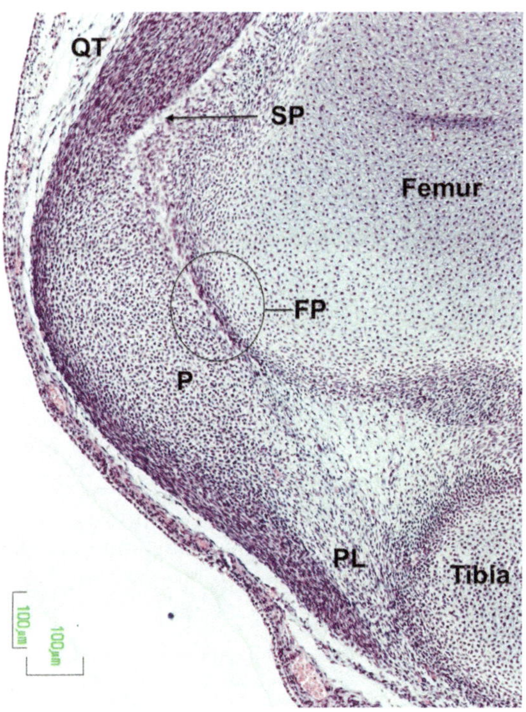

Fig. 2.4 Human fetus at 7 weeks of development (22 mm GL, 21 Carnegie stage). Sagittal section of the femoropatellar joint. Beginning of the cavitation of the femoropatellar interzone (*FP*). Patella (*P*), Patellar tendon (*PL*), Tendon of quadriceps (*QT*), Suprapatellar bursa (*SP*)

of the menisci increases at the lateral parts of the femorotibial interzone [6], with small cavities between the menisci and the condyles of the femur and tibia (Fig. 2.6) appearing laterally and at inter-layer level. On the contrary, the interzone of the superior tibiofibular joint is visible without signs of cavitation (Fig. 2.6). The cruciate ligaments are clearly visible at the intercondylar notch (Fig. 2.5), surrounded by a poorly organized mesenchyme which contains many vascular elements.

From the margins of the patella, the articular capsule surrounds the femoral condyles and attaches itself to the external surface of the menisci. The formation of the patellofemoral ligaments has also commenced (Fig. 2.5). The articular cavities of the femoropatellar, femoromeniscal and meniscotibial joints now became apparent.

After 9 weeks of development, the patella faces the femoral trochlea (Fig. 2.7). The topographical arrangement of the fibers which form the quadriceps tendon and its length on the superficial aspect of the patella become apparent, constituting the patellar tendon which reaches up to the developing tibial tuberosity (Fig. 2.7). Below the patella, deeper than the patellar ligament and ventrally to the anterior cruciate ligament, there

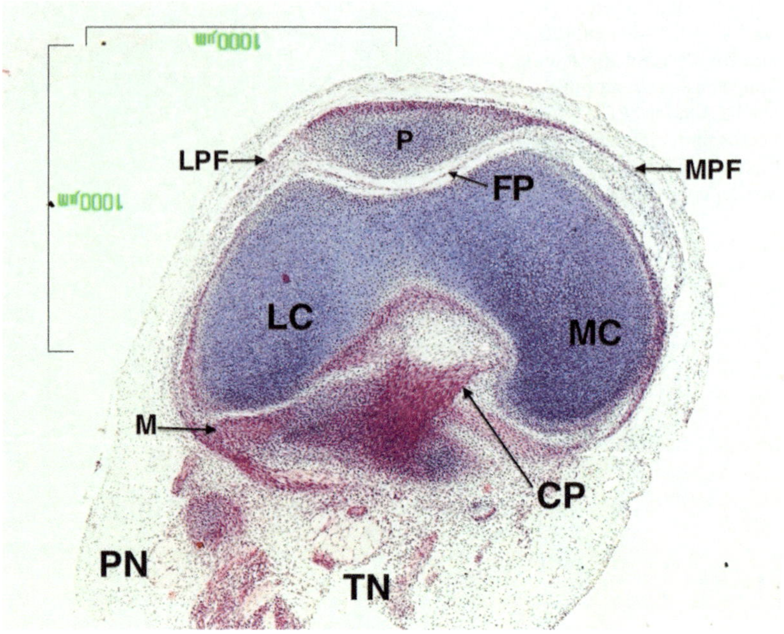

Fig. 2.5 Human fetus at 8 weeks of development (28.5 mm GL, 23 Carnegie stage). Transverse section of the femoropatellar joint. Formation of the cavity in the femoropatellar joint (*FP*). Cruciate posterior ligament (*CP*). Lateral meniscus (*M*), Lateral condyle (*LC*), Medial condyle (*MC*), Lateral patellofemoral ligament (*LPF*), Medial patellofemoral ligament (*MPF*), Patella (*P*), Common peroneal nerve (*PN*), Tibial nerve (*TN*)

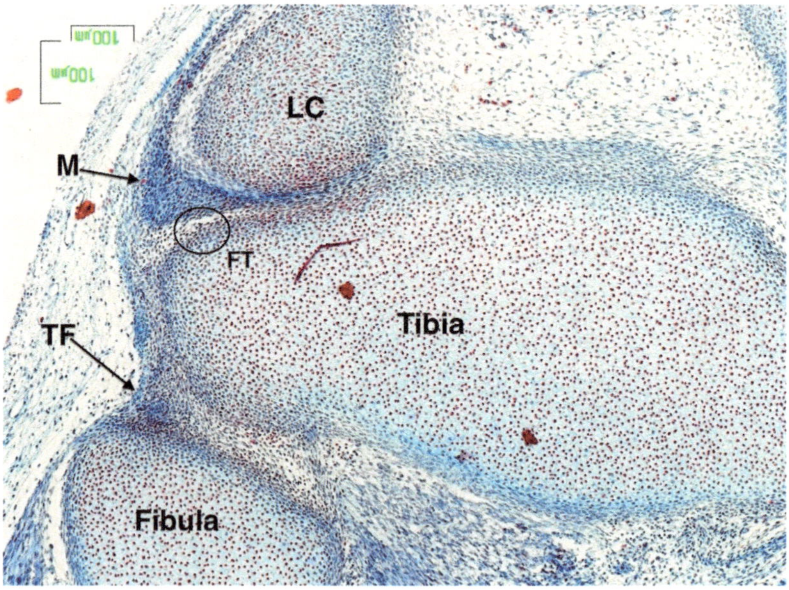

Fig. 2.6 Human fetus at 8 weeks of development (28.5 mm GL, 23 Carnegie stage). Transverse section of the knee and the tibiofibular joint (*TF*). Beginning of the formation of the menisci (*M*) and the cavitation of the femorotibial interzone (*FT*). Lateral meniscus (*M*), Lateral condyle (*LC*)

is a loose mesenchyme mass arranged as a septum forming the medial septum or mediastinum, the anlage of the fat pad ligament (Fig. 2.8).

In weeks 10 and 11, the progressive reduction of the medial wall and its further arrangement consequently forms the fat pad ligament [2, 7] with the presence of collagen fibers as well as the regression of the remaining mesenchymal tissue limited by the space between the posterior aspect and inferior border of the patella together with the prespinal portion of the tibia, both of which determine that the joint will have a similar

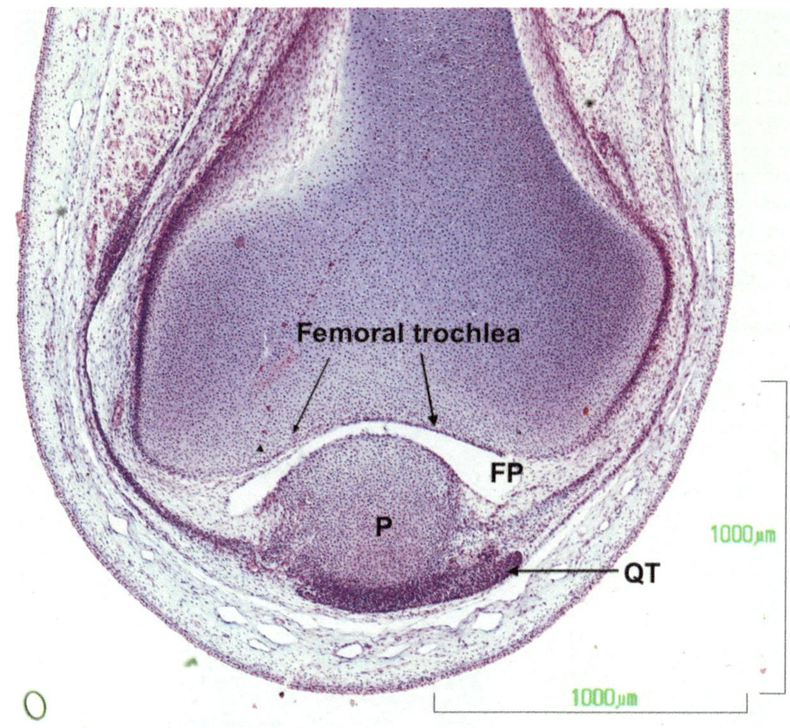

Fig. 2.7 Human fetus at 9 weeks of development (38 mm GL). Frontal section of the femoropatellar joint (*FP*). Patella (*P*), Tendon of quadriceps (*QT*)

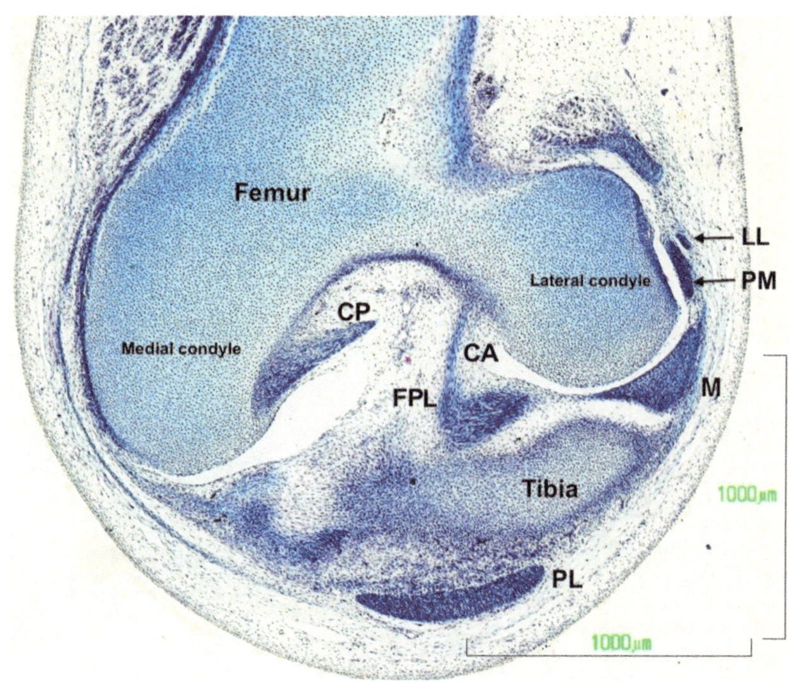

Fig. 2.8 Human fetus at 9 weeks of development (38 mm GL). Frontal section of the knee joint and the patellar tendon (*PL*). Formation of the anlage of the fat pad ligament (*FPL*). Anterior cruciate ligament (*CA*), Posterior cruciate ligament (*CP*), Fibular collateral ligament (*LL*), Popliteus muscle (*PM*), Lateral meniscus (*M*)

morphology to what it will be in adulthood. The posterior cruciate ligament attaches itself to the posterior area of the articular surface of the tibia and extends dorsoventrally trough the internal aspect of the medial condyle of the femur. The anterior cruciate ligament extends from the anterior part of the surface of the tibia to the internal aspect of the lateral condyle of the femur and in

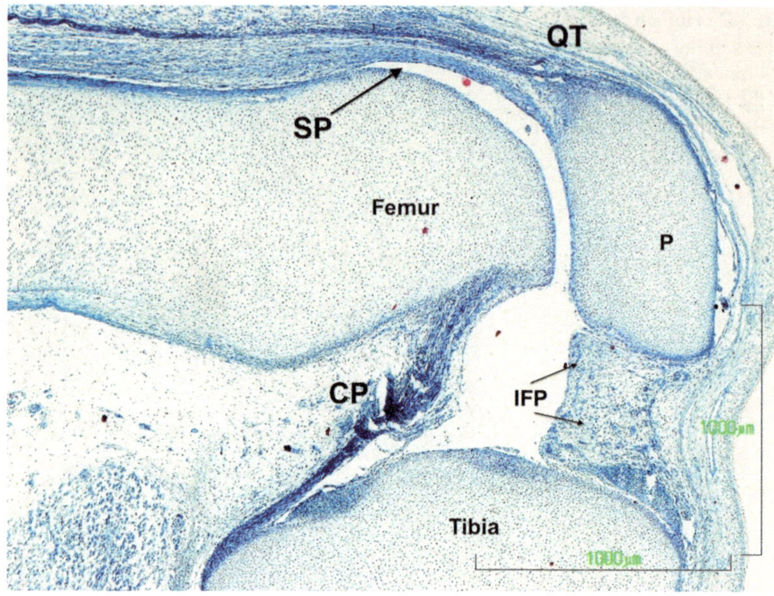

Fig. 2.9 Human fetus at 10 weeks of development (47 mm GL). Sagittal section of the knee and the femoropatellar joint. Topographical arrangement of the suprapatellar bursa (*SP*), intrapatellar fat pad (*IFP*) and the posterior cruciate ligament (*CP*). Patella (*P*), Tendon of quadriceps (*QT*)

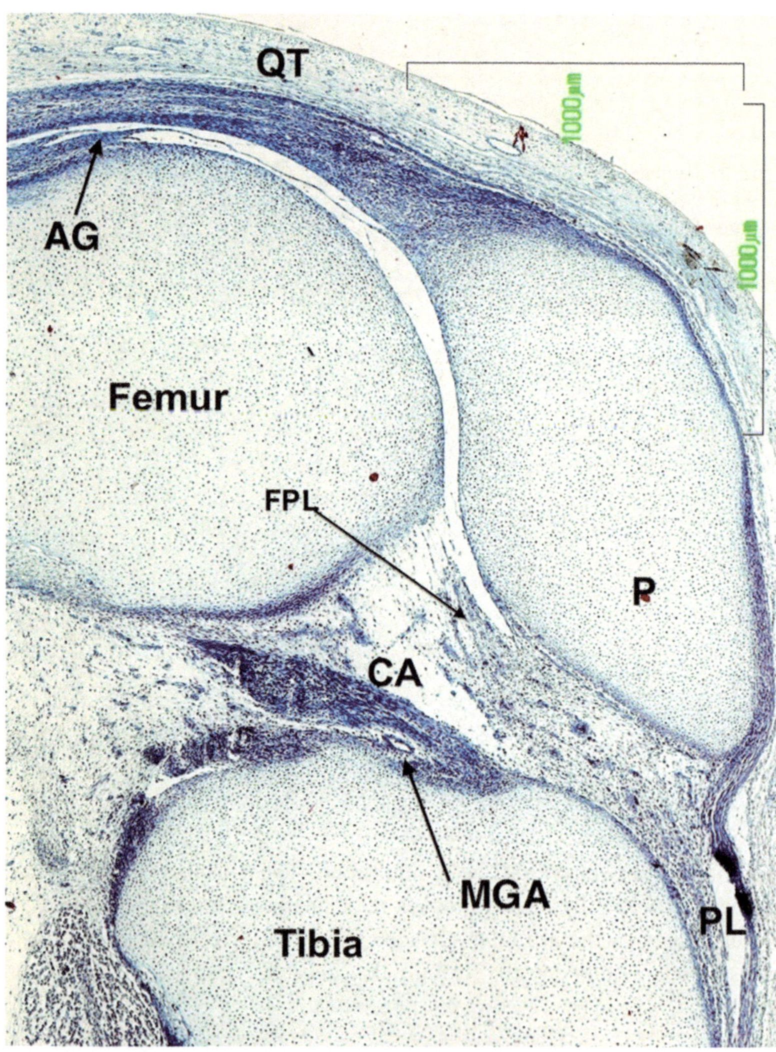

Fig. 2.10 Human fetus at 10 weeks of development (47 mm GL). Sagittal section of the knee and the femoropatellar joint. Topographical arrangement of the anterior cruciate ligament (*CA*) and fat pad ligament (*FPL*). Patella (*P*), Patellar tendon (*PL*), Articularis genus muscle (*AG*), Middle genicular artery (*MGA*), Tendon of quadriceps (*QT*)

relation to the fat pad ligament. Also, the lateral articular surface of the patella clearly starts to be larger than the medial one. The suprapatellar bursa extends to the femoral diaphysis, on which a little contingent of muscular fibers appears, the anlage of the articular muscle of knee or articularis genus muscle. However, the suprapatellar bursa is not totally completed until weeks 14–15 of development [4, 10, 11].

After 12–13 weeks, the articular cavity of the knee has reached its adult appearance since the communicating structure between the lateral meniscotibial cavity and the superior tibiofibular cavity [5], the latter previously formed at 11 weeks, disappears [1] giving way to the fully completed organization of the ligaments of the knee joint.

After 14 weeks, the ossification of the patella commences with cartilage canals penetrating from the anterior and superior surfaces. The patella increases its relative size until the sixth month of fetal life, after which it will follow the same ratio as other bones of the lower limb [8].

References

1. Andersen H. Histochemical studies on the histogenesis of the knee joint and superior tibio-fibular joint in human foetuses. Acta Anat. 1961;46:279–303.
2. Cáceres E, Caja VL. Estudio de las cavidades intraarticulares de la rodilla en los periodos embrionario y fetal humano. Ann Desarr. 1980;56:79–85.
3. Gardner E, O'Rahilly R. The early development of the knee joint in staged human embryos. J Anat. 1968; 102:289–99.
4. Gray DJ, Gardner E. Prenatal development of the human knee and superior tibiofibular joints. Am J Anat. 1950;86:235–87.
5. Haines RW. The early development of the femoro-tibial and tibio-fibular joints. J Anat. 1953;86:192–206.
6. Haines RW. The development of joints. J Anat. 1947; 81:33–55.
7. Jiménez Collado J, Guillén García P, Sobrado Pérez J. Rodilla: Morfogénesis. Anatomía aplicada. Vías de acceso. Editorial Mapfre. Madrid; 1994.
8. Koyuncu E, Cankara N, Sulak O, Özgüner G, Albay S. The morphometry of patella and patellar ligament during the fetal period. Clin Anat. 2011;24:225–31.
9. Lucien M. Dévelopment de l'articulation du genou et formation du ligament adipeux. Bull Assoc Anat. 1904;3:133–5.
10. Mérida- Velasco JA, Sánchez-Montesinos I, Espín-Ferra J, Rodríguez-Vázquez JF, Mérida-Velasco JR, Jimenez-Collado J. Development of the human knee joint. Anat Rec. 1997;248:269–78.
11. Mérida- Velasco JA, Sánchez-Montesinos I, Espín-Ferra J, Mérida-Velasco JR, Rodríguez-Vázquez JF, Jimenez-Collado J. Development of the human knee joint ligaments. Anat Rec. 1997;248:259–68.
12. O'Rahilly R, Müller F. Human embryology and teratology. 2nd ed. New York: Wiley-Liss; 1996.
13. Walmsley R. The development of the patella. J Anat. 1940;74:360–8.

Patellofemoral Anatomy

3

Pedro Guillen-Garcia, Vicente Concejero-Lopez, Jose F. Rodriguez-Vazquez, Isabel Guillen-Vicente, Marta Guillen Vicente, and Tomas F. Fernandez-Jaén

Contents

3.1 Introduction

The femororotulian joint situated on the anterior side of the knee is made up of the kneecap and the femoral trochlea as joint components. It works as a reflection pulley through the flexo-extension movements of the knee similar to the way a mechanical pulley slides, the kneecap being the footing for the transmission of forces. The femororotulian joint has been the forgotten compartment of the knee for quite a while now. Its anatomy (Fig. 3.1), biomechanics, and function are well known, but there are unfinished business:

1. The origin of pain; the true value of cartilage damage
2. The adequate diagnostic means
3. The reliability of surgical techniques

In this chapter, we will try to explain all the relevant anatomical data that explain the femororotulian biomechanics and pathology.

3.2 Morphology

According to Jiménez Collado, P Guillen Garcia, and Sobrado Perez [1], the kneecap can be studied theoretically through its anterior or superficial side, posterior side, and deep or articular side and also through its upper or base rim, lower vertex or angle, and lateral rims.

On the morphological level, the kneecap is a transverse edged trochlear arthrosis with a degree of freedom of movement. The arrangement of the

P. Guillen-Garcia (✉) • V. Concejero-Lopez
J.F. Rodriguez-Vazquez • I. Guillen-Vicente
M.G. Vicente • T.F. Fernandez-Jaén
Trauma Unit, Clinica Cemtro, Madrid, Spain
e-mail: pedroguillen612@gmail.com

A. Gobbi et al. (eds.), *The Patellofemoral Joint*,
DOI 10.1007/978-3-642-54965-6_3, © ISAKOS 2014

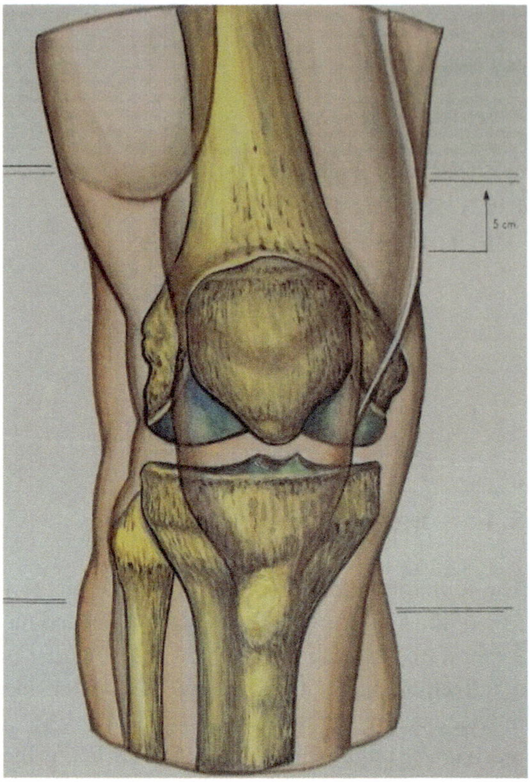

Fig. 3.1 Schematic view of the anterior side of right kneecap and boundaries (Licensed by Jimenez Collado et al. [1])

femoral trochlea and the deep articular side determines the degree of the kneecap lateral instability during the first degrees of bending and the major overload of one of its articular sides on the other. Wiseber classification is prototypical: type I, both articular sides are concave and significantly of the same amplitude; type II, lower internal concave side than the external one; and the most common of all, type III, internal convex side with a small surface borderlining patellar hypoplasia. On the other hand, the normal trochlear angle is 140°; larger angles cause a greater degree of lateral instability; lower angles are more likely to cause cartilage involvement (Fig. 3.2).

When it comes to the femur, the rotulian cartilage footing on the femoro-trochlear cartilage can be established according to the various degrees of bending [1, 3]:

- From 0° to 10°: Only the rotulian cartilage lower part contacts the trochlea – low lateral rotulian stability.
- From 10° to 30°: The rotulian cartilage lower part contacts the upper side of the trochlea – greater rotulian stability.
- From 30° to 60°: The rotulian cartilage lower part contacts the medial side of the trochlea, and a geometrical dynamic fitting of the kneecap and the trochlea occurs – the degree of biomechanical stability is just perfect.
- From 60° to 90°: The kneecap upper side rests on the lower side of the trochlear area.
- Over 90°: The kneecap enters the intercondylar notch and only the marginal parts of its sides – specially the lateral ones – make the femorotibial contact.

The ligament and tendon structures for rotulian support formed by the quadriceps tendon and its insertion in the upper rim of the kneecap; the rotulian tendon inserted in the lower rim, typically called by the anatomist rotulian ligament since it goes through from bone to bone; and the external and internal rotulian retinacula make up what we know as the cruciform elements (Fig. 3.3). We can see this cross-like arrangement of the rotulian ligament and tendon structures in a 30 mm fetus, and if tendons transmit muscle strength to the rotulian retinacula, they act as a joint guide and controllers.

Different pathologies are associated to such cross-like arrangement based on the length between the rotulian tendon and the kneecap which frames the upper or lower patella – the longer the rotulian tendon, the weaker the medial stabilizers. When alterations depend on the angle made up of the quadricipital tendon medial line and the rotulian tendon medial line, the decompensation of the force vectors increases or reduces the lateral rotulian instability. Also the greater or lesser tension variation between retinacula or reduction of one retinaculum with respect to the other can unleash factors of rotulian inclinations at the axial level and a greater or lesser degree of instability [1, 3].

The internal rotulian retinaculum is made up of three layers. The most important of all from the point of view of rotulian semiology is the intermediate layer where we find the medial patellofemoral ligament, MPFL, as the main stabilizer together with the medial patellotibial ligament, MPTL, and

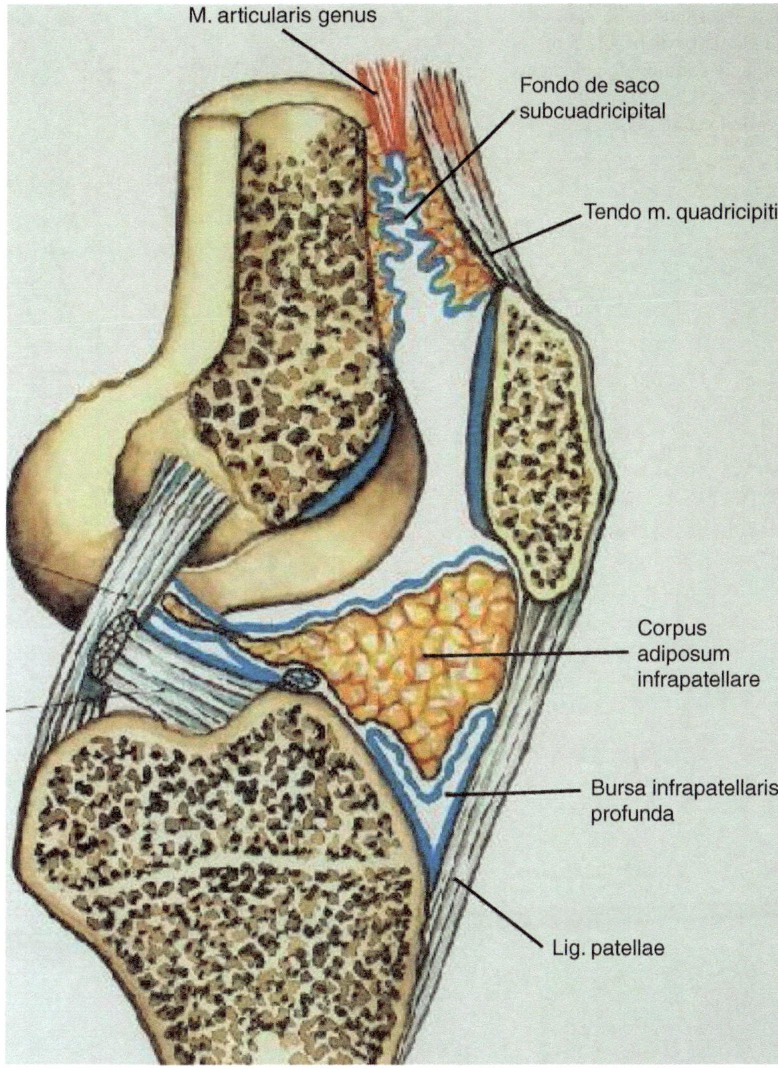

Fig. 3.2 Lateral schematic view of the knee joint. See the massive thickness of the kneecap cartilage – largest in human body (Licensed by Jimenez Collado et al. [1])

the medial patellomeniscal ligament, MPML. The anatomy of the medial patellofemoral ligament includes the insertion in the internal condyle mainly at the adductor tubercle or the existing bone sulcus level and between this and the epicondyle. Vastus medialis usually covers its proximal part. It stretches out progressively to reach the upper 2/3 of the kneecap where it inserts itself.

Some authors talk about a constant structure; others say it is present in 7 out of 20 anatomical pieces only. There is this interesting negative relationship between the length of the rotulian tendon and the width of this medial femororotulian ligament. Such a ligament is considered to be the main stabilizer of the internal side of the kneecap and the most important external translation restrictor. Its most important role is extension but it rapidly loses its function at a 20° bending [2].

It is made up of two sets of fibers – superficial and deep. The superficial layer(superficial oblique retinaculum) is in turn made up of oblique fibers in distal and anterior direction from the anterior rim of the iliotibial band to the lateral rim of the kneecap and lateral area of the rotulian tendon.

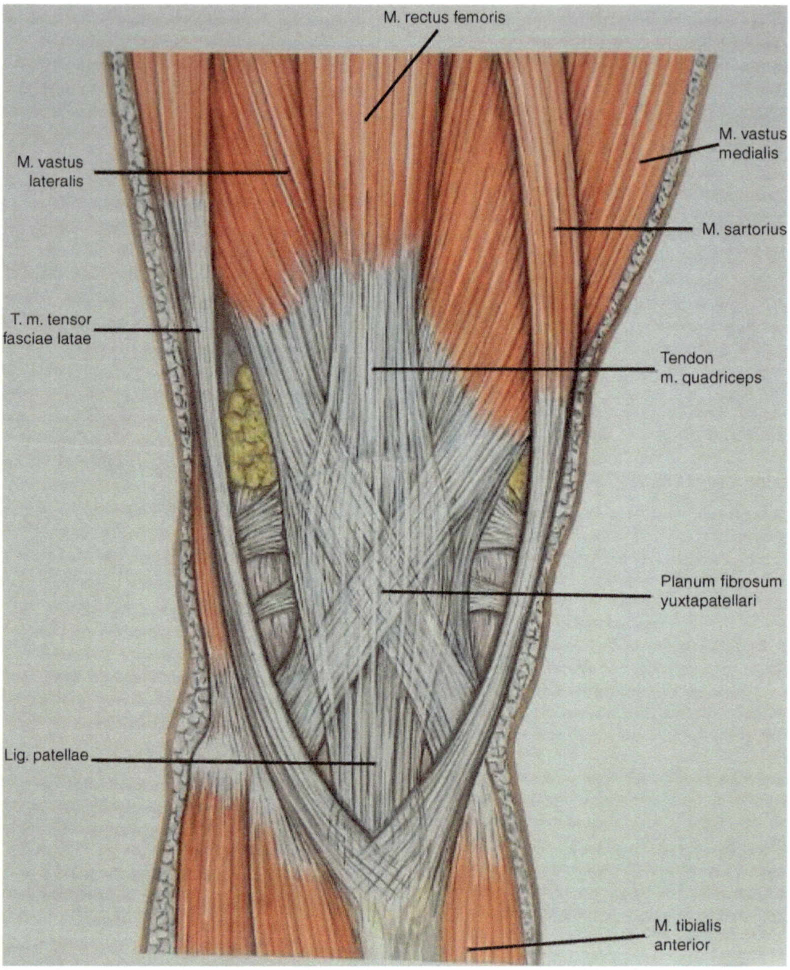

Fig. 3.3 Schematic view of the anterior distribution of ligament and tendon elements (Licensed by Jimenez Collado et al. [1])

The deep layer consists of three very differentiated structures from proximal to distal: the inconstant rotulian epicondylar band, the deep transverse retinaculum stretching from the iliotibial band deep surface to the kneecap lateral rim, and the lowest tibial-rotulian band also known as the rotulian meniscal ligament.

The fibrous juxta-rotulian external tract from the kneecap anterior side and external rim stretches to the external epicondyle.

The fibrous juxta-rotulian internal tract from the kneecap anterior side and upper half of the internal rim stretches to the internal epicondyle.

Orthopedic surgeons call these "surgical retinacula" and they have the exact same structure as the fibrous juxta-rotulian tracts aforementioned [4]. These arciform fibers meet at the kneecap anterior side, and as if they were a semiring or arch, they make up a very important band of biomechanical value, since they stabilize the kneecap position when the knee flexion-extension movements occur. If there is damage or retraction to the rotulian retinacula because of scarring, flexion-extension is limited and there is a femoropatellar displacement (subluxation or balancing).

On a muscular level, the various muscular components of the quadriceps meet in the proximal quadricipital tendon at the rotulian upper edge; when they get there, the quadricipital

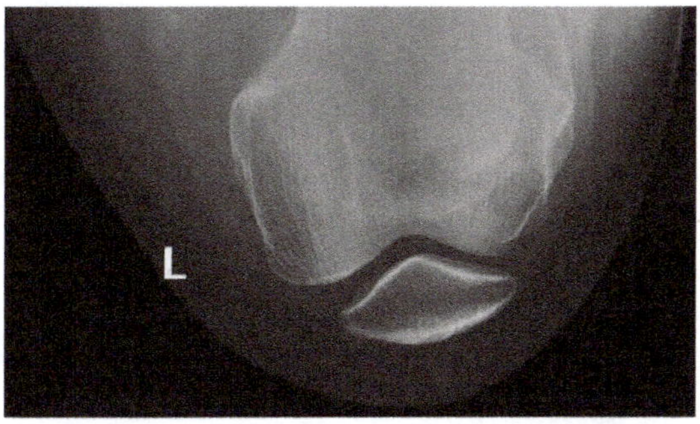

Fig. 3.4 Axial radiological image of a normal femoropatellar joint

tendon looks delaminated in three overimposed tendon levels:

- Superficial level: It is made up of the anterior rectus muscle tendon showing some double insertion when divided into two layers – short and deep fibers; the most anterior and superficial of all slide through the superficial or anterior face of the kneecap not being attached to the kneecap though distally united to the superficial fibers of the rotulian tendon that stretches to the anterior tibial tuberosity.
- Medium level: It is made up of tendons from the internal and external vastus medialis.

Before they merge into the kneecap, each tendon shows two different expansions: (1) direct or latero-rotulian expansion located at each side of the kneecap and the rotulian tendon descending vertically to meet those from fascia lata and femoral biceps muscle at the external side and internal rectus muscles, sartorius muscle, and semimembranosus tendon reflex at the internal side and (2) crossed or pre-rotulian expansion made up of oblique fibers superficially sliding through the kneecap (Fig. 3.4).

3.3 Patellar (Kneecap) Cartilage

In 1998, we initiated the study of the knee joint cartilage (KJC) through arthroscopy with an indenter that we designed and developed at the Valencian Institute of Biomechanics, Spain. Acufex developed the first prototype back in 1985 and we used it in many arthroscopies.

We measured the stiffness of the KJC through arthroscopic technique and saw nine areas inside the knee: three at the kneecap (lateral or external side, ridge of the patella, and medial or internal side), three at the femur (lateral or external femoral condyle, trochlea, and medial or internal femoral condyle), and three at other areas at the tibia (medial or internal tibial disk, medial tibial disk, and lateral or external tibial disk) [4].

We used this means to measure the stiffness of the KJC in 10-, 20-, 30-, 40-, 50-, and 60-year-old individuals and designed what we called the *Cartilaginous Model or Pattern of the Knee*. We found the cartilage of the kneecap to be the thickest cartilage of the knee, the medial or internal side of the KJC to be the softest of the kneecap, and the patellar cartilage of the external or medial side to be the toughest of the knee [4].

Investigators led by Prof. Pedro Guillén, MD, are starting to use this indenter to evolve autologous cultured chondrocyte implantation and see how tough they are, how they look like, and how striated they get. This study of the KJC with indenter led us to talk about chondropenia or bad cartilage [4].

There are two different types of chondropenia: quantitative and qualitative. In quantitative chondropenia, there is partial or total loss of the KJC, while in qualitative chondropenia KJC looks normal, a little softer maybe, and histological analysis shows changes at the extra-articular matrix and chondrocyte level. It is very important to find pathology now, given image diagnostic systems are always negative – X-ray, CAT, and MR [2–4].

On the vascular level, the kneecap is irrigated by the superior external articular artery on the kneecap external rim and rotulian ligament ahead of the kneecap articular capsule creating the so-called peri-rotulian rete articulare genus network. The internal rim running through the inferior internal articular artery stems ascending branches that anastomose with a descending branch from the superior external articular artery and horizontal branches that anastomose ahead of the rotulian tendon with the superior external articular artery and the anterior tibial recurrent artery [1]. These branches made up the peri-rotulian network.

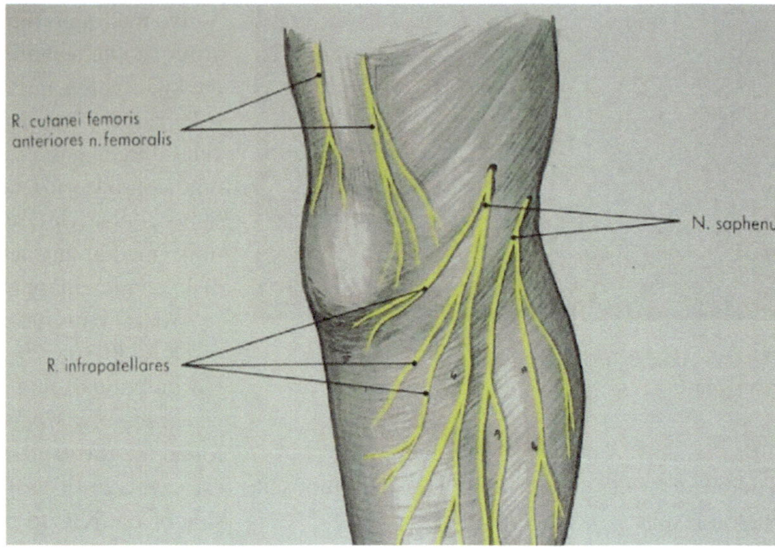

Fig. 3.5 Drawing of neural distribution of the anterior and internal side of the knee and internal saphenous branches (Licensed by Jimenez Collado et al. [1])

Let us not forget the innervation of blood vessels, whose damage means pain to the anterior side of the knee that can be easily misdiagnosed in the clinical practice.

Innervation of the kneecap external rim largely depends on the infrapatellar branch of the internal saphenous nerve that receives sensibility cutaneously from the external pararotulian area and external peri- and pararotulian areas (Fig. 3.5). Nerve damage caused by the arthroscopy internal portal contusions on the internal side of the knee induces a sort of neuropathic pain described as burning, allodynia, dysesthesia, or a reduced sensibility on the kneecap anterior side with a positive Tinel's sign [1, 3]. On the other hand, at the bone kneecap level, innervation stretches the superficial side only and a third of rotulian thickness and rotulian retinacula with no innervation on the deep or articular side; the rich innervation at this level induces some intense pain caused by direct trauma to the kneecap.

References

1. Jimenez Collado J, Guillen Garcia P, Sobrado J Rodilla. Morfogénesis, anatomíaclínica; vías de acceso, Ed. Mapfre. Madrid, Spain. 1994; ISBN: 84-7100-962-5.
2. Insall JN, Scott WN. Cirugía de la rodilla. 4ª ed. Ed. Elsevier. 2007. Madrid, Spain; ISBN:978-84-8174-952-6.
3. Concejero V, Madrigal JM. Traumatología de la rodilla. Ed. Panamericana 2002. Madrid, Spain; ISBN: 84-7903-769-5.
4. Guillen P and als. Traumatismos Articulares. Actualizacion; Editorial Mapfre. 1997. "Indentador o Medidor de Cartilago bajo control Artroscopico"; Madrid, Spain. ISBN. 84-7100-869-6.

4 Anatomy and Biomechanics of Medial Patellofemoral Ligament

Shurong Zhang, Shinya Oka, Ryosuke Kuroda, and Masahiro Kurosaka

Contents

Regarding the patellar instability and patellar dislocation, several aspects should be considered such as the anatomy of patella and trochlea, the congruency of the joint surfaces, the quadriceps muscle force transition, and surrounding tissue's mechanical features. The medial patellofemoral ligament (MPFL) is an important mechanical factor in knee stabilization. A conventional concept that the MPFL does not exist in all knees [1, 2] is challenged by recent studies which states that the MPFL is one of the medial retinacula structures extending from medial border of patella to proximal region of femoral epicondyle, contributing most lateral restraining force [3–6].

In the respect of medial patellofemoral anatomy and biomechanics, this article reviews the knowledge of MPFL that is regarded as one of the ligaments maintaining knee stability.

4.1 Medial Patellar Retinacula Complex

Patella, known as the largest sesamoid bone in the human body, is surrounded by several muscles, tendons, and ligaments. The quadriceps muscles (rectus femoris, vastus lateralis, vastus medialis, vastus intermedius) exist proximal to the patella, generating active restraint force on the patella to stabilize it in axis and medial and lateral direction [7]. The passive restraints are composed of patella ligament that is placed to the distal of patella and medial and lateral retinacular complex, which connect patella to the femur and tibia.

S. Zhang • S. Oka • R. Kuroda • M. Kurosaka (✉)
The Department of Orthopaedic Surgery,
Kobe University Graduate School of Medicine,
Kobe, Japan
e-mail: kurosaka@med.kobe-u.ac.jp

A. Gobbi et al. (eds.), *The Patellofemoral Joint*,
DOI 10.1007/978-3-642-54965-6_4, © ISAKOS 2014

The lateral retinaculum is a complex structure composed of the deep transverse retinaculum and the superficial oblique retinaculum [8]. The medial retinacular complex is divided into three primary layers [9].

Layer I is the crural fascia, which is wrapped by the sartorius and extended to the tibia in the distal part, while it is wrapped by the medial collateral ligament (MCL) and vastus medialis oblique (VMO) in the proximal part. Layer II is the superficial medial collateral ligament (MCL) and other structures. Layer III is the capsule of the knee joint and the deep MCL. Though it is difficult to identify the borderline from layer I to layer II since some structures fuse toward the same places, MPFL is placed in layer II [5, 10]. The medial patellotibial ligament (MPTL), wrapping over patella and tibia, also exists in the same layer with MPFL.

4.2 The Anatomical Footprint of MPFL

4.2.1 The Attachment on Patella

The length of MPFL attachment on the patella is around 20 mm, extending from supra-medial corner to the middle level [11, 12]. Four types of tissue can be recognized at the tendon-bone insertion site, which are bone, calcified fibrocartilage, noncalcified fibrocartilage, and tendon.

The proximal attachment of MPFL is located in the deep aspect of vastus medialis' insertion site, and its superficial fibers interdigitates with VMO [12, 13]. A proximal restraint force would be generated on the MPFL when the vastus medialis muscle contracts [14].

4.2.2 The Attachment on Femur

Smirk and Morris [15] reported various MPFL femoral attachments through a research performed on 25 knee specimens. Mainly, the MPFL attached to the posterior part of the medial femoral epicondyle (44 %), but other attachments were also detected as adductor tubercle (4 %), the adductor magnus tendon (12 %), the area just posterior to the adductor magnus tendon (20 %), or a combination of the above (4 %). It also pointed out that MPFL attached to the anterior part of the medial femoral epicondyle (16 %). Nomura et al. also found that the MPFL might attach on the distal to the adductor tubercle or superoposterior of the medial femoral epicondyle [14]. Additionally, according to the previous investigations, MPFL's femoral attachments were detected, such as on the adductor tubercle [3, 12, 16], on the anterior of medial femoral epicondyle [11, 17], or in between the adductor tubercle and anterior of the medial femoral epicondyle [10, 14]. But Baldwin reported that the MPFL femoral origin of the ligament had a mean width of 10.6 ± 2.9 mm (range, 6–15 mm) and arose strongly from bone in the groove between the adductor tubercle and the medial femoral epicondyle in 49 cadaver specimens [10]. These arguments could be explained by the structure of the MPFL that it is not a bundle of uniform fiber, but a combination of tissues from both superficial layer and deep layer.

The superficial fibers of MPFL cover the area between medial femoral epicondyle and inferior of adductor tubercle, where converging with MCL superficial fibers and adductor tendon occurs [1, 5, 13, 15, 18]. The deep layer of MPFL attaches in between the adductor tubercle and anterior medial femoral epicondyle, with the width of 10–15 mm [10, 12, 14]. A summary could be drawn from these literatures that the attachment of MPFL is slightly behind the adductor tubercle and middle of the medial femoral epicondyle (Fig. 4.1). The proximal of the femoral attachment is called transverse portion (t), where a branch of geniculate artery passes beneath. The distal part of the attachment is named oblique decussation (OD), which is aroused from MCL (Fig. 4.2).

4.3 The Biomechanical Features of MPFL

MPFL is a thin fascial band approximately 55–60 mm in length, 10–20 mm in width [1, 12, 14, 16], and withstands around 208 N of failure load [19]. It is the primary passive restraint that

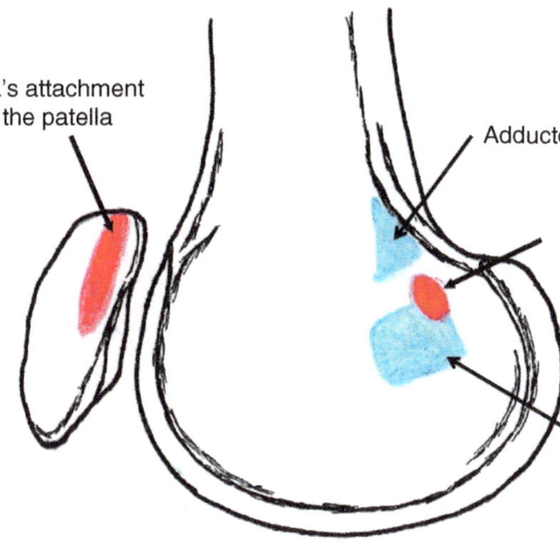

Fig. 4.1 The footprint of MPFL. MPFL extents from supra-medial corner of femur to the middle portion of patella

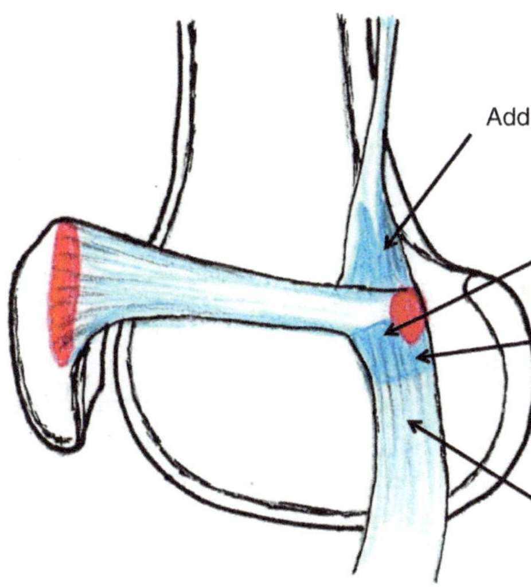

Fig. 4.2 Structure of MPFL on femur. The proximal of the femoral attachment is called transverse portion (*t*), where a branch of geniculate artery passes beneath. The distal part of the attachment is named oblique decussation (*OD*), which is aroused from MCL

resists lateral translation of the patella generating 50–60 % resistant force [1, 3]. According to an investigation by Amis et al., the patella subluxates most easily at 20°of knee flexion [20, 21]. At 20–30°of flexion, the patella will engage in femoral trochlea where lateral facet of patella will be the stabilizer. When the knee bends at 60–70°, the femoral trochlea notch becomes the patella stabilizer. Therefore, the MPFL is tight with the knee in full extension, then slack on flexion of the knee, which suggests MPFL plays an important role in drawing patella toward the trochlea in concert with VMO during early knee flexion [22]. Nomura et al. reported MPFL as the main resistance during 20–90°of knee flexion, which suggested that MPFL is also critical for stabilizing the patella in the trochlea [5]. Hautamaa et al. also emphasized that the MPFL is the major medial ligamentous stabilizer of the patella because there is a 50 % increase in lateral subluxation of the patella with an isolated release of the MPFL [4].

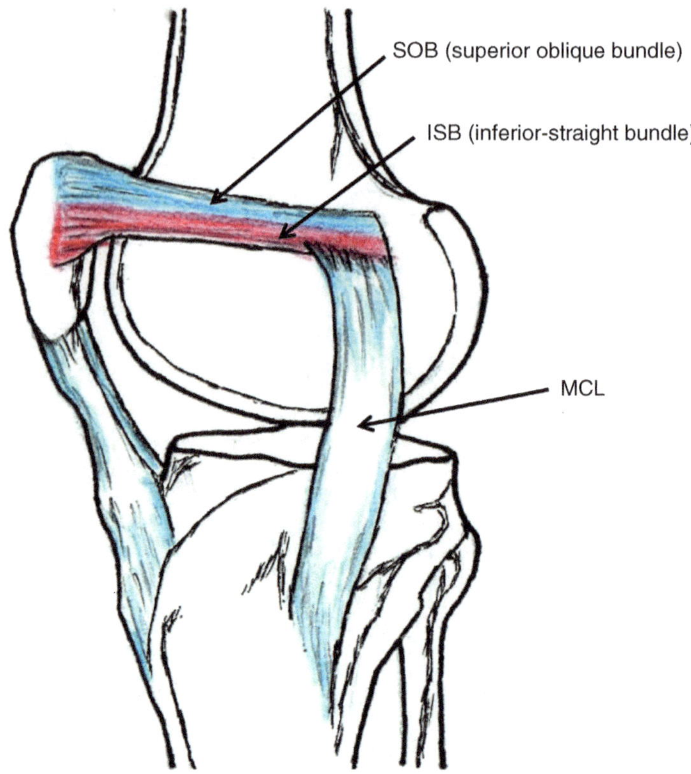

Fig. 4.3 The biomechanical features of MPFL. The patellar insertion of the MPFL is wider than its femoral insertion. The lower fiber bundle, named as inferior-straight bundle (*ISB*), attaches the medial aspect of the patella nearly horizontally, while the upper fiber bundle, named as superior-oblique bundle (*SOB*), attaches the superior-medial aspect of patella

The patellar insertion of the MPFL is wider than its femoral insertion. Amis et al. found that the proximal and distal point of attachment played different function [12]. Further, after an investigation on 12 cadaveric knees, Kang et al. reported that MPFL is formed by two relative centered fiber bundles. The lower fiber bundle, named as inferior-straight bundle (ISB), attaches the medial aspect of the patella nearly horizontally, while the upper fiber bundle, named as superior-oblique bundle (SOB), attaches the superior-medial aspect of patella. They stem from the femoral origination together and not separated entirely. SOB is not only the static structure but also serves as dynamic maintenance of patella stability, while ISB acts as the main static soft tissue restraints [23] (Fig. 4.3). According to a patellar tracking experiment on human knee specimen conducted by Victor et al., cranial and caudal parts of MPFL behave differently. The cranial parts are taut at full extension, while caudal parts are most taut at 30° flexion [24].

It is confirmed that the pathology of patella dislocation is related to the MPFL rupture, because MPFL is the major soft tissue restraint to resist lateral patellar translation. Therefore, recently MPFL reconstruction becomes a popular surgical treatment for patella instability and recurrent patella dislocation.

However, malposition and shorten ligament may increase ligament tension and patellofemoral compressive force, and these causes are regarded as parts of the main reason for MPFL reconstruction failure [25]. A better understanding on the MPFL length change pattern is important. Smirk et al. reported that the fiber connecting proximal patella to 1 cm distal to adductor tubercle behaves isometrically during 0 ~ 80° flexion, but at 120° flexion, it decreases more than 7 mm in length [15], and suggests that the optimal attachment points for the MPFL are the superior patella and just distal to the adductor tubercle, including points posterior and inferior to this [15]. This finding was supported by Victor's results that cranial parts of MFPL are isometric during 0 ~ 80° flexion and decrease 7 mm in length at 120° flexion [24].

Another research performed by Higuchi et al. regarding examination on normal volunteers' MPFL length by MRI pointed out that during 0 ~ 60° flexion, the MPFL length almost does not change, but when knee is flexed over 90°, a decreasing MPFL length pattern is detected [26]. Though multiple MPFL reconstruction techniques have been developed, which are differing in the type of graft used and method of fixation to the femur and patella, there is one point worthy to be noted that the femoral origin of MPFL reconstruction is most sensitive to reproduction of proper ligament isometry. Too proximal placement would cause tightening of the MPFL in flexion with overload of the medial patellar facet. Or a too distal placement would lead the MPFL graft to tighten in extension, causing nonphysiological patellar motion.

In addition, patellar dislocation or patellar subluxation is also concomitant with ligamentous laxity, bone malalignment, or joint rotation abnormality, which means their pathology should be very complicated. But an MPFL injury or malfunction is diagnosed in almost all of these patients, thus indicating that MPFL is a very important component in supportive retinacula complex.

In conclusion, this article gives a general review on MPFL, the most important retinacula tissue in medial patellar. However, there still remains controversy regarding its attachment structure and mechanism features. Therefore, more researches should be performed and more elaborations and elucidations are required. The consensus is that MPFL is a critical factor in knee instability and patella dislocation, and appropriate diagnosis, conservative treatment, MPFL repair, or MPFL reconstruction surgery would be a great benefit for patients.

References

1. Conlan T, Garth Jr WP, Lemons JE. Evaluation of the medial soft-tissue restraints of the extensor mechanism of the knee. J Bone Joint Surg Am. 1993;75(5):682–93.
2. Reider B, Marshall JL, Ring B. Patellar tracking. Clin Orthop Relat Res. 1981;157:143–8.
3. Desio SM, Burks RT, Bachus KN. Soft tissue restraints to lateral patellar translation in the human knee. Am J Sports Med. 1998;26(1):59–65.
4. Hautamaa PV, Fithian DC, Kaufman KR, et al. Medial soft tissue restraints in lateral patellar instability and repair. Clin Orthop Relat Res. 1998;349:174–82.
5. Nomura E, Horiuchi Y, Kihara M. Medial patellofemoral ligament restraint in lateral patellar translation and reconstruction. Knee. 2000;7(2):121–7.
6. Panagiotopoulos E, Strzelczyk P, Herrmann M, et al. Cadaveric study on static medial patellar stabilizers: the dynamizing role of the vastus medialis obliquus on medial patellofemoral ligament. Knee Surg Sports Traumatol Arthrosc. 2006;14(1):7–12.
7. Farahmand F, Senavongse W, Amis AA. Quantitative study of the quadriceps muscles and trochlear groove geometry related to instability of the patellofemoral joint. J Orthop Res. 1998;16:136–43.
8. Fulkerson JP, Gossling HR. Anatomy of the knee joint lateral retinaculum. Clin Orthop Relat Res. 1980;153:183–8.
9. Warren LF, Marshall JL. The supporting structures and layers on the medial side of the knee: an anatomical analysis. J Bone Joint Surg Am. 1979;61(1):56–62.
10. Baldwin JL. The anatomy of the medial patellofemoral ligament. Am J Sports Med. 2009;37(12):2355–61.
11. Steensen RN, Dopirak RM, McDonald 3rd WG. The anatomy and isometry of the medial patellofemoral ligament: implications for reconstruction. Am J Sports Med. 2004;32(6):1509–13.
12. Amis AA, Firer P, Mountney J, et al. Anatomy and biomechanics of the medial patellofemoral ligament. Knee. 2003;10(3):215–20.
13. Dirim B, Haghighi P, Trudell D, et al. Medial patellofemoral ligament: cadaveric investigation of anatomy with MRI, MR arthrography, and histologic correlation. AJR Am J Roentgenol. 2008;191(2):490–8.
14. Nomura E, Inoue M, Osada N. Anatomical analysis of the medial patellofemoral ligament of the knee, especially the femoral attachment. Knee Surg Sports Traumatol Arthrosc. 2005;13(7):510–5.
15. Smirk C, Morris H. The anatomy and reconstruction of the medial patellofemoral ligament. Knee. 2003;10(3):221–7.
16. Tuxøe JI, Teir M, Winge S, et al. The medial patellofemoral ligament: a dissection study. Knee Surg Sports Traumatol Arthrosc. 2002;10(3):138–40.
17. Feller JA, Feagin Jr JA, Garrett Jr WE. The medial patellofemoral ligament revisited: an anatomical study. Knee Surg Sports Traumatol Arthrosc. 1993;1(3–4):184–6.
18. Philippot R, Chouteau J, Wegrzyn J, et al. Medial patellofemoral ligament anatomy: implications for its surgical reconstruction. Knee Surg Sports Traumatol Arthrosc. 2009;17(5):475–9.
19. Mountney J, Senavongse W, Amis AA, et al. Tensile strength of the medial patellofemoral ligament before and after repair or reconstruction. J Bone Joint Surg Br. 2005;87(1):36–40.
20. Hungerford DS, Barry M. Biomechanics of the patellofemoral joint. Clin Orthop Relat Res. 1979;(144):9–15.
21. Heegaard J, Leyvraz PF, Van Kampen A, et al. Influence of soft structures on patellar three-dimensional

tracking. Clin Orthop Relat Res. 1994; 299:235–43.
22. Bicos J, Fulkerson JP, Amis A, et al. The medial patellofemoral ligament. Am J Sports Med. 2007;35(3): 484–92.
23. Kang HJ, Wang F, Chen BC, et al. Functional bundles of the medial patellofemoral ligament. Knee Surg Sports Traumatol Arthrosc. 2010;18(11):1511–6.
24. Victor J, Wong P, Witvrouw E, et al. How isometric are the medial patellofemoral, superficial medial collateral, and lateral collateral ligaments of the knee? Am J Sports Med. 2009;37(10):2028–36.
25. Elias JJ, Cosgarea AJ. Technical errors during medial patellofemoral ligament reconstruction could overload medial patellofemoral cartilage: a computational analysis. Am J Sports Med. 2006;34(9):1478–85.
26. Higuchi T, Arai Y, Takamiya H, et al. An analysis of the medial patellofemoral ligament length change pattern using open-MRI. Knee Surg Sports Traumatol Arthrosc. 2010;18(11):1470–5.

Clinical Examination of the Patellofemoral Joint

5

Sukesh Rao Sankineani, Georgios Karnatzikos, Sanyam Chaurasia, and Alberto Gobbi

Contents

5.1 Introduction

The diagnosis of patellofemoral joint disorders is challenging for orthopaedic surgeons on the account of complicated pathophysiology that underlies them. The origin of patellofemoral symptoms is generally multifactorial [1], and therefore the clinical examination should take into account the patient as a whole, and meticulous attention must be drawn to the history, morphology and clinical examination to have a definitive diagnosis and address the patient's pathology. It should be remembered that a proper therapeutic approach can only be engineered with a thorough analysis of the patients' disorder.

5.2 History

A thorough and proper history focusing on the patient's symptoms, duration and onset of problem and the effect of the problem in modifying his/her activities of daily living must be taken. A careful history will point to the presence of anterior knee pain with or without patellar instability. *Patellar instability* is described as episodes of giving away, subluxation or dislocation of the patella. While evaluating the history of pain, the clinician must ask for the exact location of pain, timing of pain (is it activity related?) and association of trauma with the onset of symptoms. Any referred pain must also be evaluated as sometimes a pain in the hip can radiate to the region immediately above or around the knee.

S.R. Sankineani (✉) • G. Karnatzikos
S. Chaurasia • A. Gobbi
Orthopaedic Arthroscopic Surgery International (O.A.S.I.) Bioresearch Foundation, Gobbi N.P.O., Milan, Italy
e-mail: sukeshrao.sankineni@gmail.com

A. Gobbi et al. (eds.), *The Patellofemoral Joint*,
DOI 10.1007/978-3-642-54965-6_5, © ISAKOS 2014

If the patient has instability, the cause of the event and confirmation that it is due to patella must be made. Often, the *feeling of giving away* of the knee may be because of quadriceps weakness, meniscal tear, ligament deficiency or other disorders around the knee. A history of previous surgery should also be taken properly to rule out the possibility of medial patellar instability after a lateral release or realignment surgery of knee. Failure to consider this possibility might lead to a misdiagnosis of persisting lateral instability and to unnecessary additional surgery, resulting in further exacerbation of the problem.

After a careful analysis of the history, a systematic clinical examination should be carried out with the patient in standing, sitting and supine position along with gait analysis.

5.3 Physical Examination

5.3.1 Standing Examination

The patient should be asked to stand barefoot in bipodal position and asked to stand facing the examiner with both feet together and pointing ahead. From the front, the *alignment* of the knee (varus or valgus deformity), *orientation* of the patellae (squinting or outfacing patellae) and patellar *height* are noted. Patients with a valgus angulation have a more predisposition to lateral subluxation; furthermore, patients with squinting patellae have a higher incidence of patellofemoral pain, and patients with outfacing patellae are associated with habitual subluxation or dislocation of knee caps.

From the side, the inclination of the pelvis, spinal curvatures particularly hyperlordosis and the position of the body with respect to the pelvis can be assessed, and from the back, the presence or absence of a scoliotic curve, pelvic tilt or foot rotation anomalies can be assessed.

5.4 Q Angle

The Q angle or quadriceps angle is a reflection of the valgus vector of the quadriceps pull acting on the patella and gives an idea of the tracking of the knee. It is usually measured in standing position and is done by drawing the first line between the centre of the patella and the anterior-superior iliac spine and the second line between the centre of the patella and the centre of the tibial tubercle. It averages 15° in normal individuals: 14° in men and 17° in women. Anatomic variants that produce either in-facing patellae or lateral displacement of the tibial tubercle can result in an increased Q angle which increases the tendency for patellofemoral pain [2, 3].

5.4.1 Gait Analysis

The patient is then asked to walk while the examiner evaluates gait to see if there is evidence of *anatalgia* localized to the hip or knee joint. Moreover, it is possible to assess the *symmetry* of gait, the *length* of the stride, the *orientation* of the patellae, the *varus* or *valgus* alignment of the knee and the pelvic tilt. A *single-leg knee-bend test* is performed wherein the patient is asked to bend his/her leg bringing the knee close to the thorax. This position helps assess the quadriceps support as well as evaluate the core stability at the hip and pronation of foot and ankle. Establishing the level of lower extremity support in any patient with patellofemoral instability or pain is important to guide physical therapy appropriately in order to improve the overall function of the lower extremity [4].

5.4.2 Seated Position

The patient is then examined in sitting position with the legs hanging from the table. On inspection, a swelling of the tibial tuberosity or the inferior pole of the patella might suggest osteochondrosis, insertional tendinitis or a partial rupture of the patellar tendon. An overview of muscular atrophy especially vastus medialis is also evident from the inspection. In addition, an assessment of patellar height can be made in this position. In this position, the patellae should face directly forward in a normal patient. In *patella alta*, the high-riding patella faces upward towards the ceiling, and in *patella baja*, the patella is lower than the normal side and seems to be drawn into the sulcus between the femoral condyles. The patellar tracking is then evaluated by asking the patient to flex and extend the knee while sitting. In the presence of instability, the patella undergoes subluxation near full extension constituting the so-called *J-sign* [5, 6].

In addition, an assessment of the *tubercle-sulcus angle* can be made at 90 degree flexion of the knee and is a variation of the Q angle described by Kolowich [7] to eliminate the effect of femoral rotation and to detect the abnormal displacement of the tibial tubercle. The angle is measured by drawing the first line passing through the centre of the patella and the centre of tibial tubercle and the second line drawn perpendicular to the transepicondylar axis. This angle is normally less than 8° in women and less than 5° in men. An increase in this angle reflects lateral displacement of the tibial tubercle and is associated with patellofemoral pain and instability.

5.4.3 Supine Position

The examiner should look for rotation of the extremities and evaluate flexion and extension of the knee to see if there is any evidence of lateral patellar tracking. *Patellar facet tenderness* [8] can be elicited in this position. To elicit it, the patella is shifted medially to expose the medial facet and then it is palpated with the other hand and observed for signs of tenderness. The procedure is repeated for the lateral facet as well. In the case of *excessive lateral pressure syndrome*, very little glide is possible and the tenderness is localized to the patellofemoral ligament. Furthermore, anteromedial knee pain can be occasionally seen due to an inflamed medial patellar plica [9] which can be palpated as a fibrous cord running between the patella and the medial femoral condyle which can be made prominent by flexing the knee.

Extensor muscles and tendons mechanism is always indicated to rule out associated soft tissue disorders. In case of quadriceps tendon rupture, a gap can be palpated along with tenderness on attempted straight-leg raise test. Similarly, palpating the patellar tendon during straight-leg raise also is a good way to check for patellar tendon rupture. In case of *jumper's knee*, tenderness is most commonly elicited in the proximal patellar tendon just inferior to the tip of patella; in addition, a soft spongy, crepitant sensation can also be felt.

5.5 Special Tests for Patellofemoral Joint

5.5.1 Passive Patellar Grind Test (Patellar Inhibition Test)

This test is the most commonly known test for eliciting patellofemoral crepitus and is performed with the patient in supine position. The patella is pressed with the palm of one hand against the femoral groove with one hand and the knee is passively flexed with the other hand. In the presence of degeneration or irregularity of the articular surface, a distinct crunching sensation is transmitted along with pain. However, the test has a low specificity because of the fact that it is elicited even in normal individuals sometimes [10]. Pain with crepitus and recurrent effusion suggest degenerative changes; pain alone suggests articular injury [11].

5.5.2 Step-Up-Step-Down Test

This test is highly sensitive for patellofemoral crepitus and is used to identify distal articular lesions which are often missed on other tests. In this test, the patient is asked to stand on a small step and is asked to step down on one side and then repeated on the other side. Any evidence of pain and crepitus is felt during the test. A patient who has an intense pain on early step down may have the distal articular lesion and may benefit from an unloading operation of the distal pole such as an anterior or anteromedial tibial tubercle transfer [12].

5.5.3 Patellar Glide Test

This test allows us to estimate the passive patellar mobility and is performed in supine position while holding the knee at 20°–30° flexion along with a relaxed quadriceps. The amount of translation is estimated between the centre of the patella and the medial and lateral epicondyles which are roughly equal in this position. In the presence of a tear of the medial or lateral structures, the translation is increased. Conversely, the translation is decreased in case of arthrofibrosis [7, 13].

5.5.4 Passive Patellar Tilt Test

This test is used to evaluate the lateral retinaculum and is performed with the knee in full extension and the quadriceps relaxed. The normal tilt is considered to be 0° and it is impossible to lift the lateral edge of the patella. In case of excessive surgical release of the lateral retinaculum, the tilt is increased and the patellar plane may be rotated internally [14, 15].

5.5.5 Engagement Sign

This test is used to assess the tracking and engagement of the patella over the proximal trochlea and is usually abnormal in cases of dysplastic trochlea, patella alta or knee recurvatum leading to patellofemoral pain. This test is performed in supine position with the knees placed in full extension, and a firm pressure is applied with a thumb over the tip of the patella and the knee is then flexed to 20°. In case of the presence of the described disorders above, pain is felt by the patient at the inferior pole of the patella when the patella engages the trochlea; furthermore, a bump can also be felt in case of dysplastic trochlea [16].

5.5.6 Apprehension Test (Fairbanks Sign)

This test is pathognomonic of clinically symptomatic patellar instability and simulates an episode of patellar dislocation under controlled conditions [17]. The patient is placed supine and the limb is abducted sufficiently to allow the knee to be flexed over the side of the table by grasping the symptomatic limb at the ankle. With the other hand, the patella is pushed laterally while the knee is flexed slowly. In patients with history of patellar subluxation or dislocation, apprehension manifesting as expressions of anxiety or quadriceps contraction to prevent knee flexion is observed. Often this test correlates with an abnormal lateral glide, a positive engagement test or patellar maltracking.

Conclusions

The diagnostic workup of patellofemoral pain should always start with a meticulous clinical examination including the overall morphology, static and dynamic equilibrium of the patient. Often, findings on clinical examination will pave for a clear diagnosis once supplemented with diagnostic studies. Along with a proper history taking, it should allow us to identify painful areas and patellofemoral instability in order to plan the management of the disorder. Most importantly, clinical exam is the best tool to determine the additional investigations needed to have a definitive diagnosis and lead to address the problem in the most effective manner.

References

1. Lankhorst NE, Bierma-Zeinstra SM, van Middelkoop M. Factors associated with patellofemoral pain syndrome: a systematic review. Br J Sports Med. 2013; 47(4):193–206. doi:10.1136/bjsports-2011-090369. Epub 2012 Jul 19.
2. Kaya D, Doral MN. Is there any relationship between Q-angle and lower extremity malalignment? Acta Orthop Traumatol Turc. 2012;46(6):416–9.
3. Christian SR, Anderson MB, Workman R, et al. Imaging of anterior knee pain. Clin Sports Med. 2006;25(4):681–702.
4. Claudon B, Poussel M, Billon-Grumillier C, Beyaert C, Paysant J. Knee kinetic pattern during gait and anterior knee pain before and after rehabilitation in patients with patellofemoral pain syndrome. Gait Posture. 2012;36(1):139–43.
5. Chapman MW, Szabo RM, Marder RA, et al. Chapman's orthopaedic surgery. Philadelphia: Lippincott Williams & Wilkins; 2001.
6. Yormak JH, Scuderi GL. Chapter 5: Physical examination of the patellofemoral joint. Patellar alignment in subjects with symptomatic knees: student raters. J Orthop Sports Phys Ther. 2001;31(7): 368–74.

7. Kolowich PA, Paulos LE. Lateral release of the patella: indications and contraindications. Am J Sports Med. 1990;18(4):359–65.
8. Calmbach WL, Hutchens M. Evaluation of patients presenting with knee pain: part II. Differential diagnosis. Am Fam Physician. 2003;68(5):917–22. Review.
9. Blok A, Weiss W, Dolata T, Szczepaniec M. Medial synovial plica. Ortop Traumatol Rehabil. 2005;7(4): 397–400.
10. Theut PC, Fulkerson JP. Anterior knee pain and patellar subluxation in the adult. In: DeLee JC, Drez Jr D, Miller MD, editors. DeLee and Drez's orthopaedic sports medicine: principles and practice. 2nd ed. Philadelphia: Saunders; 2003. p. 1772–815.
11. Tria AJ. Ligaments of the knee. Philadelphia: Churchill Livingstone; 1995.
12. Johnson D. Operative Arthroscopy. 4th ed. Philadelphia: Lippincott Williams & Wilkins; 2012.
13. Powers CM, Shellock FG, Pfaff M. Quantification of patellar tracking using kinematic MRI. J Magn Reson Imaging. 1998;8:724–32.
14. Baratz ME, Watson AD, Imbriglia JE. Orthopaedic surgery: the essentials. New York: Thieme; 1999.
15. Amis AA. Current concepts on anatomy and biomechanics of patellar stability. Sports Med Arthrosc. 2007;15(2):48–56.
16. Vallotton J, Echeverri S. Clinical examination of the patellofemoral patient. In: Zaffagnini S et al., editors. Patellofemoral pain, instability and arthritis. Berlin/Heidelberg: Springer; 2010. p. 45–6.
17. Fairbank HA. Internal derangement of the knee in children. Proc R Soc Lond. 1937;3:11.

Imaging of Patellofemoral Joint

6

Ignacio Cardona-Muñoz, Jesus I. Cardona-Medina, and Ariel de la Rosa

Contents

I. Cardona-Muoaz, MD (✉)
Centro de Ortopedia y Medicina del Deporte (Orthopaedic & Sports Medicine Center), Institute of Physical Activity and Sports Applied Sciences, University of Guadalajara, Guadalajara, Mexico
e-mail: dr.ignaciocardona@gmail.com

J.I. Cardona-Medina, MD
Centro de Ortopedia y Medicina del Deporte (Orthopaedic & Sports Medicine Center), Human Movement Sciences, University of Guadalajara, Guadalajara, Mexico

A. de la Rosa, MD
Centro de Ortopedia y Medicina del Deporte (Orthopaedic & Sports Medicine Center)
Av. Patria Sur 649 Zapopan, Jalisco 45030, Mexico

6.1 Radiographic Evaluation of the Patellofemoral Joint

Patellofemoral syndrome is the most common cause of knee pain in young active patients, and although its aetiology is multifactorial, there is an apparent consensus that the most common cause is the extensor mechanism dysfunction caused by patellofemoral malalignment [1]. Due to the complexity for determining the precise aetiology, additionally to an accurate physical examination, radiographic approach is needed to identify those anatomic factors. In most cases, standard radiographs are sufficient for evaluating those anatomic abnormalities [2, 3]. In specific or complex cases, additional studies such computerised tomography (CT), magnetic resonance (MRI), bone scan (BE) or ultrasonography may be required.

Clinical applications and interpretation of the most commonly standard radiographs and imaging studies for diagnosis and decision-making of treatment are presented.

6.1.1 Clinical Applications

From a clinical point of view the most widely nomenclature used is given by the French group that classifies clinical patellofemoral problems in three categories: (1) objective instability with anatomic factors, patients with recurrent patellofemoral dislocation; (2) potential instability, patients without dislocation but with pain and anatomic factors present; and (3) pain caused for overload

A. Gobbi et al. (eds.), *The Patellofemoral Joint*,
DOI 10.1007/978-3-642-54965-6_6, © ISAKOS 2014

of the joint for functional imbalance between the forces controlling patellar tracking, patients with pain without instability or anatomic factors. Although no clear evidence of incluencia that anatomical factors play in the causation of pain your assessment is necessary [4].

6.1.2 Standard Views

AP View

This projection must be taken with patient in natural standing position to avoid a distorted interpretation of the position of the patella [5].

By obtaining standing AP views of the knee in normal alignment for the patient, the relationship of the patella to the femur will be shown, as it exists under normal standing conditions. These should be taken in full extension and to 30° of knee flexion. A careful evaluation of the AP or PA radiograph can determine whether there is a high- or low-riding patella. Finally, patellar and condylar measurements may be taken also from an accurate evaluation of this projection. In general, a standing AP view is most appropriate, so that medial or lateral articular cartilage loss (joint space narrowing) may be detected (Fig. 6.1).

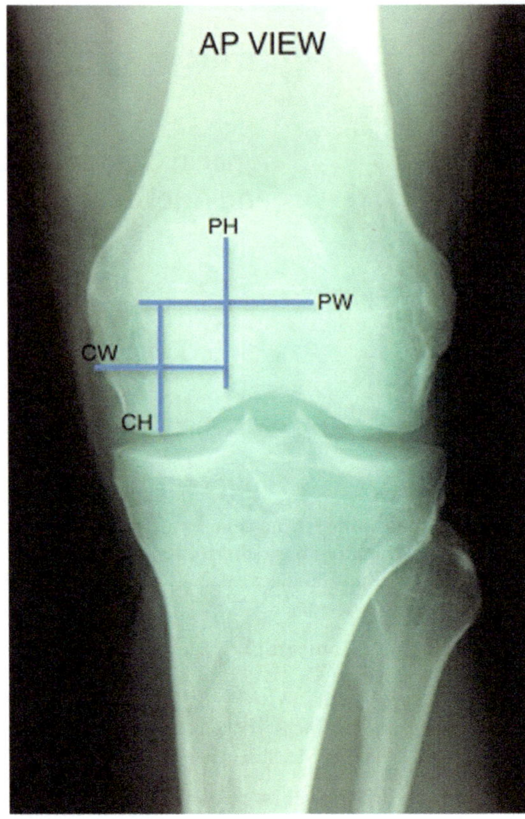

Fig. 6.1 AP view. Patellar and condylar height and width can be measured. Position of the patella. Tibial plateau dimensions and tibial spine can also be assessed. *PH* patella heigth, *PW* patella width, *CH* condylar heigth, *CW* condylar width

Lateral View

The lateral projection may be taken in the lateral decubitus position or standing with the knee flexed to 30° to place the patellar tendon under tension. The beam must be centred on the tibiofemoral joint line. Patellar height and thickness can be measured on this view. An exact lateral projection was described by Maldague and Malghem, in which the femoral condyles are overlapped. It is necessary to have precise lateral radiograms in AP and lateral view.

Through this view, the clinician can obtain fairly good knowledge of the patella.

General morphology, patellar height and thickness can be measured. Also, subchondral sclerosis, peripatellar tendon calcifications and evidence of arthrosis can be determined. Lund and Nilsson described a shallow excavation or subchondral cyst in most patients with proven chondromalacia [3].

This view provides the functional relationship of the patella to the tibia and, more important, of the patellar facets to the femur. The lateral view provides information regarding patellar rotation (tilt) with the patient in weight-bearing position. In most patients, this projection is taken at 30° of knee flexion. Thus, one can be able to get a more functional view of patellar rotation in weight bearing. Several authors reported the sensitivity and reliability of the lateral view compared to routine axial radiographs [6, 8]. The precise lateral view taken in standing position, let us evaluate patellar tilt and trochlear morphology [7] including trochlear depth. Dejour and his group studied the trochlear morphology in the lateral radiograph of the knee (Fig. 6.2).

To determine patella's height most used are Blumensaat's line and Insall-Salvati ratio, although they have been demonstrated being

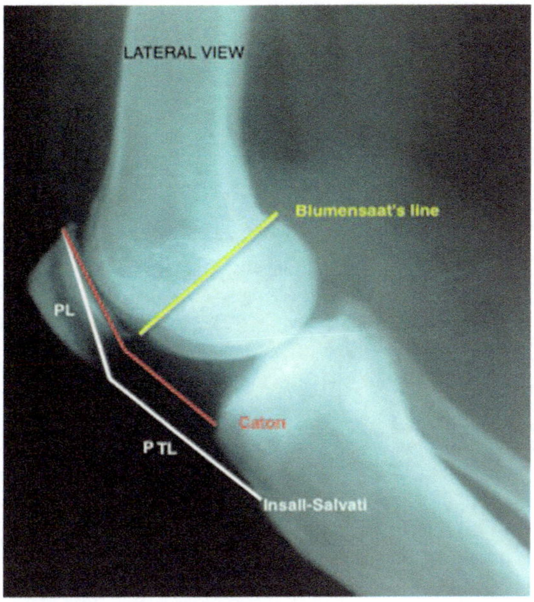

Fig. 6.2 Lateral radiograph of 30° of flexion. Blumensaat's line in *yellow*. The method of Insall-Salvati in *white*. Comparing patella length PL and patellar tendon PT. Normal ratio is 1 with a+2 or −2. Caton in *red* demonstrates the importance of considering articular length in evaluating patellar height

inaccurate. Blumensaat's line is not completely reliable because there is sometimes difficulty to identify the ventral border of the intercondylar fossa (Blumensaat's line), in addition to the wide variability of the state of the roof of the intercondylar notch of the femur. However, Insall-Salvati ratio lacks sensitivity because of variations in patella morphology (Grelsamer and meadows). Blackburne and Peel proposed a modification of this ratio using the tibial articular surface as reference. The normal reference of this ratio is 0.8. Caton et al. described a similar method.

Finally, Bernageau and Goutallier pointed out the importance of describing the patellar articular surface as it relates to the proximal central trochlea. Recently, Biedert and Albrecht [9, 10] proposed a new index for assessing the length of the lateral articular trochlea as predisposing factor for patellar instability.

Axial (Tangential) View

This view provides valuable information regarding the anatomy and understanding of functional disorders. There are several techniques for taking a tangential patellar radiograph which are taken with knee flexion ranging from 20 to 115 degrees. The most common projection is called "sunrise view" taken at 115°. Mild abnormality may be missed due to the degree of knee bending that is required in this projection.

Laurin 20° and Merchant 45° have been most used. These techniques can induce the possibility of erroneous interpretations because technically it is impossible to detect maltracking of the patella in the first 20–30° of flexion [11]. However, a good axial radiograph can be useful in evaluating patellofemoral disorders in basic approach.

With the axial projection of Merchant view, the groove angle and angle of congruency can be measured. Normally, the groove angle, which is formed by the highest points of the femoral condyles and the deepest point of the intercondylar groove, is about 138°. The congruence angle is obtained drawing two lines: the first line is the bisection of the femoral Groove angle and the second line is drawn from the lowest point of the articular margin of the patella to the deepest point of the Groove. When the deepest point of the knee cap edge lies medial to the reference line, a negative value is assigned to the angle formed; when placed side to the reference line, a positive value is assigned. In their study, the average congruence angle was −6°. It was observed that the angles of +16° or higher were associated with different patellofemoral disorders, particularly lateral subluxation of the patella. Sometimes, patellofemoral disorders may require additional tangential projections obtained with a knee flexion of 30, 60 and 90° of flexion [12].

Fulkerson [1] has found two measurements to be most helpful: the Laurin lateral patellofemoral angle (patellar tilt angle) and Merchant congruence angle (Fig. 6.3).

6.2 Computed Tomography

Computed tomography (CT) is a technique that allows imaging in different degrees of flexion using the midtransverse patella as a stable plane of reference. Omitting image overlap and distortion CT, let us define accurately the relationship between the midtransverse patella and the femoral trochlea. It is impossible to note a variety

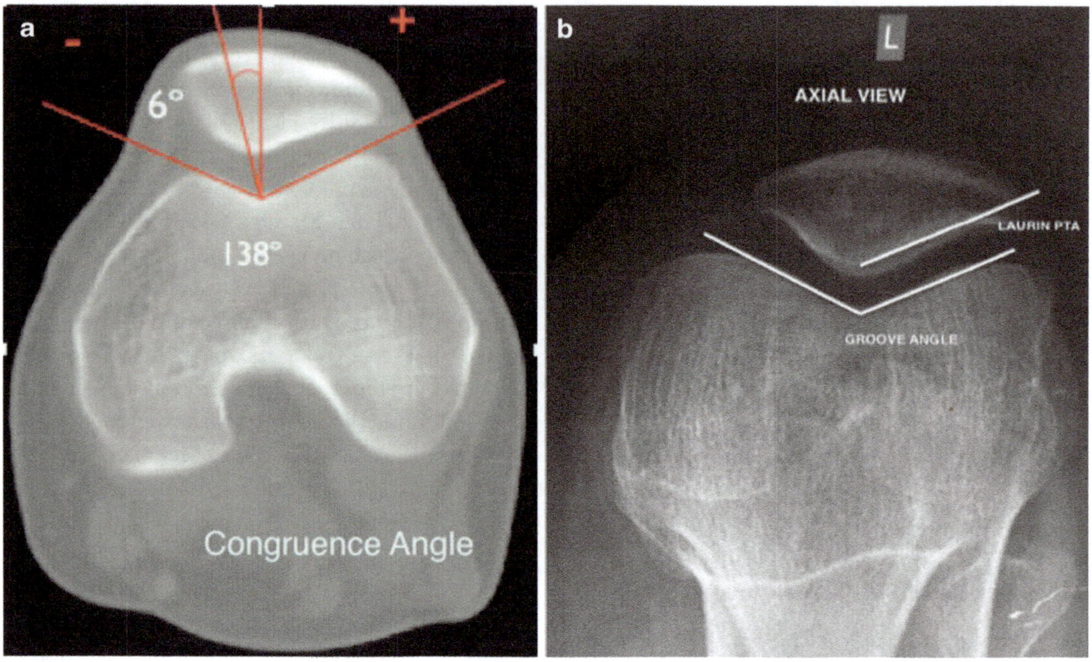

Fig. 6.3 (**a**, **b**) Congruence angle. (1) Bisect the angle of the femoral trochlea. (2) Draw a line from the apex of the femoral trochlea through the apex of the patella. The angle between these two lines is the congruence angle (CA). If the patellar apex is medial to the trochlear apex, the CA is negative. If it is lateral, the CA is positive considering patellar displacement

of different tracking patterns that are not as well defined using standard axial radiography [13].

It is also possible to make measurements with quadriceps contraction which, as described in the literature, increases sensitivity by detecting abnormal deterioration in patients with normal values at rest [13]. Evaluation of patellofemoral congruence with the quadriceps both contracted and relaxed may improve sensitivity. Moreover, selectively (Fulkerson) using the sensitive CT criteria, one can recognise that tilt may occur with or without subluxation.

6.2.1 Advantages of CT

- Allowing a true axial view of the patellofemoral joint
- Ease of patient positioning
- Ability to reliably image the patellofemoral joint in multiple degrees of flexion

6.2.2 Patient Position

Place the patient in a lateral decubitus position. Place a spacer between the knees and ankles to create an alignment similar to a relaxed stance. Scan through the patellar apices at 0, 20, 30, 40 and 50° of flexion. At each degree of flexion, ask the patient contract his or her quadriceps muscles. Assess four aspects of alignment: depth of the femoral trochlea (Fig. 6.4).

Measure the opening angle of the trochlea, which is 138° on an average. Schoettle et al. [11] reported that an increase in this angle is defined as trochlear dysplasia. It determines, in axial section, an angle between the deepest point of the trochlea and anterior uppermost points and both femoral condyles. Third medial side of the axial slide of CT is the cutting recommended for this measurement [12]. Determine the existence of trochlear dysplasia, if the angle is greater than 140° (Fig. 6.5).

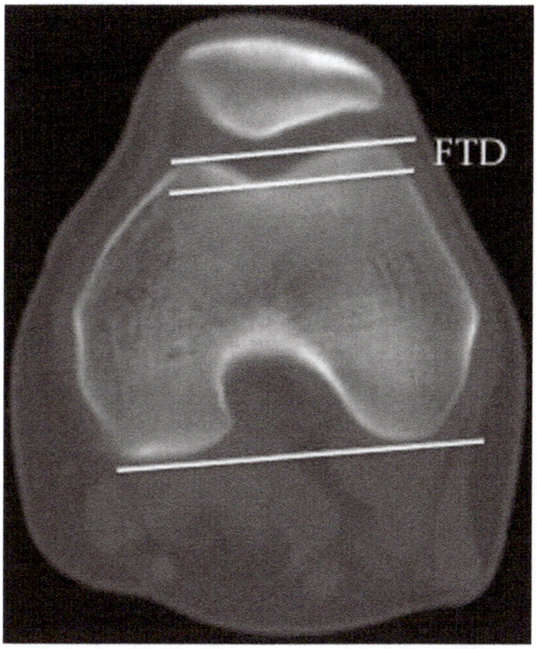

Fig. 6.4 The femoral trochlear depth (*FTD*). (1) Draw a line parallel to the femoral condyles. (2) Draw a *second* line parallel to the first at the deepest point of the femoral trochlea. (3) Draw a *third* line parallel to the highest points of the medial and lateral aspects of the trochlea. Measure between the *second* and *third* lines at the trochlear apex

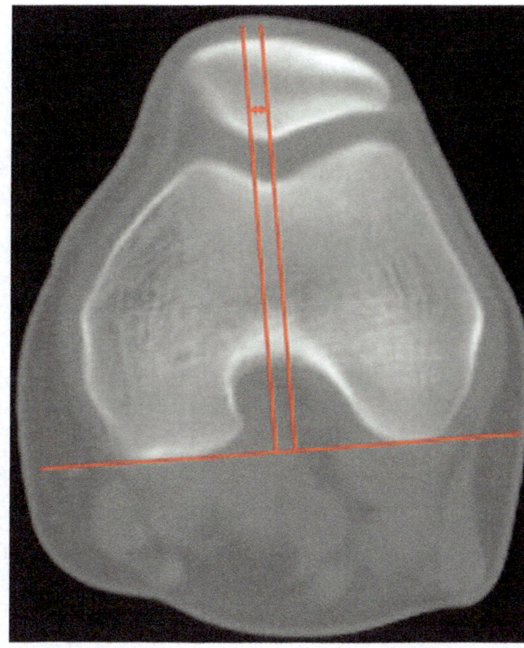

Fig. 6.5 To measure patellar displacement. (1) Draw a line parallel to the femoral condyles. (2) At 90° to the first line, draw one line intersecting the patellar apex. Then draw a line bisecting the trochlear apex. The difference between the second and third lines is the measurement of patellar subluxation. If negative, the patella is medial; if positive, the patella is lateral to the femoral trochlea

6.2.3 Evaluating the Measurements

The femoral trochlear angle or sulcus angle will normally decrease with increasing flexion. Normal femoral trochlear angle is <130°. Some studies have indicated a normal trochlear angle as high as 156°. The femoral trochlear angle and femoral trochlear depth will decrease with increasing flexion in the normal knee.

6.2.4 Significance of Measurements

Normal Values

Several studies have been performed to determine normal values in asymptomatic knees and results show a wide range with patellar displacement and patellar tilt being the most specific and sensitive indicators of patellar malalignment [13, 14]. There is no agreement on what constitutes normal anatomy of the patellofemoral joint. Each measurement has an associated cause or underlying anatomic abnormality. Shallow femoral trochlear depth and angle indicates trochlear dysplasia or sequelae of previous dislocation as well as predisposition to recurrent patellar dislocation.

An abnormal patellar tilt and angulation indicates a tight lateral retinaculum, deficient medial ligamentous support or malposition of the distal extensor mechanism (the tibial tubercle). Identification of the anatomic abnormality allows the orthopedist to select the appropriate treatment for the patient (Fig. 6.6).

The patellar displacement and patellar angulation will remain relatively constant throughout flexion in the normal knee but may increase with quadriceps contraction and decrease with progressive knee flexion. There are no absolute radiographic measurements that indicate surgical correction.

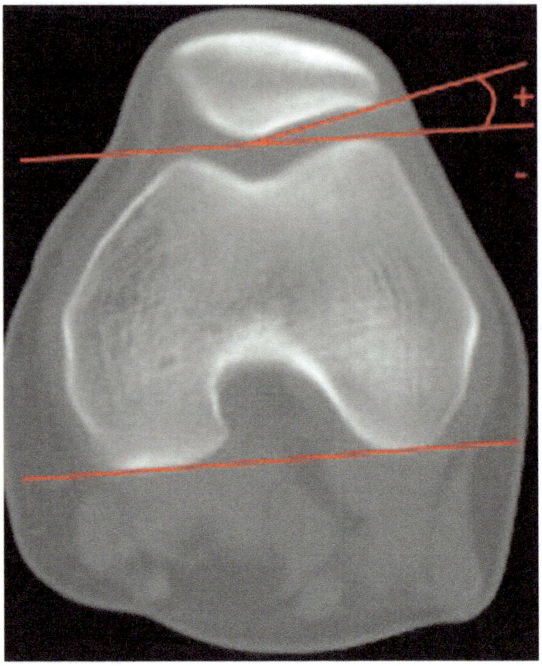

Fig. 6.6 Patellar tilt angle (PTA). (1) Draw the first line parallel to the posterior femoral condyles. (2) Draw the *second* line parallel to the first at the patellar apex. (3) The angle between the lateral patellar facet and trochlea

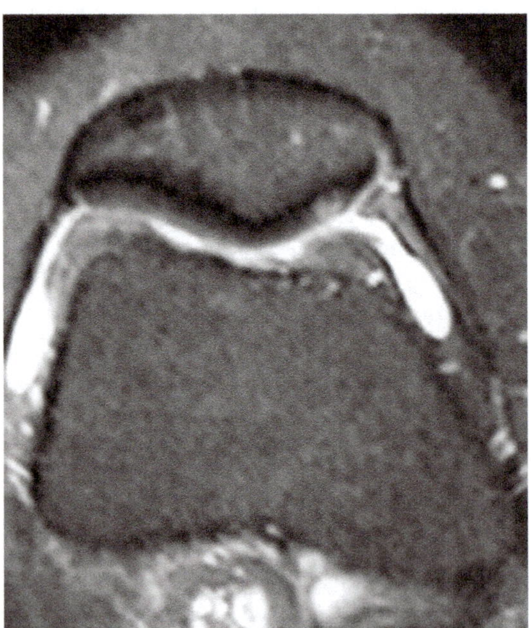

Fig. 6.7 Representative MRI sagittal T2-weighted images obtained in a subject with a 5.1 % increase in cartilage T2 joint effusion. The focal cartilage defect overlying medial facet of the patella. However, there was interval development of signal heterogeneity in the lateral facet of the patellar cartilage which is suggestive of early degeneration

6.3 Magnetic Resonance Imaging

MRI has been used as systematised approach to evaluate patellofemoral joint in complex cases and whenever there is a need specifically for soft tissue imaging [15, 16]. MRI is a common indication in young patients with knee pain. Many static and dynamic internal derangements of the patellofemoral joint in these patients lead to various secondary MRI findings. Intra-articular fractures, neoplasms and Hoffa's fat pad oedema causing maltracking may be identified [17].

It is the most effective noninvasive method currently available in the evaluation of articular cartilage injury. Many researchers agree that MRI is more useful for evaluating moderate to advanced patellar cartilage damage [18]. Patients with chondromalacia patellae have focal defects on T2-weighted images. Currently computer software for cartilage evaluation "CartiGram" improves MRI sensitivity in detecting soft but unbroken patellar articular cartilage [18, 19].

Kinematic MRI and CT studies exploit these dynamic patellofemoral relationships by imaging patients in varying degrees of knee flexion and extension [20]. However, these studies demand pre-performance optimisation, patient cooperation and significant interpreter skill. Although MRI is an increasingly reliable study its diagnostic accuracy is still not infalible [21].

6.4 Other Imaging Techniques

Arthrography is rarely necessary since MRI is available. Bone scintigraphy may be helpful in the evaluation of patients with resistant anterior knee pain that has not responded to conventional nonoperative treatment, mainly in cases with significant trauma to the patella, to rule out the possibility of major patellar disease [22, 23].

Ultrasound is an expertise hand-dependent imaging technique with a sensitivity rate of 92 %.

It is limited to peripatellar tissue evaluation [24].

Three-dimensional reconstruction (3D) using CT may be particularly helpful in difficult cases involving fracture, dysplasia or patellofemoral deformity.

References

1. Fulkerson JP. Patellofemoral pain disorders: evaluation and management. J Am Acad Orthop Surg. 1994; 2(2):124–32.
2. Madott SD, Burak JS, Math KR, Waz DM. Knee Imagin Techniques and Norml Anatomy In Insall & Scott Surgery of the Knee. W. Norman Scott. 5th ed. Churchill Livingstone; 2012;223:252.
3. Fulkerson JP. Disorders of the patellofemoral joint. Philadelphia: Lippincott Williams & Wilkins; 2004. p. 76–106.
4. Miranda VE, Muñoz-Ch S, Paolinelli GP, Astudillo AC, Wainer EM, Duboy UJ. Imaging studies of the patellofemoral joint: current reality. Rev Chile Radiol. 2010;16(3):101–15.
5. Endo Y, Shubin Stein BE, Potter HG. Radiologic assessment of patellofemoral pain in the athlete. Sports Health. 2011;3(2):195–210.
6. Mirza Tolouei F, Afshar A, Salarilak S, Sina S. CT patellar cortex tilt angle: a radiological method to measure patellar tilt. Iran J Radiol. 2005;3(1):17–21.
7. Smith TO, Davies L, Toms AP, Hing CB, Donell ST. The reliability and validity of radiological assessment for patellar instability. A systematic review and meta-analysis. Skeletal Radiol. 2011;40(4):399–414.
8. Biedert RM, Netzer P, Gal I, Sigg A, Tscholl PM. The lateral condyle index: a new index for assessing the length of the lateral articular trochlea as predisposing factor for patellar instability. Int Orthop. 2011;35: 1327–31.
9. Biedert RM, Albrecht S. The patellotrochlear index: a new index for assessing patellar height. Knee Surg Sports Traumatol Arthrosc. 2006;14(8):707–12.
10. Alemparte J, Ekdahl M, Burnier L, Hernández R, Cardemil A, Cielo R, Danilla S. Patellofemoral evaluation with radiographs and computed tomography scans in 60 knees of asymptomatic subjects. Arthroscopy. 2007;23(2):170–7.
11. Schoettle PB, Zanetti M, Seifert B, Pfirrmann CW, Fucentese SF, Romero J. The tibial tuberosity-trochlear groove distance; a comparative study between CT and MRI scanning. Knee. 2006;13(1):26–31.
12. Delgado-Martins AD, Rodríguez-Merchán EC, Ballesteros R, Luna JD. Reproducibility of patellofemoral CT scan measurements. Int Orthop. 2000;24:5–8. Springer-Verlag 2000.
13. Laprade J, Lee R. Real-time measurement of patellofemoral kinematics in asymptomatic subjects. Knee. 2005;12(1):63–72.
14. Nicolaas L, Tigchelaar S, Koeter S. Patellofemoral evaluation with magnetic resonance imaging in 51 knees of asymptomatic subjects. Knee Surg Sports Traumatol Arthrosc. 2011;19:1735–9.
15. Wittstein JR, Bartlett EC, Easterbrook J, Byrd JC. Magnetic resonance imaging evaluation of patellofemoral malalignment. Arthroscopy. 2006;22(6): 643–9.
16. Chhabra A, Subhawong TK, Carrino JA. A systematised MRI approach to evaluating the patellofemoral joint. Skeletal Radiol. 2011;40(4):375–87.
17. Jibri Z, Martin D, Mansour R, Kamath S. The association of infrapatellar fat pad edema with patellar maltracking: a case–control study. Skeletal Radiol. 2012;41(8):925–31.
18. Aquerreta-Beola JD. Diagnostic imaging of articular cartilage disease. Rev Orthop Traumatol. 2002;5:417–27.
19. Harris JD, Brophy RH, Jia G, Price B, Knopp M, Siston RA, Flanigan DC. Sensitivity of magnetic resonance imaging for detection of patellofemoral articular cartilage defects. Arthroscopy. 2012;28(11):1728–37.
20. Pan J, Pialat JB, Joseph T, Kuo D, Joseph GB, Nevitt MC, Link TM. Knee cartilage T2 characteristics and evolution in relation to morphologic abnormalities detected at 3-T MR imaging: a longitudinal study of the normal control cohort from the Osteoarthritis Initiative. Radiology. 2011;261(2):507–15.
21. LaPrade RF, Burnett 2nd QM, Veenstra MA, Hodgman CG. The prevalence of abnormal magnetic resonance imaging findings in asymptomatic knees. With correlation of magnetic resonance imaging to arthroscopic findings in symptomatic knees. Am J Sports Med. 1994;22(6):739–45.
22. Dye SF, Boll DA. Radionuclide imaging of the patellofemoral joint in young adults with anterior knee pain. Orthop Clin North Am. 1986;17(2):249–62.
23. Naslund JE, Odenbring S, Naslund U-B, Lundeberg T. Diffusely increased bone scintigraphic uptake in patellofemoral pain syndrome. Br J Sports Med. 2005;39:162–5.
24. Gemignani M, Busoni F, Tonerini M, Scaglione M. The patellar tendinopathy in athletes: a sonographic grading correlated to prognosis and therapy. Emerg Radiol. 2008;15(6):399–404.

Patellofemoral Evaluation: Do We Need an Objective Kinematic Approach?

7

Ana Leal, Rogério Pereira, Hélder Pereira, F. Samuel Silva, Paulo Flores, and João Espregueira-Mendes

Contents

7.1 The Multifactorial Issues on Patellofemoral Joint

The problematic of patellofemoral cases is set in several aspects. The anterior knee pain that is related to patellofemoral joint is the most complex and problematic pathological condition of the knee; on the other hand, it is among the less understood and one of the most neglected [1–6].

A. Leal
Clínica do Dragão – Espregueira Mendes Sports Centre – Porto FIFA Medical Centre of Excellence, F.C. Porto Dragão Stadium, 4350-415 Porto, Portugal

Mechanical Engineering Department, University of Minho, Campus de Azurém, 4800-058 Guimarães, Portugal

CT2M – Centre for Mechanical and Materials Technologies, Campus de Azurém, 4800-058 Guimarães, Portugal
e-mail: aleal@dem.uminho.pt

R. Pereira
Clínica do Dragão – Espregueira Mendes Sports Centre – Porto FIFA Medical Centre of Excellence, F.C. Porto Dragão Stadium, 4350-415 Porto, Portugal
e-mail: rp@espregueira.com

H. Pereira
Clínica do Dragão – Espregueira Mendes Sports Centre – Porto FIFA Medical Centre of Excellence, F.C. Porto Dragão Stadium, 4350-415 Porto, Portugal

3B's Research Group - Biomaterials, Biodegradables and Biomimetics, Headquarters of the European Institute of Excellence on Tissue Engineering and Regenerative Medicine, Univ. Minho, Minho, Portugal

ICVS/3B's – PT Government Associated Laboratory, Braga/Guimarães, Portugal

Orthopedic Department, Centro Hospitalar Póvoa de Varzim, Vila do Conde, Portugal
e-mail: heldermdpereira@gmail.com

F.S. Silva • P. Flores
Mechanical Engineering Department, University of Minho, Campus de Azurém, 4800-058 Guimarães, Portugal

CT2M – Centre for Mechanical and Materials Technologies, Campus de Azurém, 4800-058 Guimarães, Portugal
e-mail: fsamuel@dem.uminho.pt; pflores@dem.uminho.pt

J. Espregueira-Mendes (✉)
Clínica do Dragão – Espregueira-Mendes Sports Centre – FIFA Medical Centre of Excellence, F.C. Porto Stadium, 4350-415 Porto, Portugal

ICVS/3B's Research Group – Biomaterials, Biodegradables and Biomimetics, University of Minho, Headquarters of the European Institute of Excellence on Tissue Engineering and Regenerative Medicine, Caldas das Taipas, Guimarães, Portugal

Department of Orthopaedics, Minho University, Braga, Portugal
e-mail: jem@espregueira.com

A. Gobbi et al. (eds.), *The Patellofemoral Joint*,
DOI 10.1007/978-3-642-54965-6_7, © ISAKOS 2014

In addition, patellofemoral disorders have a high incidence, representing between 20 and 40 % of the knee problems [7] and affecting one of the most active population sectors – the young adults – with high intensity and a strong negative long-term social impact for health economy and life quality, especially considering its association with osteoarthritis [1, 2]. Actually, the correction of structural abnormalities of patellofemoral articulation, if performing a correct and meticulous diagnosis, may have a crucial role on prevention or delay of the development of patellofemoral arthrosis [8, 9], which represents unestimated costs for national health services worldwide.

Pathology related to the patellofemoral joint results from several anatomical and biomechanical factors related to osseous and soft-tissue abnormalities, such as [4, 5, 8, 10–18] (1) trochlear dysplasia, (2) quadriceps dysplasia, (3) excessive distance between the tibial tubercle and the trochlear groove, and (4) patella *alta*. These first four osseous abnormalities are directly linked to the four main factors for patellar instability, defined by Henri Dejour et al. (1994) [8]. Additionally, other situations can conduce to instability of the patellofemoral joint [4, 11, 14, 19–23]: (5) torn medial patellofemoral ligament (MPFL), (6) weakened *vastus medialis obliquus* (VMO), (7) the length of the patellar tendon, (8) excessive femoral anteversion, and (9) excessive tibial external rotation.

In sum, the involving problematic arises from several issues such as (1) high complexity of patellofemoral joint biomechanics and subsequently of its pathological cases [4–7, 24, 25]; (2) discrepancy in the definition of what is considered normal [1, 2, 8, 17, 20]; (3) ambiguous definition of patellofemoral disorders, as long as they share a common thread with and have impact on nearly all knee conditions [1–4, 6, 26]; (4) lack of correlation between symptoms and physical and radiological findings that often occurs [5, 6, 27]; and, additionally, (5) a multifactorial etiology that intensifies the misunderstanding of the physiopathology [17, 20], compromising (6) the diagnosis, which per se is actually not perfectly performed [8, 28–31]. The limitations on the etiopathogenic understanding interfere in the assessment and subsequent final diagnosis of the patellofemoral cases, hence limiting the choice of most adequate therapeutics [1, 28, 29].

7.2 Patellofemoral Joint Evaluation

Patellofemoral disorders are initially diagnosed on clinical history [10, 32]. There are important questions that must be asked and answered when evaluating a patient with anterior knee complaints, related to location, duration, timing of symptoms, past history, trauma history, occupation, and general and specific sports history, among other specifications that will contribute to analyze previous events or complications that could affect the patient [10, 33].

The subsequent step of evaluation is the most important task; however, the physical examination is often performed improperly, presenting an assortment of flaws [10, 28, 29, 31, 34, 35]. It consists of a physical assessment, in which manual strength is applied on the patella and whose validation is governed by the manifestation of pain by the patient, not showing accuracy, precision, or reproducibility of the method and being intrinsically dependent on the experience and skills of the examinant.

A wide range of clinical tests, scores, and indexes are available for patellofemoral joint evaluation [28, 29, 36, 37], from which can be highlighted specific indexes, as follows: Insall-Salvati [38], modified Insall-Salvati [39], Blackburne-Peel [40], Caton-Deschamps [41, 42], and Labelle-Laurin et al. [43], among others, established decades ago and most of them still currently applied. Nonetheless, their sensitivity and specificity, as well as reliability and validity, still remain unclear [10, 13, 28, 29], so, in general, the majority of these evaluation tools are more qualitatively than quantitatively useful. Furthermore, a usual poor interobserver reliability is known among clinicians and may be due to all of this differing examination methods worldwide [13, 29, 44]. Therefore, there is no supported accuracy and/or validity for the existent methods, and none is suitable for universal application.

The last diagnostic step, the imaging exam, allows the measurement and quantification of some indexes described above, in order to evaluate the influence of the instability factors [8, 14]. There is a wide variety of imaging techniques that can be applied on the systematic study of the patellofemoral articulation, each one of them having diverse advantages and disadvantages [45–47]. Standard radiographs, CT scan (computed tomography scan), and MRI (magnetic resonance imaging) are the most common applied techniques for the diagnosis by image [46]. However, only MRI allows for the simultaneous evaluation of all the structures that constitute the complexity of patellofemoral joint, distinguishing the different human tissues and, consequently, being the most complete imaging technique [48, 49]. Besides that, MRI presents excellent sensitivity and specificity when applied to MPFL lesions [48]. On the other hand, also CT scan is essential since it allows the quantification of small bone defects in a reliable, fast, simple, and easily reproducible way, which is crucial for surgical planning [45, 47].

In order to obtain a complete diagnosis, the aforementioned diagnostic phases should be taken as complementary procedures, never replacing each other [45]. However, it is difficult to establish a correlation between these phases, mainly because, due to physical space limitations and incompatibilities related to imaging devices, they cannot yet be simultaneously performed [1, 2, 46].

Facing the arguments presented, the majority of the cases are recurrently misdiagnosed [6, 50, 51]. As long as they cannot accurately measure the severity of the injuries, current evaluation tools also fail in predictive and indicative value of additional health complications and therapeutic or preventive strategies. Resultant diagnostic errors are serious and can even lead to unnecessary interventions [1, 5, 6]: (1) surgery instead of applying a conservative physical rehabilitation or opposing to the previous and (2) surgical procedures not addressing critical problems (e.g., MPFL insufficiency).

An appropriate diagnosis, applying dynamic, anatomical, and functional assessment and objective, reliable, and reproducible methodology, should be expected in the near future, in order to accomplish the most adequate treatment to the patient.

7.3 Anatomical and Functional Patellofemoral Joint Assessment

Several authors alert to the fact that it would be convenient to reach a better understanding of the kinematics of the patellofemoral joint and its stabilizing mechanisms, specially the contribution of the patellar ligamentous passive restraints [52]. Due to this fact, some studies have been addressing to the analysis of patellofemoral joint behavior, from anatomical to functional evaluation, in vitro [53–57] or in vivo [58, 59]. Nonetheless, most of these studies are focused on the joint kinematics thus presenting limitations on the dynamic articular evaluation and lacking on biomechanical and clinical important aspects, such as on the understanding of collateral effects of soft-tissue damage (such as on the *vastus medialis obliquus* or medial patellofemoral ligament) [60].

On the other hand, in the last decades, few patents have been applied describing medical devices for patellofemoral joint assessment, from which some examples should be taken into account; see Table 7.1. Generally, their development is based in mechanisms intended to more objectively evaluate the kinematics of this joint. In 1987, Kurt Groeben [61] published a patent supporting the invention of a device for detecting a patellar anomaly which permits quantitative measurement of the changes in position and inclination, as well as of the pressure sensitivity of the patella in a particularly simple manner. Five years later, *Medmetric Corporation* claims an invention [62], by D.M. Daniels and K.R. Watkins, of a device for the evaluation of the patellofemoral joint integrity that was able to position against the patella and measure linear displacement of the patella in response to a lateral force, which magnitude applied is correlated with the degree of leg flexion (relaxed muscles). Again, in 1998, *Medmetric Corporation* claims another invention [63], by K.R. Watkins and D.C. Fithian, of a device with exactly the same purpose and functions as the previous one, but with an upgrade, that is, a monitor which measures the displacement of the patella. Already in 2005, Castillo, Leitner, and Reese [64] describe a device for measuring a

Table 7.1 Most significant publications of mechanisms for patellofemoral joint assessment

Publication date	Publication title	Inventors	Publication type	Scheme/image	
12/11/1987	Device for detecting a patellar anomaly	Groeben, Kurt	*DE* 3615675 *A1*	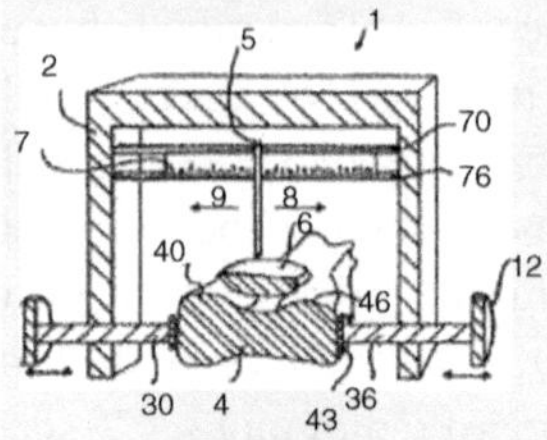	1
20/10/1992	Patella displacement measuring device	Dale M. Daniels, K. Richard Watkins (Medmetric Corporation)	*US* 5156163 *A*	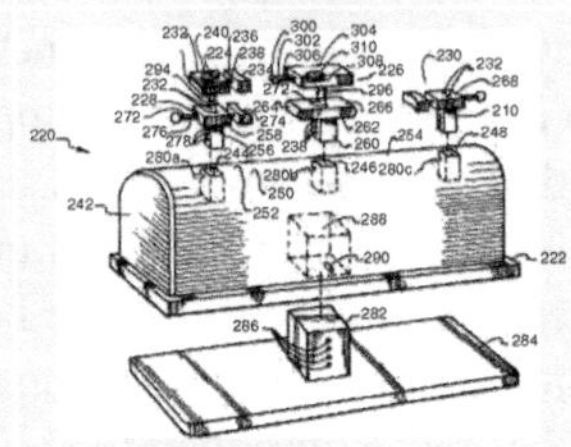	2
11/01/2000	Patella displacement tester	Donald C., Fithian, K., Richard Watkins (Medmetric Corporation)	*US* 6013039 *A*	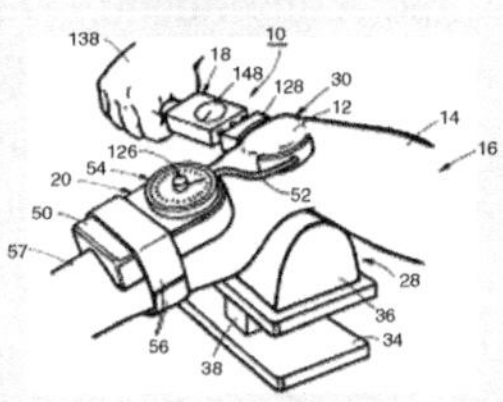	3
09/06/2005	Measuring device	James Castillo, Bernhard Leitner, Isaac Reese	*US* 2005/0124919 *A1*	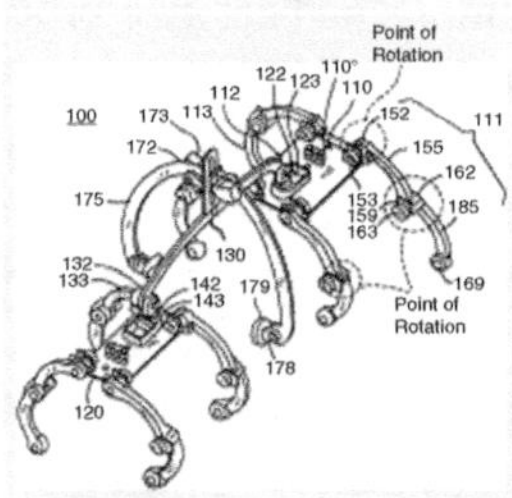	4
2006	Concurrent criterion-related validity and reliability of a clinical device used to assess lateral patellar displacement	Ota S. et al.	*Journal Of Orthopaedic and Sports Physical Therapy*	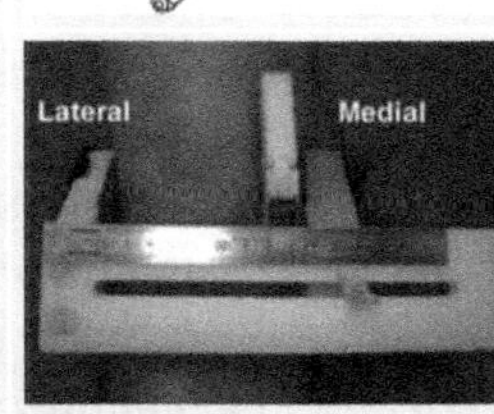	5
14/08/2008 14/08/2008 12/03/2009	System and method for diagnosing and treating patellar maltracking and malalignment	Li-Qun Zhang (London Metropolitan University)	WO 2008/097752 *A2* *US* 2008/0194997 *A1* WO 2008/097752 *A3*	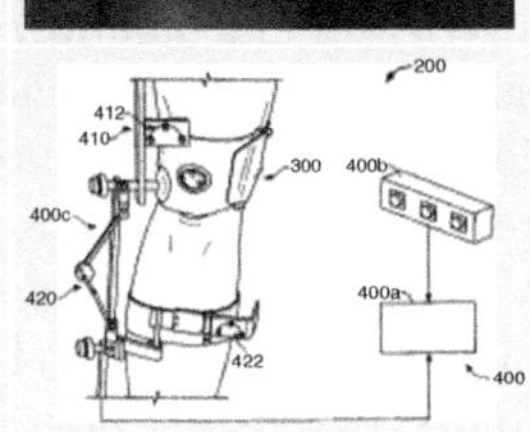	6
03/03/2010 04/03/2010 20/07/2011	Patella position measuring device	Kevin Daniel Campbell-Karn (London Metropolitan University)	*GB* 2463061 *A* *WO* 2010/023478 *A1* *EP* 2344037 *A1*	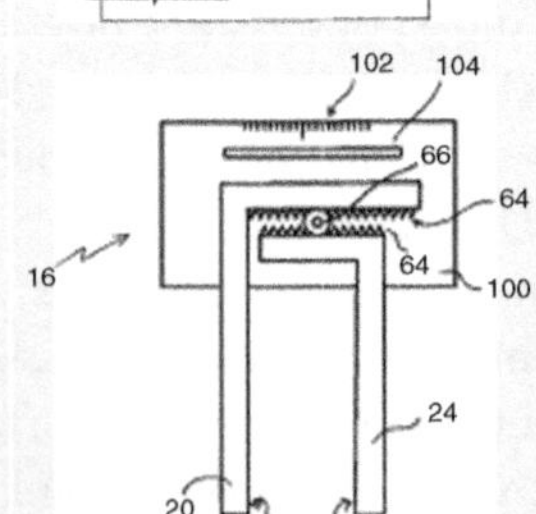	7

body part of a patient, which contains an embodiment including a reference component, with reference point, capable of being secured to a body part, and an articulated measurement arm movably coupled to the reference component that includes multiple sections, each section having a measurement point, associated with a plurality of sensors, capable of providing a plurality of data sufficient to allow determination of the position of each measurement point relative to the reference point. In the scientific context, in 2006, Ota et al. [65] have performed a study to assess the concurrent criterion-related validity and reliability of a patellofemoral arthrometer, used as an instrument to measure the patellar lateral displacement, achieving, respectively, good and excellent ICCs of agreement level (between MRI and device measures) and intra-/inter-tester reliability. Returning to patents, between 2008 and 2009, the Rehabilitation Institute of Chicago, by Li-Qun Zhang, claims the invention [66] of a system and method for in vivo, noninvasive diagnosis of patellar and scapular malalignment and maltracking that includes a patella-engaging apparatus, which has a member with a custom-contoured posterior surface that is configured to engage the anterior surface of a subject's patella. Through a plurality of markers, an analysis system optically tracks the member motion. Just 2 years later, London Metropolitan University claims an invention [67, 68], by K. D. Campbell-Karn, which respects to a device for measuring lateral position of the patella relative to the femur of an individual. It is constituted by two calipers, each one of them with a pair of jaws capable of movement toward and away from each other, in order to contact the outer edges of the patella or the femur, respectively, and includes measuring means to measure the lateral displacement of the first pair of jaws to the second one.

Nonetheless, these proposed solutions present several limitations. First of all, ferromagnetic materials are used in these devices, which make them unsafe and incompatible with MRI or CT scan devices, interfering with the image processing and representing a risk for patient safety. Therefore, its use with MRI and CT scan is impracticable and does not allow the accurate examination of soft-tissue structures and precise articular position. Moreover, on the scope of patellofemoral articular movements/kinematics, their application would be limited, once they just intend to determine patellar lateral position/inclination (1,4, 7) or motion/displacement (2,3, 5, 6), although few exceptionally consider loading (2,3). Actually, they were not widely used in the clinical practice nor scientific research, or even did not reach the market, reflecting their limitative application.

Further research and ongoing work has been focused on the development of an accurate and comprehensive assessment of the patellofemoral joint, exploring its whole kinematics; quantifying biomechanical interactions, alterations, and consequences; as well as characterizing clinical susceptibility to further damage. Through a mechanical system (ideally inert to magnetic fields), one should be able to apply a controlled loading force acting on the patella in order to move it along different axes and plans; see Fig. 7.1. The displacement caused in the joint elements would be subsequently detected and precisely measured in the images reproduced by the MRI, CT scan, or any other imaging equipment used. Furthermore, such tests could enable to increase our dynamic understanding of the patellofemoral joint under pathologic conditions (correlated to patient complaints) and to better understand its normal kinematics; see Fig. 7.2.

Such application could have the potential to find specific parameters (load, pressure, dislocation, and so on), and such data, translated in valuable information, can be useful for innumerous findings as follows:

1. Verify, by comparison between normal and pathological condition, if some of these parameters are embodied as diagnostic indicators.
2. Detect standard values for these parameters, to try to define a boundary between normal and pathological condition.
3. Identify values which can have clinical correlation to be considered factors of pathological risk (e.g., for the case of subluxation).
4. Establish correlations between patients' complaints and imaging findings, to try to find sustainable explanations for current misunderstandings.

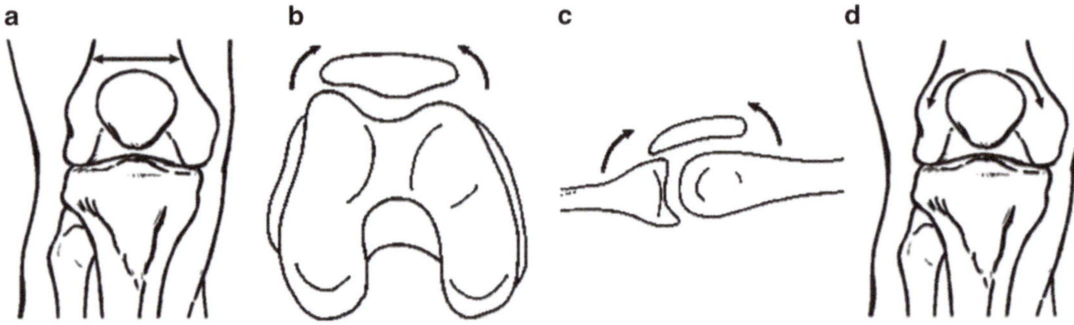

Fig. 7.1 Patellofemoral joint movements along different axes and planes: (**a**) medial-lateral translation; (**b**) internal-external tilt; (**c**) anterior-posterior tilt; (**d**) internal-external rotation

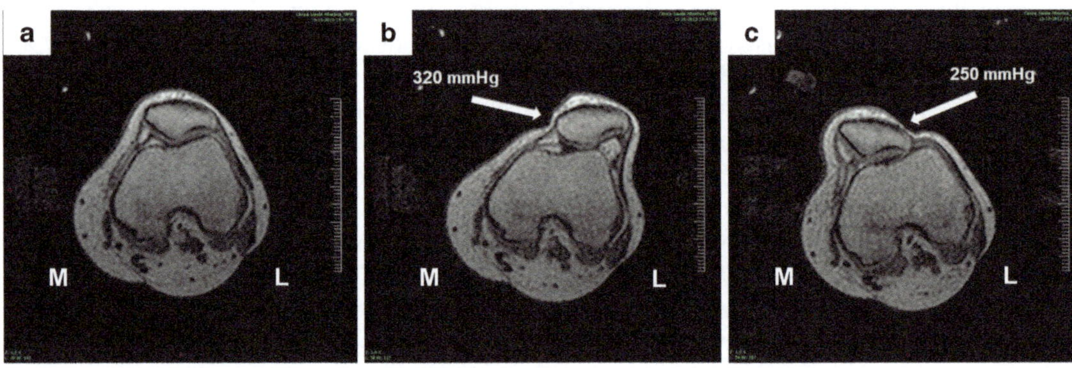

Fig. 7.2 MRI images of the patellofemoral joint on axial view: (**a**) patellar neutral position with no loading; (**b**) patellar lateral translation under loading; (**c**) patellar medial translation under loading *M* medial and *L* lateral, sides of the leg.

By using and identifying them as criteria, the results obtained may constitute important clinical information for treatment, rehabilitation, and prevention purposes.

7.4 Final Remarks

Owing to its complexity, many researchers have been studying the problematic of the patellofemoral cases for a better understanding and evaluation, aiming to demystify and to globalize concepts and methodologies. Nonetheless, in spite of all the efforts, no published work has presented yet a complete solution for the specific purpose within this discussion.

The approach under development consists not just in the development of a new methodology but also in a subsequent definition of diagnosis criteria. The methodology will ideally perform an objective measurement of the patellofemoral instabilities, through a kinematic evaluation of this joint simultaneously with imaging support. Thereafter, diagnosis criteria could be identified and will be possible to standardize, as result of a battery of clinical tests performed worldwide. Beyond the paramount importance, these are completely innovative aspects; both device/methodology and standard criteria are not yet available nowadays for patellofemoral joint evaluation. New paradigms may be established as conceptual framework guiding further development of research and accounting for a great scientific breakthrough.

References

1. Sanchis-Alfonso V. Anterior knee pain and patellar instability. London: Springer; 2006.
2. Grelsamer RP. Patellar nomenclature: the tower of Babel revisited. Clin Orthop Relat Res. 2005;436: 60–5.

3. Arendt EA, Dejour D. Patella instability: building bridges across the ocean a historic review. Knee Surg Sports Traumatol Arthrosc. 2012;21(2):279–93.
4. Merchant AC. Classification of patellofemoral disorders. Arthroscopy. 1988;4(4):235–40.
5. Fithian DC, Neyret P, Servien E. Patellar instability: the Lyon experience. Curr Orthop Pract. 2008;19(3): 328–38.
6. Zaffagnini S, Dejour D, Arendt EA. Patellofemoral pain, instability, and arthritis: clinical presentation. Springer: Imaging and Treatment; 2010.
7. Zaffagnini S, Giordano G, Bruni D, Maria G, Muccioli M, Marcacci M. Pathophysiology of lateral patellar dislocation. In: Zaffagnini S, et al, editors. Patellofemoral Pain, Instability, and Arthritis: Clinical Presentation. Springer: Imaging and Treatment. 2010;17–27.
8. Dejour H, Walch G, Nove-Josserand L, Guier C. Factors of patellar instability: an anatomic radiographic study. Knee Surg Sports Traumatol Arthrosc. 1994;2(1):19–26.
9. Arendt EA. Anatomy and malalignment of the patellofemoral joint: its relation to patellofemoral arthrosis: patellofemoral arthritis and arthroplasty, Clin Orthop Relat Res. 2005;436:71–5.
10. Kantaras AT, Selby J, Johnson DL. History and physical examination of the patellofemoral joint with patellar instability. Oper Tech Sports Med. 2001;9(3): 129–33.
11. Ficat P, Hungerford D. Disorders of the patellofemoral joint. Williams & Wilkins: Baltimore; 1977.
12. Lancourt JE, Cristini JA. Patella alta and patella infera. Their etiological role in patellar dislocation, chondromalacia, and apophysitis of the tibial tubercle. J Bone Joint Surg Am. 1975;57(8):1112–5.
13. Seil R, Müller B, Georg T, Kohn D, Rupp S. Reliability and interobserver variability in radiological patellar height ratios. Knee Surg Sports Traumatol Arthrosc. 2000;8(4):231–6.
14. Diederichs G, Issever AS, Scheffler S. MR imaging of patellar instability: injury patterns and assessment of risk factors. Radiographics. 2010;30(4):961–81.
15. Lippac S, Dejour D, Els M, Dornac D, Ring C, Lippacher S, Elsharkawi M, Dornacher D, Dreyhaupt J, Reichel H, Nelitz M. Observer agreement on the Dejour trochlear dysplasia classification: a comparison of true lateral radiographs and axial magnetic resonance images. Am J Sports Med. 2012;40(4):837–43.
16. Nelitz M, Lippacher S, Reichel H, Dornacher D. Evaluation of trochlear dysplasia using MRI: correlation between the classification system of Dejour and objective parameters of trochlear dysplasia. Knee Surg Sports Traumatol Arthrosc. 2014;22(1):120–7.
17. Panni AS, Cerciello S, Maffulli N, Di Cesare M, Servien E, Neyret P. Patellar shape can be a predisposing factor in patellar instability. Knee Surg Sports Traumatol Arthrosc. 2011;19(4):663–70.
18. Fucentese SSF, von Roll A, Koch PPP, Epari DDR, Fuchs B, Schottle PB, von Roll A. The patella morphology in trochlear dysplasia–a comparative MRI study. Knee. 2006;13(2):145–50.
19. Arendt EA. MPFL reconstruction for PF instability. The soft (tissue) approach. Orthop Traumatol Surg Res. 2009;95(8 Suppl 1):S97–100.
20. Neyret P, Robinson AHN, Le Coultre B, Lapra C, Chambat P. Patellar tendon length–the factor in patellar instability? Knee. 2002;9(1):3–6.
21. Escala JS, Mellado JM, Olona M, Giné J, Saurí A, Neyret P. Objective patellar instability: MR-based quantitative assessment of potentially associated anatomical features. Knee Surg Sports Traumatol Arthrosc. 2006;14(3):264–72.
22. Thaunat M, Erasmus PJ. Recurrent patellar dislocation after medial patellofemoral ligament reconstruction. Knee Surg Sports Traumatol Arthrosc. 2008; 16:40–3.
23. Philippot R, Chouteau J, Wegrzyn J, Testa R, Fessy MH, Moyen B. Medial patellofemoral ligament anatomy: implications for its surgical reconstruction. Knee Surg Sports Traumatol Arthrosc. 2009;17(5):475–9.
24. Espregueira-Mendes JDC, Pessoa P. O Joelho. Porto: Lidel; 2006.
25. Completo A, Fonseca F. Fundamentos de Biomecânica – Músculo, Esquelética e Ortopédica. Portugal: Porto; 2011.
26. Arendt EA, Fithian DC, Cohen E. Current concepts of lateral patella dislocation. Clin Sports Med. 2002; 21(3):499–519.
27. Fucentese S, von Roll A, Koch P, Epari D. The patella morphology in trochlear dysplasia – a comparative MRI study. Knee. 2006;13(2):145–50.
28. Smith TO, Davies L, O'Driscoll ML, Donell ST. An evaluation of the clinical tests and outcome measures used to assess patellar instability. Knee. 2008;15(4): 255–62.
29. Smith TO, Clark A, Neda S, Arendt EA, Post WR, Grelsamer RP, Dejour D, Almqvist KF, Donell ST, Fields A. The intra- and inter-observer reliability of the physical examination methods used to assess patients with patellofemoral joint instability. Knee. 2012;19(4):404–10.
30. Leal A, Pereira H, Silva FS, Flores P, Espregueira-Mendes J. Clinical diagnosis of patellofemoral disorders: the role of bioengineering. In: Proceedings of the 3rd Portuguese Bioengineering meeting. IEEE; Braga. 2013.
31. Leal A, Pereira H, Silva FS, Flores P, Espregueira-Mendes J. On the development of advanced methodologies to assist on the diagnosis of human articulations pathologies: a biomechanical approach. In: 5° Congresso Nacional de Biomecânica. 2013.
32. Fulkerson JP, Hungerford DS. Biomechanics of the patellofemoral joint. Disord Patellofemoral Joint. 1990;4:24–42.
33. Halbrecht J. Patellar problems in athletes. In: Johnson DH, Pedowitz RA, editors. Practical orthopaedic sports medicine & arthroscopy. 2nd ed. Philadelphia: Lippincott Williams & Wilkins; 2007. p. 680–709.
34. Smith TO, Hunt NJ, Donell ST. The reliability and validity of the Q-angle: a systematic review. Knee Surg Sports Traumatol Arthrosc. 2008;16(12): 1068–79.

35. Johnson DH, Pedowitz RA. Practical orthopaedic sports medicine & arthroscopy patellar problems in athletes, vol. 1. 1st ed. Philadelphia: Lippincott Williams & Wilkins; 2007. p. 1–50.
36. Post WR. Clinical evaluation of patients with patellofemoral disorders. Arthroscopy. 1999;15(8):841–51.
37. Fredericson M, Yoon K. Physical examination and patellofemoral pain syndrome. Am J Phys Med Rehabil. 2006;85(3):234–43.
38. Insall J, Salvati E. Patella position in the normal knee joint. Radiology. 1971;101:101–4.
39. Grelsamer RP, Meadows S. The modified Insall-Salvati ratio for assessment of patellar height. Clin Orthop Relat Res. 1992;282:170–6.
40. Blackburne J, Peel T. A new method of measuring patellar height. J Bone Joint Surg Br. 1977;59-B(2):241–2.
41. Caton J, Deschamps G, Chambat P, Lerat JL, Dejour H. Patella infera. Apropos of 128 cases. Rev Chir Orthop Reparatrice Appar Mot. 1982;68(5):317–25.
42. Caton J. Method of measuring the height of the patella. Acta Orthop Belg. 1989;55(3):385–6.
43. Laurin CA, Dussault R, Levesque HP. The tangential x-ray investigation of the patellofemoral joint: x-ray technique, diagnostic criteria and their interpretation. Clin Orthop Relat Res. 1979;144:16–26.
44. Smith TO, Cogan A, Patel S, Shakokani M, Toms AP, Donell ST. The intra- and inter-rater reliability of X-ray radiological measurements for patellar instability. Knee. 2013;20(2):133–8.
45. Sanchis-alfonso V, Montesinos-Berry E, Serrano A, Martínez-Sanjuan V. Evaluation of the patient with anterior knee pain and patellar instability. In: Anterior knee pain and patellar instability. London: Springer; 2011. p. 105–23.
46. Tavernier T, Dejour D. Imagerie du genou: quel examen choisir? J Radiol. 2001;1:387–405.
47. Walch G, Dejour H. Radiology in femoro-patellar pathology. Acta Orthop Belg. 1989;55(3):371–80.
48. Sanders TG, Morrison WB, Singleton BA, Miller MD, Cornum KG. Medial patellofemoral ligament injury following acute transient dislocation of the patella: MR findings with surgical correlation in 14 patients. J Comput Assist Tomogr. 2001;25(6):957–62.
49. Stoller DW. MR imaging of the knee. In: Magnetic resonance imaging in orthopaedics and sports medicine. Philadelphia: Lippincott Williams & Wilkins; 2007.
50. Dejour H, Walch G. Pathologie fémoro-patellaire. In: 6ème Journée Lyonnaise de Chirurgie du Genou. Sauramps Medical; Lyon. 1987.
51. Dejour D, Bonnin M, Servien E, Fayard JM, Al E. La Patella. In: 15éme Journées Lyonnaises de Chirurgie du Genou. Sauramps Medical; Lyon. 2012
52. Zaffagnini S, Colle F, Lopomo N, Sharma B, Bignozzi S, Dejour D, Marcacci M. The influence of medial patellofemoral ligament on patellofemoral joint kinematics and patellar stability. Knee Surg Sports Traumatol Arthrosc. 2013;21(9):2164–71.
53. Amis AA, Senavongse W, Bull AMJ. Patellofemoral kinematics during knee flexion-extension: an in vitro study. J Orthop Res. 2006;24(12):2201–11.
54. Belvedere C, Catani F, Ensini A, Moctezuma de la Barrera JL, Leardini A. Patellar tracking during total knee arthroplasty: an in vitro feasibility study. Knee Surg Sports Traumatol Arthrosc. 2007;15(8):985–93.
55. Ostermeier S, Holst M, Bohnsack M, Hurschler C, Stukenborg-Colsman C, Wirth C-J. In vitro measurement of patellar kinematics following reconstruction of the medial patellofemoral ligament. Knee Surg Sports Traumatol Arthrosc. 2007;15(3):276–85.
56. Philippot R, Boyer B, Testa R, Farizon F, Moyen B. The role of the medial ligamentous structures on patellar tracking during knee flexion. Knee Surg Sports Traumatol Arthrosc. 2012;20(2):331–6.
57. Nomura E, Horiuchi Y, Kihara M. Medial patellofemoral ligament restraint in lateral patellar translation and reconstruction. Knee. 2000;7(2):121–7.
58. Yamada Y, Toritsuka Y, Horibe S, Sugamoto K, Yoshikawa H, Shino K. In vivo movement analysis of the patella using a three-dimensional computer model. J Bone Joint Surg Br. 2007;89(6):752–60.
59. Nha KW, Papannagari R, Gill TJ, Van de Velde SK, Freiberg AA, Rubash HE, Li G. In vivo patellar tracking: clinical motions and patellofemoral indices. J Orthop Res. 2008;26(8):1067–74.
60. Diederichs G, Köhlitz T, Kornaropoulos E, Heller MO, Vollnberg B, Scheffler S. Magnetic resonance imaging analysis of rotational alignment in patients with patellar dislocations. Am J Sports Med. 2013; 41(1):51–7.
61. Groeben K. Device for detecting a patellar anomaly. DE 36156751987, 12/11/1987 Germany.
62. Daniels DM, Watkins KR. Patella displacement measuring device. US 5156163 A1992, 20/10/1992 US.
63. Fithian DC, Watkins KR. Patella displacement tester. US 6013039 A2000, 11/01/2000 US.
64. Castillo J, Leitner B, Reese I. Measuring device. US 2005/124919 A12005, 09/06/2005 US.
65. Ota S, Ward SR, Chen Y, Tsai Y, Powers CM. Concurrent criterion-related validity and reliability of a clinical device used to assess lateral patellar displacement. J Orthop Sports Phys Ther. 2006;36(9):645–52.
66. Zhang L-Q. System and method for diagnosing and treating patellar maltracking and malalignment. WO 2008/097752 A22008, 14/08/2008 US.
67. Campbell-Karn KD. Patella position measuring device. WO 2010/023478 A12010, 03/03/2010 GB.
68. Campbell-Karn KD. Patella position measuring device. EP 2344037 A12011, 03/03/2010 GB.

Gait Analysis in Patellofemoral Disorders

8

Sofia Xergia, Evangelos Pappas, and Anastasios D. Georgoulis

Contents

S. Xergia
Department of Orthopaedic Surgery, Orthopaedic Sports Medicine Center of Ioannina, University of Ioannina, Ioannina, Greece

E. Pappas
Department of Physiotherapy, Faculty of Health Sciences, University of Sydney, Lidcombe, Australia

Department of Orthopaedic Surgery, Orthopaedic Sports Medicine Center of Ioannina, University of Ioannina, Ioannina, Greece

A.D. Georgoulis (✉)
2nd Orthopaedic Clinic & Orthopaedic Sports Medicine Centre, University of Ioannina, Ioannina, Greece
e-mail: georgoulis@osmci.gr

A. Gobbi et al. (eds.), *The Patellofemoral Joint*,
DOI 10.1007/978-3-642-54965-6_8, © ISAKOS 2014

8.1 Introduction

Patellofemoral pain syndrome (PFPS) is one of the most common diagnoses among young, physically active populations, affecting 1 in 4 athletes with more than 70 % being between 16 and 25 years old [1–4]. Many researchers have tried to identify predisposing factors for this ailment. The general consensus is that the etiology of patellofemoral pain remains enigmatic and multifactorial and may include intrinsic (skeletal alignment, soft tissue imbalance, biomechanical influences) and extrinsic (environment, equipment) risk factors [2, 5, 6].

Lateral tracking of the patella has been hypothesized to contribute to the development of patellofemoral pain [7–9]. In healthy individuals, during knee flexion the contact area between the femur and the patella moves along both the trochlear groove and the articular surface of the patella. In early flexion, the contact is at the distal and lateral edge of the patellar articular surface, and during flexion it moves proximally. The lateral facet of the trochlear groove counteracts subluxation of the patella during extension and early flexion. The patella engages the groove at 20° of flexion. Thus, between 20° of flexion and full extension, the function of the patella depends on factors other than the bony architecture of the patellofemoral joint [2]. This relationship may be disturbed due to a variety of anatomical factors such as dysplasia of the femoral condyles, patella alta, genu valgum, increased femoral anteversion, lower extremity

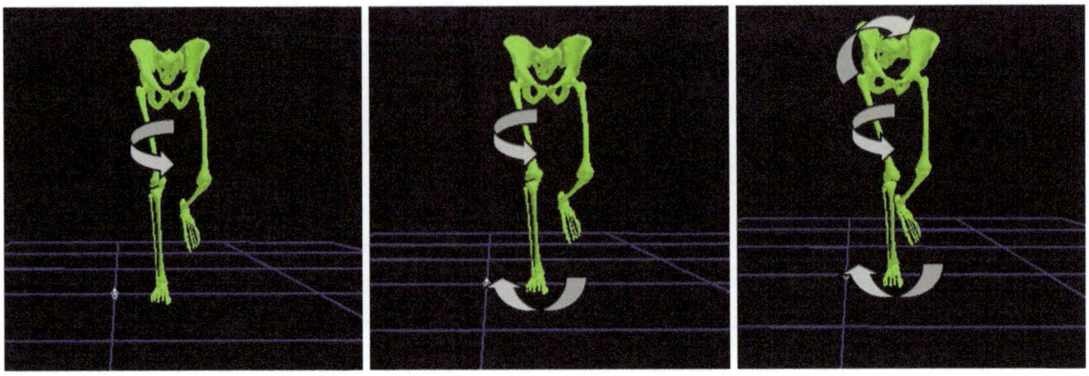

Fig. 8.1 3D skeletal model of an adult male performing a single-limb hop task. A combination of rearfoot eversion, knee valgus, hip adduction, and internal rotation is observed

malalignment with an increased Q angle, generalized hyperlaxity, and a laterally positioned tibial tuberosity [10]. For example, patients with patella alta were shown to have greater lateral displacement and tilt as well as decreased contact area when compared to a control group [11]. All these factors should be considered in the clinical investigation for the etiology of PFPS. Several studies have proposed an anatomical-biomechanica explanation for the etiology of PFPS. External tibial rotation, rearfoot motion, internal femoral rotation (Fig. 8.1), and imbalance of the femoral muscles have been suggested to contribute to PFPS [12–21]. With the use of kinematic, kinetic, and electromyographical (EMG) equipment, investigators have attempted to find differences between healthy subjects and those with PFPS that may shed light in the etiology of the disease. However, these findings should be interpreted with caution as it is difficult to know if they pre-existed the disease or developed as compensatory mechanisms. The following sections summarize the findings of the studies that investigated biomechanical factors around the knee, distal to the knee, and proximal to the knee. Finally, we summarize prospective studies that have a higher potential to identify predisposing factors.

8.2 Factors Around the Knee Joint

As the quadriceps exerts direct forces on the patella, it has been suggested that imbalances between the medially and laterally directed muscular forces applied to the patella by the vastus medialis and the vastus lateralis muscles, respectively, can contribute to PFPS. No consensus has been achieved regarding the relation between vastus muscles activity and PFPS or abnormal movement of the patella [12–17, 19–21]. Interestingly, Powers et al. [13] found that patients with patellofemoral pain exhibit decreased activity of all vastus muscles during level walking and ramp ambulation and proposed the "quadriceps femoris muscle avoidance gait." Salsich et al. [22] examined ascending and descending in patients with patellofemoral pain. They identified reduced cadence while descending and decreased peak knee extensor moment during both ascending and descending, which is also suggestive of quadriceps avoidance. Quadriceps avoidance has also been identified in stair climbing [1, 22]. It has been advocated that it could be due to a forward trunk lean, which brings the center of pressure closer to the knee joint. In addition, no consensus exists regarding changes in knee flexion in patients with patellofemoral pain [1, 5, 6, 22–25], while it has been shown that joint pressure is greater while descending than while ascending, which can explain the greater amount of pain reported by the patients during this activity [1].

8.3 Factors Distal to the Knee

Rearfoot eversion is coupled to medial rotation of the tibia during the weight acceptance portion of the stance phase in running [26, 27]. In addition

Nawoczenski et al. [28] studied the coupling pattern between tibial medial and lateral rotation and calcaneal eversion and inversion throughout the stance phase in healthy individuals during treadmill running. They concluded that individuals with high arched feet showed magnitudes of rotations favoring tibial medial and lateral rotation over calcaneal eversion and inversion. In lieu of these findings, Dierks et al. [29] examined the relationship between arch structure and knee kinematics in runners after a prolonged run. However, no difference in arch height was noted between the runners with patellofemoral pain and the control group, while no association was observed between arch height and peak knee adduction angle during running.

Grenholm et al. [30] studied stair descending in women with patellofemoral joint pain and showed greater plantar flexion in the swing leg in preparation for foot placement, which was interpreted as a compensatory mechanism. Recently, it has been experimentally demonstrated that patellofemoral stress increases with increasing shoe heel height [31]. Clinicians should consider educating their female patients with PFPS on reducing the time spent in high-heel shoes. In a systematic review by Barton et al. [32], it was concluded that both delayed timing and excessive rearfoot eversion motion are linked to PFPS.

8.4 Factors Proximal to the Knee

Several scientists have also studied the other end of kinetic chain and the relationship between patella disorders and hip and pelvis motion. Lee et al. [33] suggested that with internal femoral rotation, the lateral articular surface of the trochlea impinges upon the lateral articular facets of the retropatellar surface and thus pushes the patella medially. In a previous study Lee et al. [34] demonstrated the relationship that exists between femoral rotation and patellofemoral pressure. However, significant increases in patellofemoral pressure were noted in femoral rotation greater than 20°. Therefore it is likely that there is an association between patellofemoral joint disorders and alterations of the biomechanics of the hip joint.

Mascal et al. [35] presented two patients with patellofemoral pain who were treated with an exercise program targeting the pelvis, hip, and trunk musculature. Before treatment, they both exhibited excessive hip adduction during weight acceptance, excessive internal hip rotation in early midstance, and notable contralateral pelvic drop during midstance while walking. Hip adduction, internal rotation (in the stance limb), and contralateral pelvic drop were noted in the step-down task as well. All these factors, as well as symptoms, were improved after treatment focusing on recruitment and endurance training of the hip pelvis and trunk musculature. Weakness in hip abduction and external rotation has been identified in females with patellofemoral pain using isometric strength measurement [36]. Willson and Davis [37] showed that in women with patellofemoral pain during single-leg squat, there is medial displacement of the knee which is associated with increased hip adduction and knee external rotation. Souza et al. [38] also showed that there is increased internal femoral rotation in the same population during running, drop jump, and step-down movements. They also noted that this population is characterized by decreased hip muscle strength. However, they recorded increased gluteus maximus muscular activity. This was interpreted as recruitment of a weakened muscle perhaps in an effort to stabilize the hip joint.

Powers et al. [39] performed dynamic imaging of the patellofemoral joint using MRI in six female patients with lateral subluxation of the patella during knee extension (non-weight-bearing condition) and squatting (weight-bearing condition). Abnormal patellofemoral kinematic patterns were noted during terminal knee extension in both tasks. In addition, lateral patellar displacement was more pronounced during non-weight-bearing knee extension as compared to weight-bearing knee extension. This study demonstrated that in the non-weight-bearing task, the patella rotates on the femur, while in the weight-bearing task, the femur rotates underneath the patella, which emphasizes the role of femoral rotation on patellar malalignment.

Dierks et al. [29] examined the relationship between hip strength and hip kinematics in runners with PFPS before and after a prolonged run. They showed that there is a decrease in hip abductor and external rotator strength at the end of the run. In addition, this decrease in hip abductor strength is associated with an increase in hip adduction angle. Salsich et al. [22] showed that there are alterations in the activity of hip extensors in patients with PFPS during stair ambulation, which according to the authors seems to be a compensatory mechanism. Finally, it is important to note that there is a correlation between the biomechanics of the three large joints of the lower extremity in patients with PFPS. In a recent study, Barton et al. [40] found that tibial internal rotation and hip adduction are both linked to rearfoot eversion.

In summary, abnormal hip biomechanics are encountered in patients with patellofemoral pain. Specifically, increased hip adduction and internal rotation have been observed during walking, running, drop jump, and step-down. In addition, muscle strength measurements have demonstrated weak hip abductors and external rotators.

8.5 Prospective Studies

Prospective studies require baseline testing of a large number of healthy subjects and following them prospectively to identify those who develop PFPS. Thus, they are difficult to conduct as they require large financial and personnel resources. A meta-analysis of these studies was recently published [41] and identified quadriceps weakness as a predictor of PFPS. Other factors that were identified by single studies as predictors were weakness of the knee flexors and hip adductors, flexibility of lower extremity musculature, genu varum alignment, navicular drop, and knee valgus angle during a landing task. A more recent study [42] screened 400 healthy female runners at baseline and followed them for 2 years. It was found that athletes who went on to develop PFPS exhibited greater hip adduction than those who did not develop PFPS, thus confirming what previous case control studies have suggested.

8.6 Conclusion and Future Directions

The proper treatment of PFPS either conservative or surgical requires the knowledge and understanding of static and dynamic factors that contribute to the stability of patellofemoral joint. Kinetic, kinematic, and EMG studies have already been performed and specific mechanisms have been reported. Specifically, "quadriceps femoris muscle avoidance" has been described in patients with patellofemoral pain during level walking, ramp ambulation, and stair climbing. It has been advocated that it could be due to a forward trunk lean, which brings the center of pressure closer to the knee joint. Another important finding of these studies is the involvement of the hip joint. Specifically, increased hip adduction and internal rotation have been observed in patients with patellofemoral pain during walking, running, drop jump, and step-down, while muscle strength measurements have demonstrated weak hip abductors and external rotators.

However the field of biomechanical evaluation in patellofemoral disorders has not been fully explored. For instance, no consensus has been achieved regarding the relation between vastus muscles activity and the onset of pain or the abnormal movement of the patella. Thus, further studies are required to recognize the changes that exist in patellofemoral pain and patella instability and to discover whether these alterations have a causative or a compensatory role. In addition, prospective studies will reveal predisposing factors with greater certainty. The understanding of the mechanisms that lead to patellofemoral disorders as well as the results of studies that compare different surgical and rehabilitation techniques or just conservative measures will lead us to the ideal treatment of each patient.

References

1. Brechter JH, Powers CM. Patellofemoral joint stress during stair ascent and descent in persons with and without patellofemoral pain. Gait Posture. 2002;16(2): 115–23.
2. Canale ST, Beaty JH, Campbell WC. Campbell's operative orthopaedics. St. Louis: Mosby; 2012.
3. DeHaven KE, Lintner DM. Athletic injuries: comparison by age, sport, and gender. Am J Sports Med. 1986;14(3):218–24.
4. Devereaux M, Lachmann S. Patello-femoral arthralgia in athletes attending a Sports Injury Clinic. Br J Sports Med. 1984;18(1):18–21.
5. Crossley KM, et al. Knee flexion during stair ambulation is altered in individuals with patellofemoral pain. J Orthop Res. 2004;22(2):267–74.
6. Crossley KM, et al. Physical therapy improves knee flexion during stair ambulation in patellofemoral pain. Med Sci Sports Exerc. 2005;37(2):176–83.
7. Fulkerson JP, Shea KP. Disorders of patellofemoral alignment. J Bone Joint Surg Am. 1990;72(9): 1424–9.
8. Grana WA, Kriegshauser LA. Scientific basis of extensor mechanism disorders. Clin Sports Med. 1985;4(2):247–57.
9. Moller BN, Moller-Larsen F, Frich LH. Chondromalacia induced by patellar subluxation in the rabbit. Acta Orthop Scand. 1989;60(2):188–91.
10. Tompkins M, et al. What's new in sports medicine. J Bone Joint Surg Am. 2011;93(8):789–97. doi: 10.2106/JBJS.J.01723.
11. Ward SR, Terk MR, Powers CM. Patella alta: association with patellofemoral alignment and changes in contact area during weight-bearing. J Bone Joint Surg Am. 2007;89(8):1749–55.
12. Powers CM. Patellar kinematics, part I: the influence of vastus muscle activity in subjects with and without patellofemoral pain. Phys Ther. 2000;80(10): 956–64.
13. Powers CM, Landel R, Perry J. Timing and intensity of vastus muscle activity during functional activities in subjects with and without patellofemoral pain. Phys Ther. 1996;76(9):946–55; discussion 956–67.
14. Moller BN, et al. Isometric contractions in the patellofemoral pain syndrome. An electromyographic study. Arch Orthop Trauma Surg. 1986;105(1):24–7.
15. Wild Jr JJ, Franklin TD, Woods GW. Patellar pain and quadriceps rehabilitation. An EMG study. Am J Sports Med. 1982;10(1):12–5.
16. Karst GM, Willett GM. Onset timing of electromyographic activity in the vastus medialis oblique and vastus lateralis muscles in subjects with and without patellofemoral pain syndrome. Phys Ther. 1995;75(9):813–23.
17. Souza DR, Gross MT. Comparison of vastus medialis obliquus: vastus lateralis muscle integrated electromyographic ratios between healthy subjects and patients with patellofemoral pain. Phys Ther. 1991;71(4):310–6; discussion 317–20.
18. Turner MS. The association between tibial torsion and knee joint pathology. Clin Orthop Relat Res. 1994;302:47–51.
19. Sheehy P, et al. An electromyographic study of vastus medialis oblique and vastus lateralis activity while ascending and descending steps. J Orthop Sports Phys Ther. 1998;27(6):423–9.
20. Reynolds L, et al. EMG activity of the vastus medialis oblique and the vastus lateralis in their role in patellar alignment. Am J Phys Med. 1983;62(2):61–70.
21. Lee TQ, Morris G, Csintalan RP. Dysplasia of the quadriceps mechanism: hypoplasia of the vastus medialis muscle as related to the hypermobile patella syndrome. Surg Clin North Am. 1975;55(1):199–226.
22. Salsich GB, Brechter JH, Powers CM. Lower extremity kinetics during stair ambulation in patients with and without patellofemoral pain. Clin Biomech (Bristol, Avon). 2001;16(10):906–12.
23. Powers CM, et al. Are patellofemoral pain and quadriceps femoris muscle torque associated with locomotor function? Phys Ther. 1997;77(10): 1063–75; discussion 1075–8.
24. Greenwald AE, et al. A biomechanical and clinical evaluation of a patellofemoral knee brace. Clin Orthop Relat Res. 1996;324:187–95.
25. Nadeau S, Gravel D, Hébert LJ, Arsenault AB, Lepage Y. Gait study of patients with patellofemoral pain syndrome. Gait Posture. 1997;5(1):21–7.
26. Stergiou N, Bates BT, James SL. Asynchrony between subtalar and knee joint function during running. Med Sci Sports Exerc. 1999;31(11):1645–55.
27. Nigg BM, Cole GK, Nachbauer W. Effects of arch height of the foot on angular motion of the lower extremities in running. J Biomech. 1993;26(8): 909–16.
28. Nawoczenski DA, Saltzman CL, Cook TM. The effect of foot structure on the three-dimensional kinematic coupling behavior of the leg and rear foot. Phys Ther. 1998;78(4):404–16.
29. Dierks TA, et al. Proximal and distal influences on hip and knee kinematics in runners with patellofemoral pain during a prolonged run. J Orthop Sports Phys Ther. 2008;38(8):448–56. doi:10.2519/jospt.2008.2490. Epub 2008 Aug 1.
30. Grenholm A, Stensdotter AK, Hager-Ross C. Kinematic analyses during stair descent in young women with patellofemoral pain. Clin Biomech (Bristol, Avon). 2009;24(1):88–94. doi:10.1016/j.clinbiomech.2008.09.004. Epub 2008 Nov 4.
31. Ho KY, Blanchette MG, Powers CM. The influence of heel height on patellofemoral joint kinetics during walking. Gait Posture. 2012;36(2):271–5.
32. Barton CJ, et al. Kinematic gait characteristics associated with patellofemoral pain syndrome: a systematic review. Gait Posture. 2009;30(4):405–16. doi:10.1016/j.gaitpost.2009.07.109. Epub 2009 Aug 3.
33. Lee TQ, Morris G, Csintalan RP. The influence of tibial and femoral rotation on patellofemoral contact area and pressure. J Orthop Sports Phys Ther. 2003;33(11):686–93.

34. Lee TQ, et al. The influence of fixed rotational deformities of the femur on the patellofemoral contact pressures in human cadaver knees. Clin Orthop Relat Res. 1994;302:69–74.
35. Mascal CL, Landel R, Powers CM. Management of patellofemoral pain targeting hip, pelvis, and trunk muscle function: 2 case reports. J Orthop Sports Phys Ther. 2003;33(11):647–60.
36. Ireland ML, et al. Hip strength in females with and without patellofemoral pain. J Orthop Sports Phys Ther. 2003;33(11):671–6.
37. Willson JD, Davis IS. Lower extremity mechanics of females with and without patellofemoral pain across activities with progressively greater task demands. Clin Biomech (Bristol, Avon). 2008;23(2):203–11. Epub 2007 Oct 17.
38. Souza RB, Powers CM. Differences in hip kinematics, muscle strength, and muscle activation between subjects with and without patellofemoral pain. J Orthop Sports Phys Ther. 2009;39(1):12–9. doi:10.2519/jospt.2009.2885.
39. Powers CM, et al. Patellofemoral kinematics during weight-bearing and non-weight-bearing knee extension in persons with lateral subluxation of the patella: a preliminary study. J Orthop Sports Phys Ther. 2003;33(11):677–85.
40. Barton CJ, et al. The relationship between rearfoot, tibial and hip kinematics in individuals with patellofemoral pain syndrome. Clin Biomech (Bristol, Avon). 2012;27(7):702–5. doi:10.1016/j.clinbiomech.2012.02.007. Epub 2012 Mar 20.
41. Pappas E, Wong-Tom WM. Prospective predictors of patellofemoral pain syndrome: a systematic review with meta-analysis. Sports Health. 2012;4(2): 115–20.
42. Noehren B, Hamill J, Davis I. Prospective evidence for a hip etiology in patellofemoral pain. Med Sci Sports Exerc. 2013;45(6):1120–4.

Prevention of Patellofemoral Injuries

9

Michael T. Benke, Christopher M. Powers, and Bert R. Mandelbaum

Contents

9.1 Introduction

Patellofemoral knee injuries account for one in four of all knee injuries in athletes [1], up to 40 % of all physical therapy visits for knee pain [2, 3], and 10 % of total visits to physical therapy clinics [3]. Patellofemoral pain is the most common injury in runners [4] and the most common source of knee pain in adolescents [5].

Patellofemoral pain (PFP) is an overuse injury, characterized by retropatellar or peripatellar pain that is made worse by running, prolonged sitting, squatting, jumping, or climbing stairs. The etiology of PFP has previously been credited to lateral patellar maltracking owing to vastus medialis obliquus (VMO) muscle weakness. It is thought that abnormal patella tracking causes increased lateral compressive patellofemoral joint stress [6]. As such, interventions have focused on influencing patellar motion with taping and bracing, patella mobilization, and strengthening of the VMO [7].

While it is generally accepted that conservative treatment of PFP is helpful [8–10], recurrence rates are high and range between 25 and 91 % [1, 11–14]. This suggests that the etiology of PFP is not being addressed with current treatment approaches. More recent literature has focused on the hip musculature and their dynamic effect on patellofemoral mechanics. Given that the patella articulates with the distal femur, abnormal hip motions likely contribute to PFP [7].

M.T. Benke (✉) • B.R. Mandelbaum
Santa Monica Orthopaedic and Sports Medicine Group, Santa Monica, CA, USA
e-mail: mtbenke@gmail.com

C.M. Powers
Musculoskeletal Biomechanics Research Laboratory, Division of Biokinesiology and Physical Therapy, University of Southern California, Los Angeles, CA, USA

A. Gobbi et al. (eds.), *The Patellofemoral Joint*,
DOI 10.1007/978-3-642-54965-6_9, © ISAKOS 2014

9.2 Biomechanics

Considering the wide scope of patellofemoral pain and the impact it has on athletic performance and quality of life, prevention is imperative. Having a sound knowledge of patellofemoral biomechanics is a key part in understanding and prescribing rehabilitation and prevention programs.

As mentioned, previous theory that PFP originates from patellar maltracking stemmed from kinematic studies performed under non-weight-bearing conditions with a fixed femur [15–17]. In such a state, the patella tilts and translates lateral on the distal femur [18]. However, under weight-bearing conditions, it has been shown using dynamic MRI that the internal rotation of the femur results in a more significant contribution to relative patellar tilt and translation [18]. Lee and colleagues [19] have reported that 5–10° of internal rotation of the femur increases patellofemoral joint stress considerably.

The Q-angle, or the angle formed between the resultant quadriceps vector and the patellar tendon vector, largely determines the propensity for the patella to track laterally. While the Q-angle is typically assessed statically, segmental motions of the pelvis, hip, and knee during dynamic tasks can significantly increase the Q-angle and the laterally directed force acting on the patella. This concept has been termed the "dynamic Q-angle" [20, 21]. Chen and Powers [20, 21] reported that subjects with PFP have greatly increased dynamic Q-angles during stair descent compared with asymptomatic controls (39° versus 24°). This increased dynamic Q-angle in turn increases the contact pressures on the patella. In a cadaveric study, a 10° increase in the Q-angle resulted in a 45 % increase in peak contact pressure on the lateral aspect of the patellofemoral joint [22].

Looking at the lower extremity in the coronal and transverse planes, adduction and internal rotation of the hip move the knee joint center of rotation medially. When the foot is fixed, this causes abduction of the tibia and resultant dynamic knee valgus [7] (Fig. 9.1). Multiple studies have linked weak hip musculature to increased knee valgus [23–26], and this susceptible position of knee valgus has been implicated in patellofemoral joint and ACL injury [7, 27, 28].

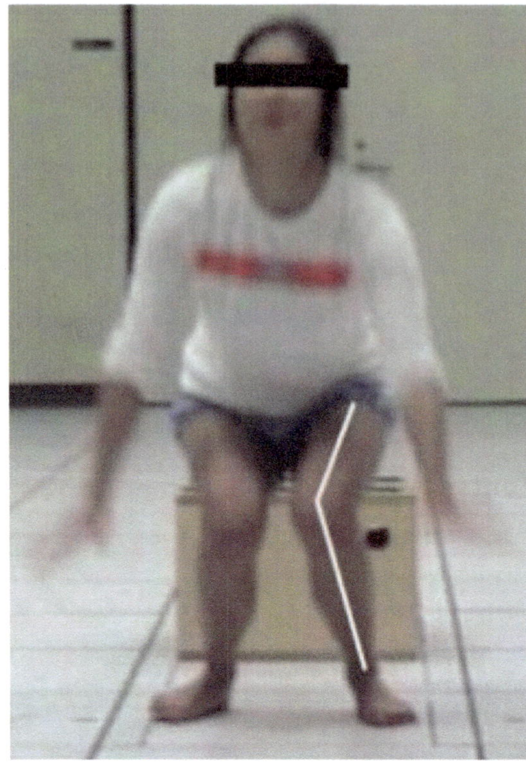

Fig. 9.1 Dynamic knee valgus. Excessive hip adduction and internal rotation results in dynamic knee valgus (Reproduced with permission from Powers [7])

Poor proximal neuromuscular control and weakness of hip musculature, in particular the hip abductors and external rotators, lead to poor lower-extremity control in the coronal and transverse planes. As the hip collapses into excessive adduction and internal rotation, the knee falls into dynamic valgus [18, 29–31].

In the sagittal plane, trunk position greatly influences muscular demands across the knee and resultant forces across the patellofemoral joint. During a drop-jump maneuver, leaning the trunk forward moves the center of mass—and therefore the ground reactive vector—anteriorly, closer to the knee joint and further from the hip joint. This decreases the knee flexion moment and demand on the knee extensors while increasing the hip flexion moment and demand on the hip extensors [7]. Landing with the trunk more erectly moves the ground reactive vector posteriorly, thereby increasing the knee flexion moment and demand on the knee extensors and

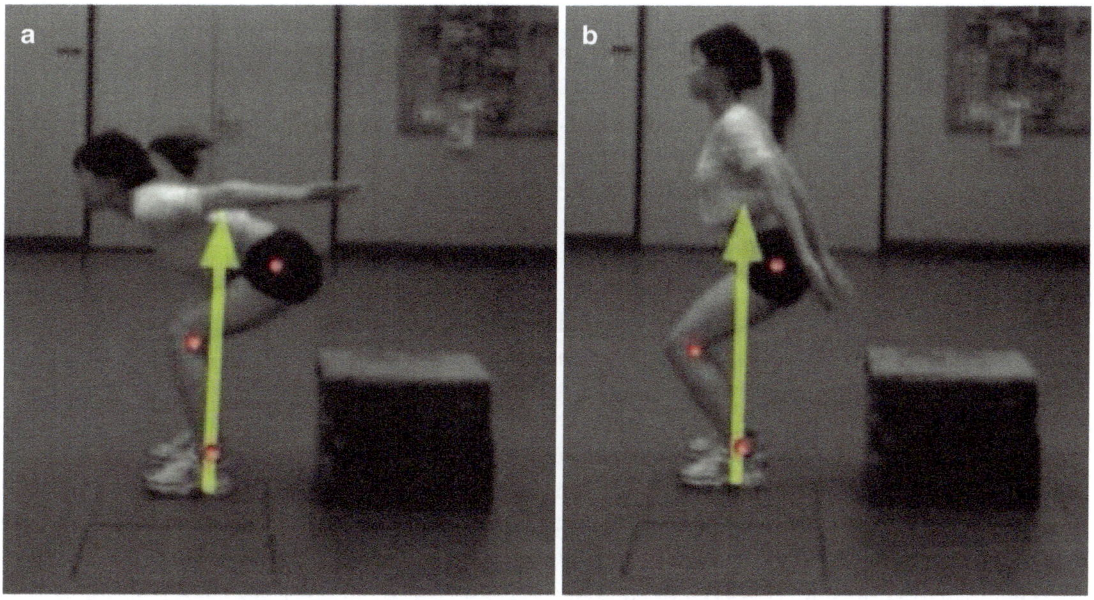

Fig. 9.2 Sagittal plane motion. (**a**) Landing with the trunk anterior increases the moment at the hip while decreasing the moment at the knee. (**b**) Landing with the trunk posterior increases the moment at the knee while decreasing the moment at the hip (Reproduced with permission from Powers [7])

decreasing the hip flexion moment and demand on the hip extensors [7, 32] (Fig. 9.2).

Patients with weak hip extensors commonly will compensate by leaning the trunk posteriorly [33]. This increases the knee flexion moment and the demand on the quadriceps, which may lead to quadriceps strain, patellar tendinopathy, patellofemoral compression, and ACL strain [7].

9.3 Clinical Findings

Multiple clinical studies have supported the theory that PFP is linked to poor proximal muscular control. In a study of 210 NCAA Division I athletes, Nadler et al. [34] showed that hip extensor weakness was significantly more common in athletes with recent patellofemoral and lower-extremity injury. In a prospective study, Leetun et al. [35] showed that collegiate athletes who developed lower-extremity injuries over a season had significantly greater weakness in hip abduction and hip extension. Likewise, Niemuth et al. [36] and Cichanowski et al. [37] found a relationship between hip muscle weakness and patellofemoral injury in athletes.

Ireland et al. [29] reported weakness of hip abductors and external rotators in 15 females with anterior knee pain compared to asymptomatic age-matched controls. Robinson et al. investigated if females with PFP were more likely to demonstrate hip abduction, external rotation, or extension weakness than a similar, asymptomatic, age-matched control group. The authors calculated limb strength as a percentage of the uninvolved extremity in order to better compare to controls. It was reported that females with PFP demonstrated less hip abduction, external rotation, and extension strength than the controls [38].

Bolgla et al. [39] found hip abductor and external rotator weakness in 18 female participants with PFP compared to asymptomatic controls. They did not, however, find altered hip and knee kinematics as theorized. Symptomatic individuals did not exhibit increased hip internal rotation, hip adduction, and knee valgus during functional activities. The authors do note that the task tested may not have been challenging enough to elucidate kinematic dysfunction [39].

Souza and Powers [40] looked at hip muscle strength as well as hip kinematics during more strenuous tasks including running, a drop jump, and a step-down maneuver in 21 subjects with PFP compared to well controls. These authors found hip abductor and extensor weakness as well as increased peak hip internal rotation, thereby supporting kinematic dysfunction of the hip and knee in this population.

Other studies, however, have found no difference between hip muscle strength and PFP [41, 42]. Piva et al. found no difference in hip external rotation and abduction when comparing between individuals with and without PFP. No comparison was made to the uninvolved extremity, making the comparison strength differences between individuals difficult. Cowan et al. [42] found no difference in hip external rotation and abduction strength between individuals with and without PFP, although they did note alteration of hip muscle function as shown by EMG evidence of delayed gluteus medius firing.

In a prospective study, Thijs et al. [43] found no difference in isometric hip muscle strength in 77 female runners that did and did not develop PFP over a 10-week period. The authors contend that there is no causal relationship found between proximal strength and PFP. This study, however, looked only at peak isometric strength. It may have been prudent to look at eccentric hip muscle function or endurance strength testing as these measures may be more relevant to overuse syndromes such as PFP [7, 43].

Despite some disagreement in the literature regarding a common finding of proximal weakness and dysfunction in patients with PFP, a systematic review performed by Prins et al. [44] found that there is strong evidence to support that females with PFP possess diminished strength of the hip abductors, external rotators, and extensors.

One must be careful when drawing conclusions about cause and effect of muscular weakness on PFP, as weakness may be a by-product of disuse secondary to pain. However, several studies have shown clinical success and pain relief following hip-focused rehabilitation [45–48].

9.4 Rehabilitation and Prevention Strategies

While certainly multifactorial, with such a large and growing body of biomechanical and clinical literature supporting the hypothesis that PFP has significant proximal influences, a strong case can be made for hip and pelvic-based rehabilitation for patellofemoral dysfunction.

This approach is supported by recent clinical studies that have shown that hip-focused rehabilitation is beneficial for persons with PFP.

Earl and Hoch [45] showed that an 8-week course of hip and core strengthening significantly reduced pain, improved function, and improved the overall lower-extremity biomechanics in a group of 19 females with PFP.

Mascal et al. [46] reported on two female patients with PFP recalcitrant to traditional quadriceps-based rehabilitation. Each patient was found to have hip abductor, external rotator, and extensor weakness as well as kinematic dysfunction shown by hip internal rotation and adduction during dynamic tasks. After a 14-week program of hip, pelvis, and trunk strengthening and stabilization, the subjects had resolution of symptoms with improvement in strength and kinematics.

Fukuda et al. [49], in a recent randomized controlled clinical trial, looked at 70 females with PFP. Patients were randomized into three groups: a control group that received no treatment, a knee rehabilitation group, and a knee and hip rehabilitation group. Treatment consisted of three sessions per week for 4 weeks. Both treatment groups showed improvements in lower-extremity function scores, anterior knee pain scores, and single-limb hop testing compared to the control group. However, the improvements in pain and function were greater in the knee and hip exercise group.

Nakagawa et al. [50] showed similar findings in a randomized controlled pilot study comparing hip and quadriceps strengthening to isolated quadriceps strengthening.

In another randomized controlled trial by Khayambashi et al. [51], the authors showed that strengthening of hip abductors and external rotators improved pain and health status compared to a control group in females with PFP. This study was unique in that it looked at isolated strengthening of hip musculature as opposed to combined hip and knee exercises. The authors show that rehabilitation of the hip alone can be effective in treating PFP.

Our institution routinely focuses on optimizing a "hip strategy" when treating patients with PFP. The concept of "hip strategy" combines hip strength (abductors, external rotators, extensors), biomechanical alignment, kinematics, and neuromuscular control.

Two key components are incorporated into our proximally based intervention program to address knee injury: pelvic stability and dynamic hip control [7]. Anterior pelvic tilt can result from hip flexor tightness or weak posterior rotators of the pelvis (hip extensors, hamstrings, abdominals). This may then lead to compensatory lumbar lordosis and posterior shift of trunk and center of mass. This in turn increases the knee flexion moment and puts more stress on knee extensors and the forces acting on the patellofemoral joint. This also decreases the hip flexion moment and decreases demand on the hip extensors, thus perpetuating hip extensor weakness and further anterior pelvic tilt. To prevent this from happening, dynamic pelvic and core stability is addressed through rehabilitation.

The second principle in PFP rehabilitation and prevention is dynamic hip control to prevent hip adduction and internal rotation [7]. The gluteus medius functions mainly in the coronal plane as a hip abductor, thereby stabilizing the femur and pelvis during single stance. The posterior fibers assist with hip extension and external rotation, but not with any major significance [52]. The gluteus maximus, however, is the most powerful hip extensor and external rotator of the hip. Working in multiple planes, it is ideally suited to minimize the at-risk position of dynamic knee valgus.

Additionally, improving gluteus maximus function in the sagittal place decreases the requirement of compensatory quadriceps activity to absorb the load of impact; this in turn decreases the forces across the patellofemoral joint [7, 53, 54].

Finally, it should be noted that the same protocol used for PFP rehabilitation and prevention is implemented following ACL reconstruction at our institution. Optimal hip strategy in order to prevent femoral internal rotation and hip adduction responsible for dynamic knee valgus is felt to be crucial prior to return to sport.

References

1. Devereaux MD, Lachmann SM. Patellofemoral arthralgia in athletes attending a sports injury clinic. Br J Sports Med. 1984;18:18–21.
2. DeHaven KE, Lintner DM. Athletic injuries: comparison by age, sport, and gender. Am J Sports Med. 1986;14:218–24.
3. Kannus P, Natri A, Paakkala T, Jarvinen M. An outcome study of chronic patellofemoral pain syndrome: seven-year follow-up of patients in a randomized, controlled trial. J Bone Joint Surg Am. 1999;81(3):355–63.
4. Taunton JE, Ryan MB, Clement DB, McKenzie DC, Lloyd-Smith DR, Zumbo BD. A retrospective case-control analysis of 2002 running injuries. Br J Sports Med. 2002;36:95–101.
5. Baxter M. Knee pain in the paediatric athlete. Paediatr Med. 1986;1:211–8.
6. Fulkerson JP, Arendt EA. Anterior knee pain in females. Clin Orthop. 2000;372:69–73.
7. Powers CM. The influence of abnormal hip mechanics on knee injury: a biomechanical perspective. J Orthop Sports Phys Ther. 2010;40(2):42–51.
8. Bizzini M, Childs J, Piva S, Delitto A. Systematic review of the quality of randomized controlled trials for patellofemoral pain syndrome. J Orthop Sports Phys Ther. 2003;33:4–20.
9. Crossley K, Bennell K, Green S, McConnell J. A systematic review of physical interventions for patellofemoral pain syndrome. Clin J Sport Med. 2001;11(2):103–10.
10. Heintjes E, Berger M, Bierma-Zeinstra S, Bernsen RM, Verhaar JA, Koes BW. Exercise therapy for patellofemoral pain syndrome (review). Cochrane Database Syst Rev. 2003;(4):CD003472.

11. Hoch A, Pepper M, Akuthota V. Stress fractures and knee injuries in runners. Phys Med Rehabil Clin N Am. 2005;16:749–77.
12. Nimon G, Murray D, Sandow M, Goodfellow J. Natural history of anterior knee pain: a 14- to 20-year follow-up of nonoperative management. J Pediatr Orthop. 1998;18(1):118–22.
13. Stathopulu E, Baildam E. Anterior knee pain: a long-term follow-up. Rheumatology (Oxford). 2003;42(2):380–2.
14. Witvrouw E, Lysens R, Bellemans J, Peers K, Vanderstraeten G. Open versus closed kinetic chain exercises for patellofemoral pain: a prospective, randomized study. Am J Sports Med. 2000;28(5):687–94.
15. Brossmann J, Muhle C, Bull CC, Schroder C, Meichert UH, Zieplies J, Spielman RF, Heller M. Evaluation of patellar tracking in patients with suspected patellar malalignment: cine MR imaging vs. arthroscopy. Am J Roentgenol. 1994;162:361–7.
16. Laprade J, Culham E. Radiographic measures in subjects who are asymptomatic and subjects with patellofemoral pain syndrome. Clin Orthop Relat Res. 2003;414:172–82.
17. MacIntyre NJ, Hill NA, Fellows RA, Ellis RE, Wilson DR. Patellofemoral joint kinematics in individuals with and without patellofemoral pain syndrome. J Bone Joint Surg Am. 2006;88:2596–605.
18. Powers CM, Ward SR, Fredericson M, Guillet M, Shellock FG. Patellar kinematics during weight-bearing and non-weight-bearing movements in persons with patellar subluxation. J Orthop Sports Phys Ther. 2003;33(11):677–85.
19. Lee TQ, Morris G, Csintalan RP. The influence of tibial and femoral rotation on patellofemoral contact area and pressure. J Orthop Sports Phys Ther. 2003; 33:686–93.
20. Chen YJ, Powers CM. The dynamic Q-angle: a comparison of persons with and without patellofemoral pain [abstract]. Proceedings of the North American Congress on Biomechanics. Ann Arbor: MI; 2008.
21. Chen YJ, Scher I, Powers CM. Quantification of patellofemoral joint reaction forces during functional tasks: a subject specific, three-dimensional model. J Appl Biomech. 2010;26(4):415–23.
22. Huberti HH, Hayes WC. Patellofemoral contact pressures. The influence of Q-angle and tendofemoral contact. J Bone Joint Surg Am. 1984;66:715–24.
23. Claiborne TL, Armstrong CW, Gandhi V, Pincivero DM. Relationship between hip and knee strength and knee valgus during a single leg squat. J Appl Biomech. 2006;22:41–50.
24. Hollman JH, Ginos BE, Kozuchowski J, Vaughn AS, Krause DA, Youdas JW. Relationships between knee valgus, hip-muscle strength, and hip-muscle recruitment during a single-limb step-down. J Sport Rehabil. 2009;18:104–17.
25. Jacobs CA, Uhl TL, Mattacola CG, Shapiro R, Rayens WS. Hip abductor function and lower extremity landing kinematics: sex differences. J Athl Train. 2007;42:76–83.
26. Willson JD, Ireland ML, Davis I. Core strength and lower extremity alignment during single leg squats. Med Sci Sports Exerc. 2006;38:945–52.
27. Powers CM. The influence of altered lower-extremity kinematics on patellofemoral joint dysfunction: a theoretical perspective. J Orthop Sports Phys Ther. 2003; 33(11):639–46.
28. Hewett TE, Myer GD, Ford KR, Heidt Jr RS, Colosimo AJ, McLean SG, van den Bogert AJ, Paterno MV, Succop P. Biomechanical measures of neuromuscular control and valgus loading of the knee predict anterior cruciate ligament injury risk in female athletes: a prospective study. Am J Sports Med. 2005;33:492–501.
29. Ireland ML, Willson JD, Ballantyne BT, Davix IM. Hip strength in females with and without patellofemoral pain. J Orthop Sports Phys Ther. 2003;33:671–6.
30. Riegger-Krugh C, Keysor J. Skeletal malalignments of the lower quarter: correlated and compensatory motions and postures. J Orthop Sports Phys Ther. 1996;23:164–70.
31. Wilk KE, Davies GJ, Mangine RE, Malone TR. Patellofemoral disorders: a classification system and clinical guidelines for nonoperative rehabilitation. J Orthop Sports Phys Ther. 1998;28:307–22.
32. Blackburn JT, Padua DA. Sagittal-plane trunk position, landing forces, and quadriceps electromyographic activity. J Athl Train. 2009;44:174–9.
33. Perry J. Gait analysis: normal and pathological function. Thorofare: Slack Inc; 1992.
34. Nadler SF, Malanga GA, DePrince M, Stitik TP, Feinberg JH. The relationship between lower extremity injury, low back pain, and hip muscle strength in male and female collegiate athletes. Clin J Sport Med. 2000;10:89–97.
35. Leetun DT, Ireland ML, Willson JD, Ballantyne BT, Davis IM. Core stability measures as risk factors for lower extremity injury in athletes. Med Sci Sports Exerc. 2004;36:926–34.
36. Niemuth PE, Johnson RJ, Myers MJ, Thieman TJ. Hip muscle weakness and overuse injuries in recreational runners. Clin J Sport Med. 2005;15:14–21.
37. Cichanowski HR, Schmitt JS, Johnson RJ, Niemuth PE. Hip strength in collegiate female athletes with patellofemoral pain. Med Sci Sports Exerc. 2007;39(8):1227–32.
38. Robinson RL, Nee RJ. Analysis of hip strength in females seeking physical therapy treatment for unilateral patellofemoral pain syndrome. J Orthop Sports Phys Ther. 2007;37(5):232–8.
39. Bolgla LA, Malone TR, Umberger BR, Uhl TL. Hip strength and hip and knee kinematics during stair descent in females with and without patellofemoral pain syndrome. J Orthop Sports Phys Ther. 2008;38(1):12–8.
40. Souza RB, Powers CM. Differences in hip kinematics, muscle strength, and muscle activation between subjects with and without patellofemoral pain. J Orthop Sports Phys Ther. 2009;39(1):12–9.
41. Piva SR, Goodnite EA, Childs JD. Strength around the hip and flexibility of soft tissues in individuals with and without patellofemoral pain syndrome. J Orthop Sports Phys Ther. 2005;35(12):793–801.

42. Cowan SM, Crossley KM, Bennell KL. Altered hip and trunk muscle function in individuals with patellofemoral pain. Br J Sports Med. 2009;43(8):584–8.
43. Thijs Y, Pattyn E, Van Tiggelen D, Rombaut L, Witvrouw E. Is hip muscle weakness a predisposing factor for patellofemoral pain in female novice runners? A prospective study. Am J Sports Med. 2011;39:1877–82.
44. Prins MR, van der Wurff P. Females with patellofemoral pain syndrome have weak hip muscles: a systematic review. Aust J Physiother. 2009;55(1):9–15.
45. Earl JE, Hoch AZ. A proximal strengthening program improves pain, function, and biomechanics in women with patellofemoral pain syndrome. Am J Sports Med. 2011;39:154–63.
46. Mascal CL, Landel R, Powers C. Management of patellofemoral pain targeting hip, pelvis, and trunk muscle function: 2 case reports. J Orthop Sports Phys Ther. 2003;33:647–60.
47. Boling MC, Padua DA, Marshall SW, Guskiewicz K, Pyne S, Beutler A. A prospective investigation of biomechanical risk factors for patellofemoral pain syndrome. The joint undertaking to monitor and prevent ACL injury (JUMP-ACL) cohort. Am J Sports Med. 2009;37:2108–16.
48. Tyler TF, Nicholas SJ, Mullaney MJ, McHugh MP. The role of hip muscle function in the treatment of patellofemoral pain syndrome. Am J Sports Med. 2006;34:630–6.
49. Fukuda TY, Rossetto FM, Magalhaes E, Bryk FF, Lucareli PR, De Almeida Carvalho NA. Short-term effects of hip abductors and lateral rotators strengthening in females with patellofemoral pain syndrome: a randomized controlled clinical trial. J Orthop Sports Phys Ther. 2010;40(11):736–42.
50. Nakagawa TH, Muniz TB, Baldon Rde M, Dias Maciel C, de Menezes Reiff RB, Serrao FV. The effect of additional strengthening of hip abductor and lateral rotator muscles in patellofemoral pain syndrome: a randomized controlled pilot study. Clin Rehabil. 2008;22:1051–60.
51. Khayambashi K, Mohammadkhani Z, Ghaznavi K, Lyle MA, Powers CM. The Effects of isolated hip abductor and external rotator muscle strengthening on pain, health status, and hip strength in females with patellofemoral pain: a randomized controlled trial. J Orthop Sports Phys Ther. 2012;42(1):22–9.
52. Neumann DA. Kinesiology of the musculoskeletal system. St Louis: Mosby Inc; 2002.
53. Pollard CD, Sigward SM, Powers CM. Limited hip and knee flexion during landing is associated with increased frontal plane motion and moments. Clin Biomech. 2010;25:142–6.
54. Zhang SN, Bates BT, Dufek JS. Contributions of lower extremity joints to energy dissipation during landings. Med Sci Sports Exerc. 2000;32:812–9.

Patellofemoral Instability 10

Massimo Petrera, Tim Dwyer, and Alberto Gobbi

Contents

M. Petrera (✉)
University of Toronto Orthopaedic Sports Medicine, Women's College Hospital, Toronto, ON, Canada
e-mail: massimopetrera@gmail.com

T. Dwyer
University of Toronto Orthopaedic Sports Medicine, Women's College and Mount Sinai Hospital, Toronto, ON, Canada

A. Gobbi
Orthopaedic Arthroscopic Surgery International (O.A.S.I.) Bioresearch Foundation, Gobbi N.P.O., Milan, Italy
e-mail: gobbi@cartilagedoctor.it

10.1 Introduction

Patellofemoral instability is a condition characterized by repeated dislocation or subluxation of the patella secondary to minimal trauma. This condition is known to be more common in females, and it is often bilateral. The reported incidence of patella dislocation in the population aged between 10 and 17 years old is reported at 29 per 100,000 [1]. After a first-time dislocation, the rate of recurrence following nonoperative treatment ranges between 15 and 44 % [2, 3].

The patella most commonly dislocates laterally – when medial dislocation occurs, it is usually iatrogenic, either as a result of excessive lateral release or overtightening of medial structures. The etiology of patella instability is multifactorial and is most commonly thought of in terms of bony abnormalities (patella alta, trochlear dysplasia, patella hypoplasia, lower limb malalignment, rotational abnormalities) or soft tissue pathology (vastus medialis obliquus atrophy, medial retinaculum laxity, injuries to the medial patellofemoral ligament (MPFL), tight lateral structures, or ligamentous laxity) [4–6]. Treatment plans seeking to address patella instability must therefore be based on the understanding of how the pathoanatomy contributes to instability.

A. Gobbi et al. (eds.), *The Patellofemoral Joint*,
DOI 10.1007/978-3-642-54965-6_10,

10.2 Anatomy and Biomechanics of Instability

Normal patellar tracking is a process that relies on both static and dynamic stabilizers, requiring a complex balance of ligamentous structures, bone morphology, lower extremity alignment, and muscle activation [4–7]. Static constraints include the bony contours of the femur, a normal rotational profile (proximal femoral anteversion and external tibial torsion), and integrity of the MPFL (which acts as constant static checkrein to patella). The quadriceps muscle, in particular the vastus medialis obliquus (VMO), represents the main dynamic stabilizer; an imbalance between VMO and vastus lateralis obliquus may contribute to instability.

The trochlear groove has a complex geometry in terms of depth and shape – the lateral facet of the trochlear groove is normally more prominent anteriorly, preventing lateral displacement of the patella. Any degree of trochlear dysplasia (flattening of the sulcus) or corresponding patella hypoplasia decreases the inherent stability of the patellofemoral joint. Limb malalignment also plays an important role in patellar tracking, with an increased Q-angle (the angle formed by a first line drawn from the anterior superior iliac spine (ASIS) to the center of patella and a second line drawn from center of patella to tibial tubercle) representing a risk factor for patellar instability. The Q-angle averages 14° in males and 17° in females [7]; this value is increased in presence of genu valgum, increased femoral anteversion, increased external tibial torsion, or a lateralized tibial tuberosity.

The MPFL is an important soft issue restraint to lateral translation of the patella during the early phase of flexion, acting as the main stabilizer from 0° to 30° of knee flexion. In this range the MPFL is tensioned in full extension, serving to guide the patella into engagement with the trochlea during the early stages of flexion. Any degree of patella alta prevents this early engagement of the patella in the trochlear groove, contributing to lateral subluxation. Between 30° and 60° of flexion, the patella moves medially and becomes centered in groove, becoming deeply engaged in the trochlear groove between 60° and 90°. At >90° the patella tilts such that the medial facet articulates with the medial femoral condyle (MFC), and at flexion of 135° or greater, the odd facet of the patella contacts the lateral border of MFC [6].

10.3 Clinical Examination

10.3.1 History

A thorough history taking and an accurate physical examination are essential tools for a correct diagnosis. Age at the first dislocation, history of trauma, the direction of instability, and the frequency of subsequent dislocations should be recorded. Usually, if the event was atraumatic or related to a minor twisting, it is likely that the patient has predisposing conditions (soft tissue or bony abnormalities) that will contribute to recurrent dislocations. Traumatic mechanisms usually involve either a direct blow to the patella or a mechanism of valgus and external rotation of tibia. With increasing severity of injury, significant swelling may occur secondary to a MPFL injury or due to hemarthrosis associated with a chondral or osteochondral injury.

10.3.2 Physical Examination

Patients should be examined while standing, walking, sitting, and lying (in both supine and prone positions). Signs of generalized ligamentous laxity can be assessed objectively using the Beighton scale [8]. This scale assigns one point for the ability to touch the floor with the palms while having the knees fully extended and one point each side for the ability to achieve hyperextension of the knee beyond −10°, hyperextension of the elbow beyond −10°, extension of the fifth metacarpophalangeal (MCP) joint beyond 90° and for the ability to extend the thumb to touch the forearm with the wrist flexed. Hypermobility is defined by a score higher than 4 (out of a maximum of 9).

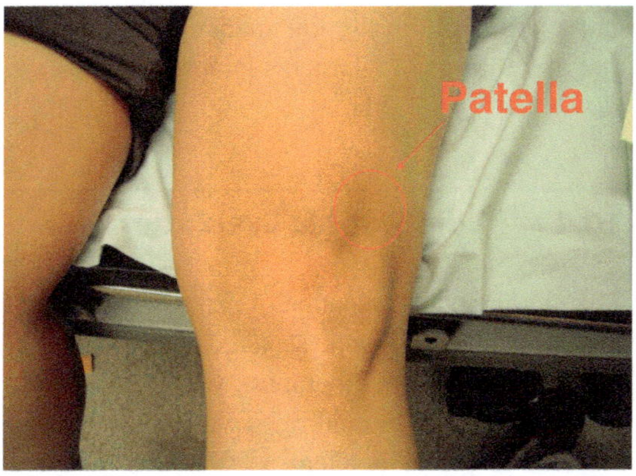

Fig. 10.1 J-sign: lateral subluxation of patella is seen as the knee from a flexed position approaches full extension (Image used with the permission of www.boneschool.com)

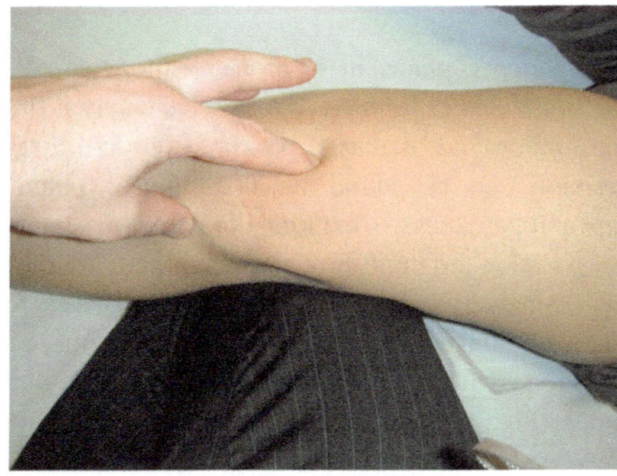

Fig. 10.2 Patellar hypermobility. With the knee in 30° of flexion, a lateral force has been applied – the patella has been moved laterally by greater than 50 % of its width (Image used with the permission of www.boneschool.com)

Standing/Gait

The patient's malalignment, especially genu valgum, is best seen in standing position. In some patients, "squinting patella" (kneeing-in posture) is noted, which can be a sign of increased femoral anteversion, increased Q-angle, excessive external tibial torsion, or hyperpronation of the foot. James et al. called the combination of increased proximal femoral anteversion and excessive external tibial torsion "miserable malalignment" syndrome [9]. Patients with patella alta may display "grasshopper eyes," so called when the patella is noted to sit in a higher and more lateral position than normal. An in-toeing gait is often indicative of increased femoral anteversion; this generates an external rotation moment about the knee with a resultant lateral force on the patella [10].

Sitting

Patellar tracking can be dynamically assessed by having the patient sit over the edge of the examining table and moving the knee through an active range of motion – comparison with the non-affected limb is useful as long as the condition is not bilateral. A positive J-sign is seen with lateral subluxation of patella as the knee approaches full extension, indicating a degree of maltracking (Fig. 10.1). Active knee extension may also elicit crepitus, suggestive of chondral lesions or early degenerative changes.

Supine

Placing the patient supine allows inspection for VMO wasting. With the knee in extension, palpation is used to quantify the degree of effusion and (in the setting of acute dislocation) areas of tenderness associated with MPFL injury either over the medial aspect of the patella or the medial epicondyle (Bassett's sign). Compression of the patella while the patient contracts the quadriceps muscle (Clarke's test or patellar grind) can produce anterior knee pain in the presence of excessive lateral pressure syndrome or cartilage lesions. The apprehension test (also known as Fairbank's test) is also performed with the patient supine – as knee flexion is increased, the patella is pushed laterally by the examiner; a positive test is the presence of apprehension and is suggestive of patella instability.

Patellar hypermobility is assessed with the knee in 30° of flexion (Sage mobility test) (Fig. 10.2); movement of the patella laterally by greater than 50 % of the patella width is associated with inadequate medial restraints. The patellar tilt test is also performed with the knee in this position, by elevating the lateral edge of patella from the lateral condyle (Fig. 10.3). It is normally possible to tilt the lateral edge of the patella up to at least 20° above the horizontal – in the setting of a tight lateral retinaculum, this is not possible. The Q-angle is also measured clinically with the knee in 30° of flexion in order to ensure the patella is engaged in the groove.

Prone

The patient is placed in the prone position to assess for rotational deformity, as excessive external tibial torsion and increased proximal femoral anteversion can be found in a setting of patella instability. Excessive external tibial torsion is measured using the thigh-foot angle, with greater than 30° considered to be excessive, while increased femoral anteversion is diagnosed when the internal rotation of the hip is greater than the external rotations (Fig. 10.4). The prone position also allows measurement of the heel-to-buttock distance – decreased flexion of the knee may indicate quadriceps contracture, while an involuntary hip flexion and a pelvic tilt during this maneuver are suggestive of tightness of the rectus femoris (Ely test).

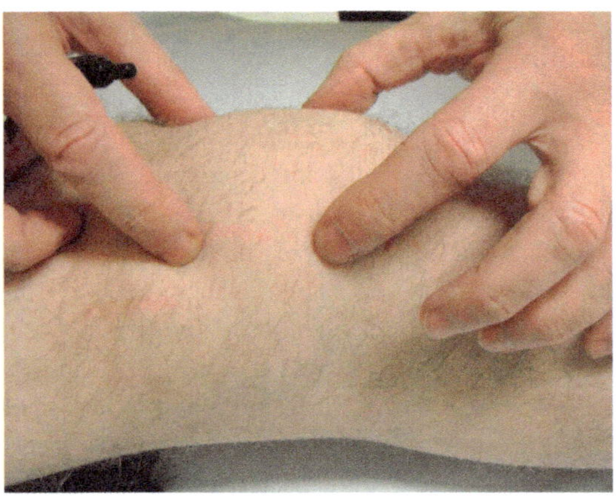

Fig. 10.3 The patellar tilt test. With the knee in 30° of flexion, the lateral edge of patella is elevated – inability to elevate to 20° or more is associated with a tight lateral retinaculum (Image used with the permission of www.boneschool.com)

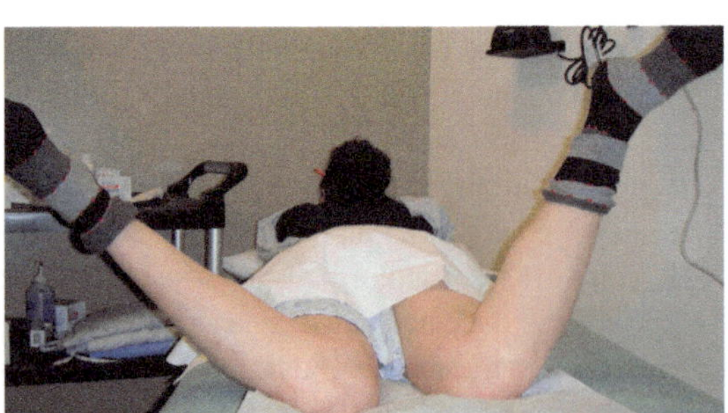

Fig. 10.4 Increased femoral anteversion on the left side, associated with increased internal rotation of the hip (Image used with the permission of www.boneschool.com)

10.4 Radiographic Evaluation

The radiographic evaluation of patients with patella instability begins with standard anteroposterior, lateral and axial radiographs, while lower limb malalignment can be confirmed with the use of long-leg radiographs. The lateral view is used to assess for the presence of patella alta. While many different techniques have been described, the Blackburne-Peel ratio is probably the most accurate for measurement of patella height [11] and is expressed as the ratio between the perpendicular distance from the lower articular margin of patella to tibial plateau and the length of patella articular surface – patella alta is defined by a ratio >1 (Fig. 10.5). Trochlear dysplasia can also be assessed on true lateral views by looking for a crossover sign, whereby the deepest part of the trochlear groove crosses the anterior aspect of the femoral condyles. The Dejour classification of trochlear dysplasia is reported in Table 10.1 [12].

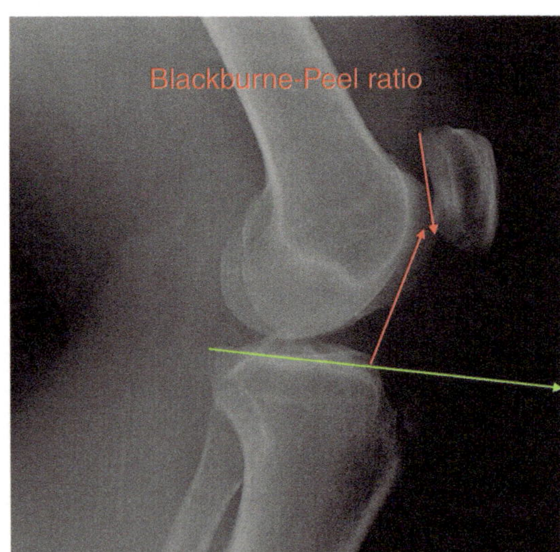

Fig. 10.5 Patella alta is defined by a Blackburne-Peel ratio >1 - ratio between the perpendicular distance from the lower articular margin of patella (*upward pointing red arrow*) to tibial plateau (*green arrow*), and the length of patella articular surface (*downward pointing red arrow*) - (Image used with the permission of www.boneschool.com)

Table 10.1 Classification of trochlea dysplasia

Dejour classification	
Type A	Crossover sign, fairly shallow trochlea (sulcus angle >140°)
Type B	Crossover sign, flat trochlea and supratrochlear spur
Type C	Crossover sign with double contour, medial hypoplasia
Type D	Type C features + vertical link between facets ("cliff pattern")

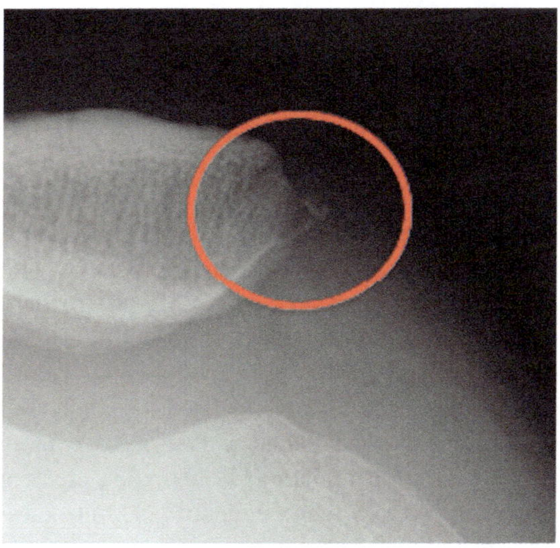

Fig. 10.6 MPFL bony avulsion (*red circle*)

Axial radiographs are used to diagnose patellar tilt, patellar subluxation, and trochlea dysplasia; bony avulsions of the MPFL can also be detected with this view (Fig. 10.6). While many variations of the axial view have been described, the most widely accepted is the Merchant view, taken with the patient supine, the knee flexed at 45°, and x-ray beam inclined 30° downward [13]. The sulcus angle is measured by the intersection of the highest points of the medial and lateral femoral condyles and the lowest point of the intercondylar sulcus [14]. A sulcus angle >140° indicates trochlear dysplasia (Fig. 10.7). Patella subluxation can be quantified by measuring the congruence angle, which reflects the relationship between the central ridge of the patella and the intercondylar sulcus. The congruence angle is the angle between the bisector of the sulcus angle and a line drawn to the central ridge of the patella; a normal angle is defined as < −16°, whereby the central ridge is medial to the bisector line. If the congruence angle is lateral, then the angle is positive (Fig. 10.8).

Magnetic resonance imaging (MRI) and computed tomography have also a role in selected cases. MRI is useful for detecting associated chondral lesions and MPFL tears; one study demonstrated 85 % sensitivity and 70 % accuracy in detecting MPFL tears [15]. Axial computed tomography slices are used to provide three-dimensional images

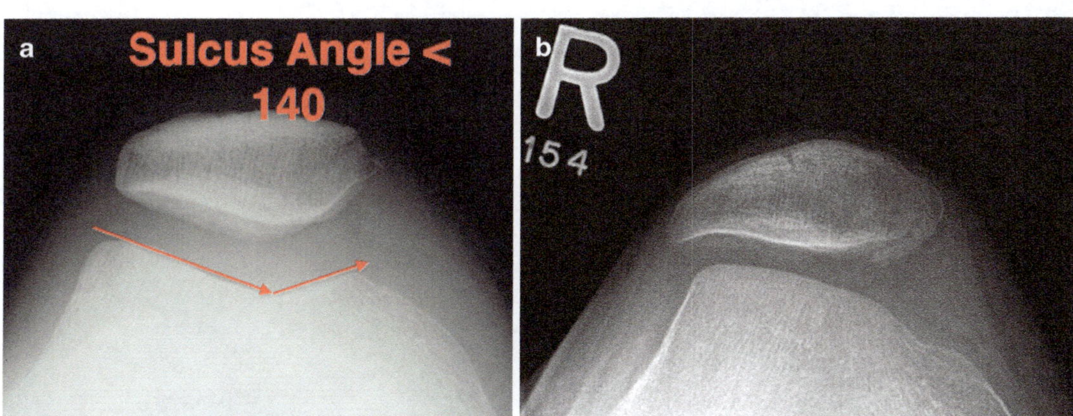

Fig. 10.7 (**a**) Normal sulcus angle of the femoral trochlea (<140°). The sulcus angle is measured by the intersection of the highest points of the medial and lateral femoral condyles and the lowest point of the intercondylar sulcus (as delineated by the *red arrows*). (**b**) Trochlear dysplasia (Image used with the permission of www.boneschool.com)

of the patellofemoral joint, defining the morphology of the trochlear groove and helping to diagnose both subluxation and lateral tilt of patella. However, many use CT to diagnose lateralization of the tibial tuberosity – by superimposing the axial slices of the trochlea and the tibial tuberosity the distance between tibial tuberosity and trochlear groove can be calculated (TT-TG); a TT-TG distance >20 mm is abnormal and is often associated with patellar instability secondary to malalignment (Fig. 10.9).

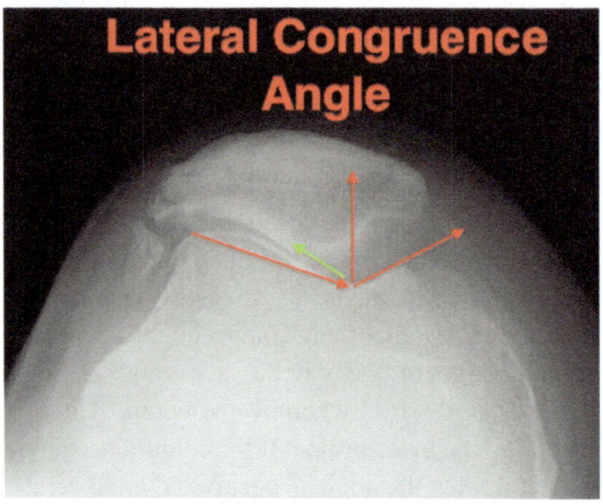

Fig. 10.8 Lateral subluxation of the patella, with the central ridge of the patella lateral to the intercondylar sulcus. The congruence angle is the angle between the bisector of the sulcus angle (*red arrows*), and a line drawn to the central ridge of the patella (*green arrow*). The central ridge should lie at or medial to the bisector of the trochlear groove. (Image used with the permission of www.boneschool.com)

10.5 Treatment

10.5.1 Nonoperative

Because of the favorable natural history of patella dislocations, nonoperative treatment is usually indicated for first-time dislocations and for patients with infrequent episodes of subluxation or dislocation. Fithian et al. reported that the recurrence rate for first-time dislocators was 17 %; however patients with subsequent dislocations had a risk of recurrence as high as 50 % [1]. In the acute phase, the immediate goals are to decrease the effusion, increase range of motion, and stimulate VMO activity.

The rehabilitation program in patient with patella instability focuses on VMO strengthening, improving core stability, restoring full ROM, and improving of proprioception. With regard to strengthening exercises, there is increased evidence that closed-chain exercises have more efficacy than open chain [16, 17], while treating weakness of the core musculature can limit excessive medial femoral rotation, which can contribute

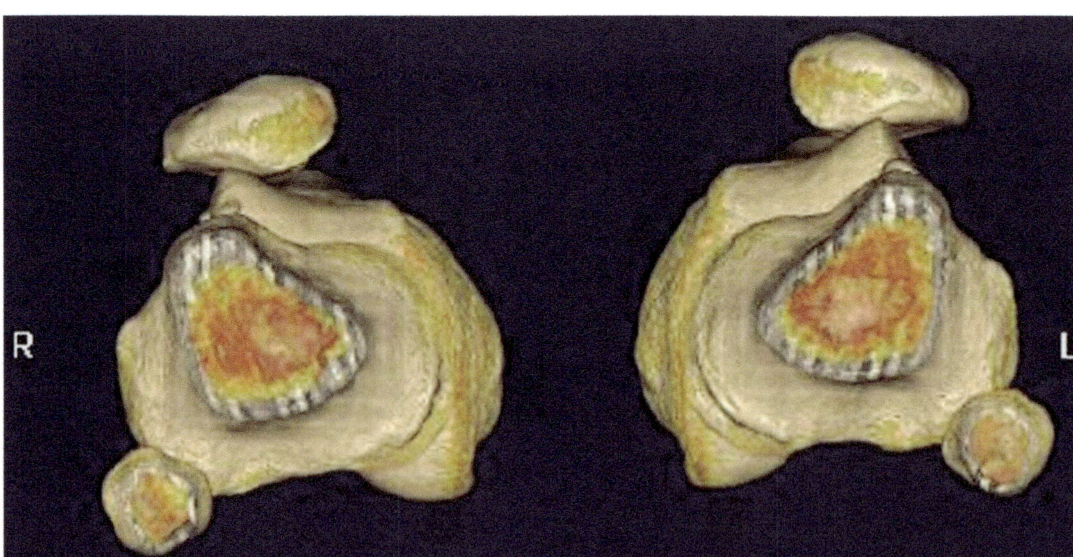

Fig. 10.9 Lateralization of the tibial tuberosity as shown in 3D axial computed tomography slices (Image used with the permission of www.boneschool.com)

to patellar instability [10]. Stretching exercises aimed at the iliotibial band and lateral retinaculum should also be part of the rehabilitation program.

The use of taping or a patellar brace is controversial; however these are widely utilized, in an attempt to apply a restraining force against lateral displacement of the patella [18–21]. It is important that braces are not be considered an alternative to the rehabilitation program.

10.5.2 Operative Treatment

Patients with predisposing factors including malalignment, bony abnormalities, and soft issue deficiencies are more prone to recurrent dislocation and can sometimes require operative intervention. Many surgical procedures have been described and will be presented in detail in the following chapters [22–30]. Broadly speaking, these procedures can be divided into soft tissue procedures (medial repair, medial imbrication, lateral release, MPFL reconstruction) and bony procedures (tibial tuberosity transfers, rotational osteotomies, trochleoplasty). Distal realignment procedures are indicated in patients with TT-TG distance >20 mm (medialization of the tibial tuberosity) or patella alta (distalization of the tibial tuberosity) or both [27–29, 31]. These procedures are often combined with a medial imbrication or MPFL reconstruction. While isolated MPFL reconstruction may be indicated in patients with normal TT-TG distance and normal patellar height [23, 24, 31], isolated lateral release is only indicated in patients with excessive lateral pressure syndrome and has no indication in the treatment of patella instability [30, 31]. In extreme circumstances, rotational abnormalities such as increased proximal femoral anteversion or excessive external tibial torsion may require surgical intervention.

In conclusion, there is no gold standard procedure for the surgical treatment of patellar instability, with a lack of good evidence to support one technique over any other. Patient examination should be thorough, focusing not only on patellofemoral joint but also on gait, lower extremity alignment, and rotational alignment. By combining examination findings with appropriate investigations, the etiology of patella instability should be clear to surgeons planning any procedures. We advocate customized treatments that correct the particular pathoanatomy of the patellofemoral joint.

References

1. Fithian DC, Paxton EW, Stone ML, Silva P, Davis DK, Elias DA, White LM. Epidemiology and natural history of acute patellar dislocation. Am J Sports Med. 2004;32:1114–21.
2. Hawkins RJ, Bell RH, Anisette G. Acute patellar dislocations. The natural history. Am J Sports Med. 1986; 14:117–20.
3. Cofield RH, Bryan RS. Acute dislocation of the patella: results of conservative treatment. J Trauma. 1977;17:526–31.
4. Andrish J. The biomechanics of patellofemoral stability. J Knee Surg. 2004;17:35–9.
5. Conlan T, Garth WP, Lemons JE. Evaluation of the medial soft-tissue restraints of the extensor mechanism of the knee. J Bone Joint Surg Am. 1993;75:682–93.
6. Goodfellow J, Hungerford DS, Zindel M. Patello-femoral joint mechanics and pathology. 1. Functional anatomy of the patello-femoral joint. J Bone Joint Surg Br. 1976;58:287–90.
7. Aglietti P, Insall JN, Cerulli G. Patellar pain and incongruence. I: measurements of incongruence. Clin Orthop Relat Res. 1983;176:217–24.
8. Beighton PH, Horan F. Orthopedic aspects of the Ehlers-Danlos syndrome. J Bone Joint Surg Br. 1969; 51:444–53.
9. James SL. Running injuries to the knee. J Am Acad Orthop Surg. 1995;3:309–18.
10. Lee TQ, Yang BY, Sandusky MD, McMahon PJ. The effects of tibial rotation on the patellofemoral joint: assessment of the changes in situ strain in the peripatellar retinaculum and the patellofemoral contact pressures and areas. J Rehabil Res Dev. 2001;38:463–9.
11. Berg EE, Mason SL, Lucas MJ. Patellar height ratios. A comparison of four measurement methods. Am J Sports Med. 1996;24:218–21.
12. Dejour D, Le Coultre B. Osteotomies in patello-femoral instabilities. Sports Med Arthrosc. 2007;15: 39–46.
13. Merchant AC, Mercer RL, Jacobsen RH, Cool CR. Roentgenographic analysis of patellofemoral congruence. J Bone Joint Surg Am. 1974;56:1391–6.
14. Dejour H, Walch G, Nove-Josserand L, Guier C. Factors of patellar instability: an anatomic radiographic study. Knee Surg Sports Traumatol Arthrosc. 1994;2:19–26.
15. Sanders TG, Morrison WB, Singleton BA, Miller MD, Cornum KG. Medial patellofemoral ligament injury following acute transient dislocation of the patella: MR findings with surgical correlation in 14 patients. J Comput Assist Tomogr. 2001;25:957–62.

16. Escamilla RF, Fleisig GS, Zheng N, Barrentine SW, Wilk KE, Andrews JR. Biomechanics of the knee during closed kinetic chain and open kinetic chain exercises. Med Sci Sports Exerc. 1998;30:556–69.
17. Stensdotter AK, Hodges PW, Mellor R, Sundelin G, Hager-Ross C. Quadriceps activation in closed and in open kinetic chain exercise. Med Sci Sports Exerc. 2003;35:2043–7.
18. Cowan SM, Bennell KL, Hodges PW. Therapeutic patellar taping changes the timing of vasti muscle activation in people with patellofemoral pain syndrome. Clin J Sport Med. 2002;12:339–47.
19. Nikku R, Nietosvaara Y, Kallio PE, Aalto K, Michelsson JE. Operative versus closed treatment of primary dislocation of the patella. Similar 2-year results in 125 randomized patients. Acta Orthop Scand. 1997;68:419–23.
20. Maenpaa H, Lehto MU. Patellar dislocation. The long-term results of nonoperative management in 100 patients. Am J Sports Med. 1997;25:213–7.
21. McConnell J. Rehabilitation and nonoperative treatment of patellar instability. Sports Med Arthrosc. 2007;15:95–104.
22. Ahmad CS, Stein BE, Matuz D, Henry JH. Immediate surgical repair of the medial patellar stabilizers for acute patellar dislocation. A review of eight cases. Am J Sports Med. 2000;28:804–10.
23. Farr J, Schepsis AA. Reconstruction of the medial patellofemoral ligament for recurrent patellar instability. J Knee Surg. 2006;19:307–16.
24. Steiner TM, Torga-Spak R, Teitge RA. Medial patellofemoral ligament reconstruction in patients with lateral patellar instability and trochlear dysplasia. Am J Sports Med. 2006;34:1254–61.
25. Donell ST, Joseph G, Hing CB, Marshall TJ. Modified Dejour trochleoplasty for severe dysplasia: operative technique and early clinical results. Knee. 2006;13:266–73.
26. Verdonk R, Jansegers E, Stuyts B. Trochleoplasty in dysplastic knee trochlea. Knee Surg Sports Traumatol Arthrosc. 2005;13:529–33.
27. Trillat A, Dejour H, Couette A. Diagnosis and treatment of recurrent dislocations of the patella. Rev Chir Orthop Reparatrice Appar Mot. 1964;50:813–24. French.
28. Fulkerson JP, Becker GJ, Meaney JA, Miranda M, Folcik MA. Anteromedial tibial tubercle transfer without bone graft. Am J Sports Med. 1990;18:490–7.
29. Barber FA, McGarry JE. Elmslie-Trillat procedure for the treatment of recurrent patellar instability. Arthroscopy. 2008;24:77–81.
30. Lattermann C, Toth J, Bach Jr BR. The role of lateral retinacular release in the treatment of patellar instability. Sports Med Arthrosc. 2007;15:57–60.
31. Colvin AC, West RV. Patellar instability. J Bone Joint Surg Am. 2008;90:2751–62.

11 Lateral Patellar Dislocation: Pathomechanism and Treatment

Yukiyoshi Toritsuka, Yuzo Yamada, Norimasa Nakamura, and Konsei Shino

Contents

Y. Toritsuka (✉)
Department of Orthopaedic Sports Medicine, Kansai Rosai Hospital, 3-1-69 Inabaso, Amagasaki, Hyogo 660-8511, Japan
e-mail: toritsuka@kanrou.net

Y. Yamada
Department of Orthopaedic Surgery, Japan Community Health Care Organization Osaka Hospital, 4-2-78 Fukushima, Fukushima, Osaka 553-0003, Japan

N. Nakamura
Department of Orthopaedic Surgery, Osaka University Graduate School of Medicine, 2-2 Yamadaoka, Suita, Osaka 565-0871, Japan

K. Shino
Osaka Yukioka College of Health Science, 1-1-41 Sojiji, Ibaraki, Osaka 567-0801, Japan

11.1 Introduction

Patellar dislocation occurs as a result of trauma or during sports activities among those with various predisposing factors. Many reports indicate a similar incidence of first-time dislocation in both males and females [1]. Dislocation generally occurs in a young population with an average age of around 20 during sports activities [2]. Disruptions of soft tissue restraints, including the medial patellofemoral ligament (MPFL), which normally prevent lateral patellar translation, almost inevitably occur [3, 4], and associated osteochondral fractures affect the prognosis [5].

11.2 Pathomechanism

11.2.1 MPFL Disruption

The MPFL originates at the adductor tubercle just distal to the insertion of the adductor magnus tendon just behind the medial epicondyle, the insertion site of the MCL and inserts into the proximal two-thirds of the medial margin of the patella, where its fibers fuse with the vastus medialis [6] (Fig. 11.1). MPFL disruption occurs most frequently at its femoral side [7], and tears in this position are prone to lead to re-dislocation by reducing the obliquity of the vastus medialis obliquus (VMO) muscle vector and its medial pull [8]. In contrast, tears at the patellar side are less frequent because of the ligament's wider

A. Gobbi et al. (eds.), *The Patellofemoral Joint*,
DOI 10.1007/978-3-642-54965-6_11, © ISAKOS 2014

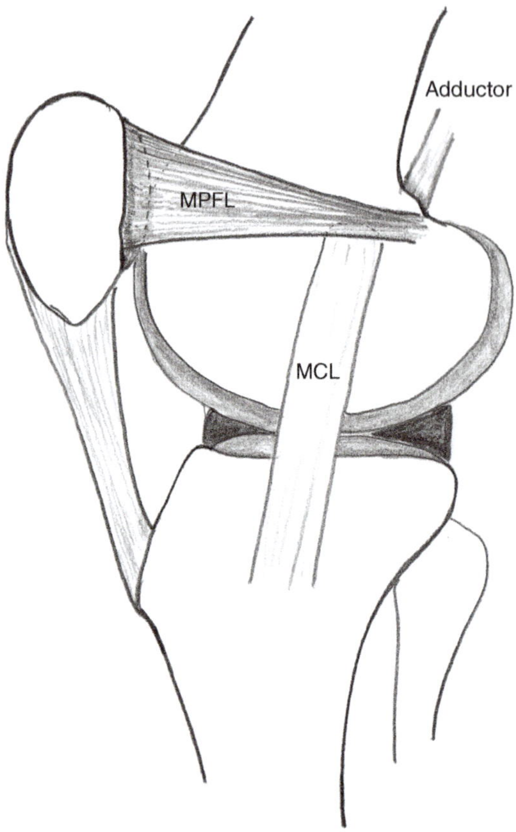

Fig. 11.1 The medial patellofemoral ligament

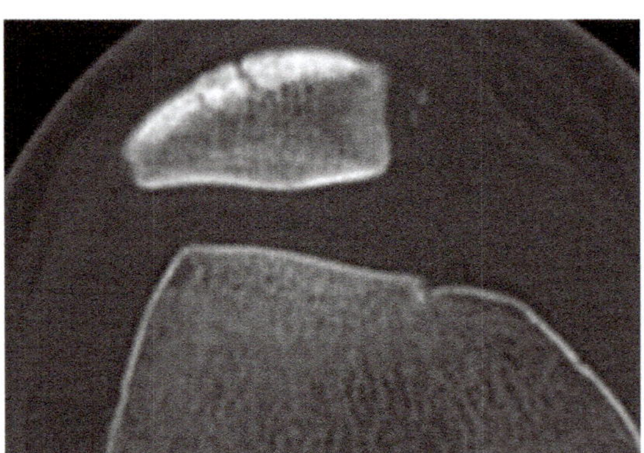

Fig. 11.2 Avulsion fracture of the MPFL

insertion, while avulsion fractures of the medial structures including retinaculum, capsule, or MPFL are not so rare [9, 10] (Fig. 11.2).

11.2.2 Osteochondral Fracture

Osteochondral fracture occurs either during dislocation or relocation. The medial facet of the patella is sheared off the lateral condyle of the femur during relocation, while the lateral condyle is wedged by the sharp lateral margin of the patella or sledge-hammered by the blunt patellar articular surface during dislocation [1, 11]. These injuries tend to occur in patients with fewer predisposing factors [12]. In order to prevent progression to osteoarthrosis, repairable fragments should be fixed as soon as possible.

11.2.3 Predisposing Factors

Patellar dislocation occurs in patients with predisposing factors such as a large Q angle, patella alta, or shallow femoral sulcus. Patellae in patients with these factors tend to laterally dislocate for the following reasons: (1) The greater Q angle causes an increase in the laterally directed traction force exerted on the patella by the quadriceps, (2) patella alta leads to insufficient restraint of the femoral sulcus because the patella does not go into the trochlea at shallow knee flexion angles, (3) the shallow femoral sulcus is less able to structurally restrain the lateral dislocation of the patella.

11.3 Treatment

First-time dislocators are treated both conservatively and surgically. Conservative treatment consists of techniques such as management of activity level, orthosis, and muscle strength exercises for the VMO, while surgical treatment includes various procedures intended to prevent re-dislocation such as tibial tubercle transfers, trochleoplasties, and MPFL reconstructions with or without lateral retinacular release. Generally a first-time dislocator without an osteochondral fracture is conservatively treated; however, this results in a re-dislocation rate of 15–44 % and limitation of strenuous sports activities in 60 % of patients without re-dislocation [1, 2].

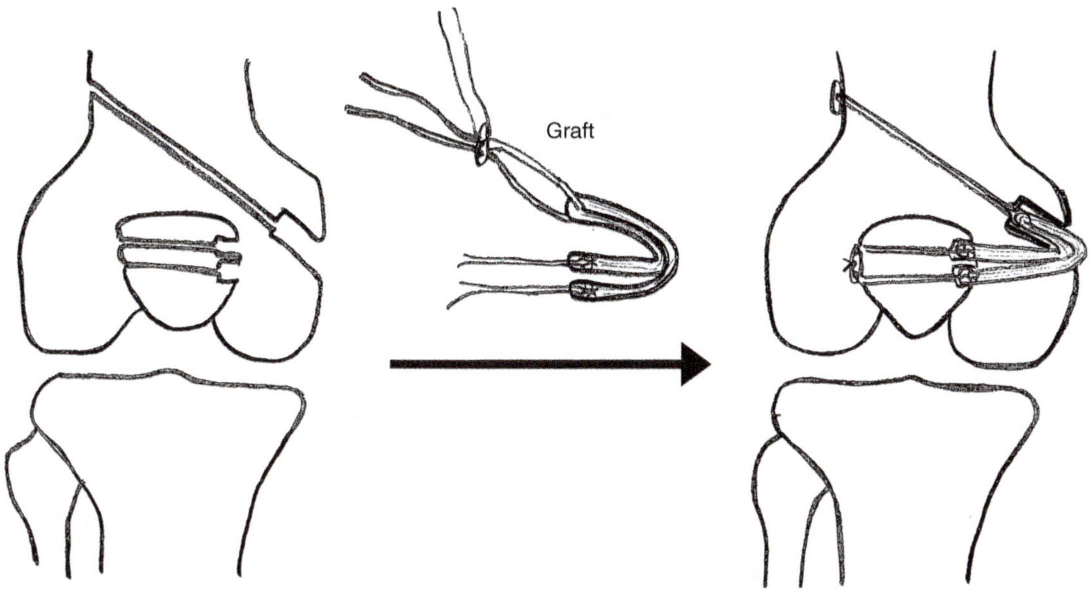

Fig. 11.3 Dual tunnel MPFL reconstruction

Traditional surgical treatments include tibial tubercle transfers to reduce the Q angles and trochleoplasties to deepen the femoral groove. However, both the advantages and disadvantages of each procedure should be understood, and individually appropriate procedures should be applied without generating secondary problems. While decreasing the Q angle by tibial tubercle transfer results in decreased lateral facet load, it results in increased loads elsewhere in the joint, leading to early arthrosis [13, 14]. In fact, Nakagawa et al. recently reported development of patellofemoral arthrosis after tibial tubercle transfer in spite of a satisfactory short-term clinical outcome [15]. Trochleoplasties make it possible to stabilize the patella by deepening the trochlea or elevating the lateral condyle but generate patellofemoral arthrosis from incongruent articulation and increased contact force [14]. Knoch et al. reported that trochleoplasty is a reasonable procedure but may not prevent subsequent development of osteoarthritis [16]. In contrast, experimental studies indicated that patellofemoral contact pressure was not significantly altered by MPFL reconstructions, and clinically satisfactory long-term results have also been reported so far [17–19]. Thus, MPFL reconstructions can be considered less invasive and safer procedures that prevent dislocation of the patella, avoiding secondary problems without any aggressive correction of the predisposing anatomies.

11.3.1 MPFL Reconstruction (Dual Tunnel MPFL Reconstruction) [20]

A schema of this procedure is shown in Fig. 11.3.

Graft Preparation

The harvested autogenous semitendinosus tendon is folded and utilized as a single double-looped tendon graft. A TightRope® (Arthrex Inc., Naples, FL) is used on the loop end to attach the femoral side, and two #2 FiberWire™ (Arthrex Inc.) sutures are placed using the Krackow suture technique on each free end to attach the patellar side (Fig. 11.4).

Operative Procedure

Prior to surgery, hypermobility of the patella is evaluated under anesthesia by Kolowich's grading [21]. Next, arthroscopic evaluation is performed, with special attention to cartilaginous damage to the patella, lateral femoral condyle and osteochondral fracture. In addition, patellar tracking is observed through full extension to 60° of flexion from the lateral suprapatellar portal.

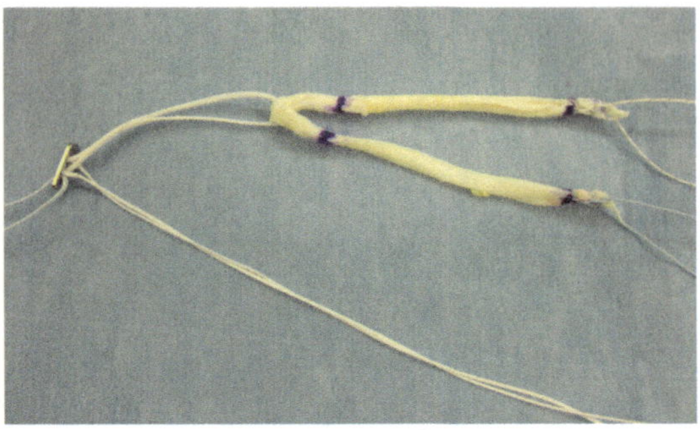

Fig. 11.4 A semitendinosus tendon graft

Following evaluation, a 5-cm transverse incision is made from 1 cm medial to the medial edge of the patella to just posterior to the medial epicondyle. The medial retinaculum is exposed and incised from the medial margin of the patella to the adductor tubercle to expose the MPFL insertion sites. A 2.4-mm guide wire is inserted from the femoral insertion site to the lateral cortex of the femur [7] and over-drilled using a 5–6-mm cannulated reamer according to the diameter of the graft. Two guide wires are transversely inserted, one from the proximal one third and one from the center of the medial edge of the patella to the lateral border under X-ray control (Fig. 11.5) and over-drilled using a 4.5-mm cannulated reamer up to 1 cm in depth. Both free ends of the graft are introduced into the patellar sockets and fixed on the lateral side

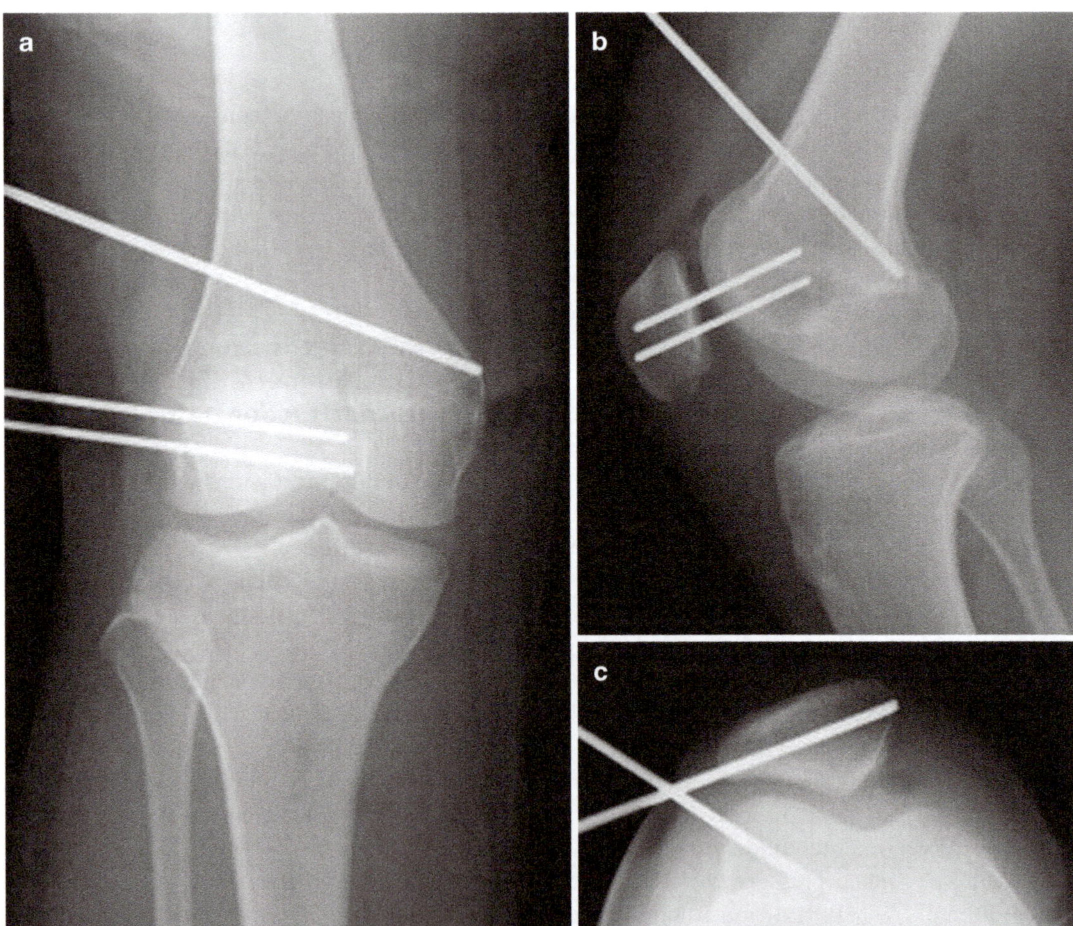

Fig. 11.5 X-ray control during surgery. (**a**) AP view, (**b**) lateral view, (**c**) axial view

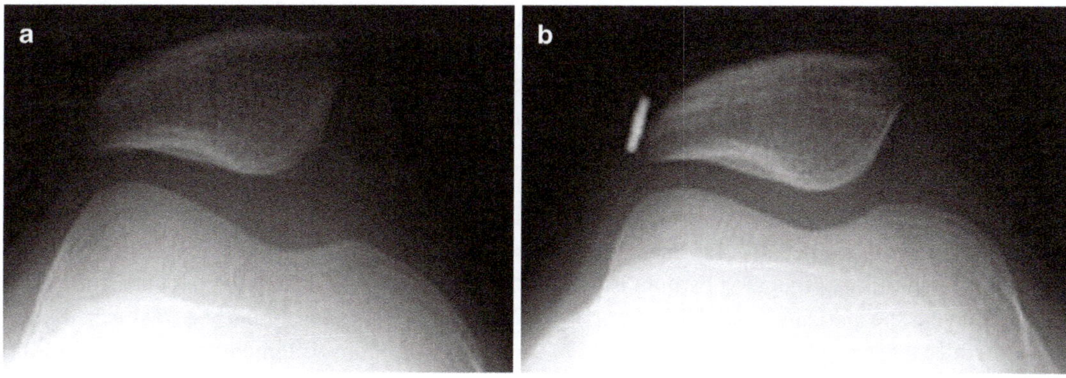

Fig. 11.6 Pre- vs. post-op radiographs: axial view. Lateral tilt of the patella was reduced while maintaining congruence of the patellofemoral joint. (**a**) pre-op, (**b**) post-op

of the patella by tying the sutures over an Endobutton™(Smith & Nephew Endoscopy, Andover, MA); then the loop end of the graft is introduced into the femoral tunnel. Femoral fixation is achieved at 45° of knee flexion under proper tension by tightening the TightRope®; fixing the graft at this flexion angle is assumed to be appropriate because the patella is already stabilized on the femoral groove (Fig. 11.6). Patellar stability is checked in near extension as well to avoid too tight fixation of the graft. The incised medial retinaculum is sutured together over the graft. When the surgeon does not want to expose a torn or elongated MPFL, a sub-retinacular tunnel is created for passage of the graft to avoid invasion of the VMO, followed by grafting as above.

Postoperative Regimen

The knee is immobilized for a week with a brace, followed by CPM exercise. Active/assisted ROM exercise is started at 2 weeks. Partial weight bearing is allowed at 3 weeks and full weight bearing is started at 5 weeks. Running is allowed at 3 months followed by a return to previous sporting activity at 6 months

Advantages of Our Procedure

Our procedure has the following theoretical advantages: First, this technique enables us to reconstruct the wide original patellar insertion of the MPFL. Secondly, better bone-tendon healing is expected by widening the contact area between the bone tunnel and the graft. Thirdly, it can reduce the potential risk of patellar fracture by creating shallow/short hollows with small-diameter tunnels compared to a larger transverse bone tunnel. Fourth, a stronger initial fixation is expected by pull-out fixation compared to suture to the periosteum or the suture anchor technique [22]. Finally, though not quantitative, it enables us to control the patellar position by adjusting the initial tension of the graft at the time of its fixation. Thus, our technique makes it possible to reliably fix a folded semitendinosus tendon graft under controlling laxity of the patella and to safely mimic the morphological characteristics of the natural MPFL leading to better graft-bone healing.

11.4 Discussion

11.4.1 Our Recommendations at the Time of the MPFL Reconstruction

Reconstruction or Repair

A disrupted MPFL should be reconstructed rather than repaired except in cases of avulsion fracture of the patella, because we believe that a repaired MPFL is not always robust enough to prevent re-dislocation.

Choice of Graft Material

We use the semitendinosus tendon rather than the gracilis tendon as a graft. Since Mountney et al.

reported that the MPFL has an average tensile strength of 208 N, the semitendinosus tendon, which has a strength of 1216 N, might be considered too strong [22, 23]. However we do not believe that stronger grafts do any harm unless they are too thick and violate the surrounding tissues. It should be noted that the grafts are required to oppose the residual predisposing factors after operation. In addition, a longer substitute is preferable because the patients' patellae tend to be more laterally located [24, 25].

Lateral Retinacular Release

Lateral retinacular release should not be universally performed as an additional procedure. It should be carefully applied only in those with a tight retinaculum, since it contributes to lateral patellar stability [26].

Additional Ligament Reconstruction

We usually reconstruct just the MPFL, the most important medial stabilizer. Drez et al. reported reconstruction of the medial patellotibial ligament (MPTL) using a gracilis tendon graft [27]. However, we consider it unnecessary to reconstruct the MPTL at the cost of additional harvesting of the gracillis tendon, because it is just one of the secondary medial stabilizers. We believe that MPFL reconstruction alone is sufficient to control patellar dislocation [5, 28].

11.4.2 Treatment Algorithm for Patellar Dislocation at Our Institute

When choosing MPFL reconstructions, the following three points should be taken into consideration: whether or not the growth plate remains, whether the dislocation is primary or recurrent, and whether osteochondral fractures are present. The treatment algorithm for patellar dislocation at our institute is shown in Fig. 11.7.

In patients with closed physes, conservative treatments are applied for first-time dislocators without osteochondral fracture. Patients with osteochondral fracture should be surgically treated using bio-absorbable pins to fix large fragments, while small fragments might be excised. When MPFL reconstruction is performed at the same time, great care should be taken to avoid promoting arthrofibrosis. In order to prevent arthrofibrosis, a staged surgery might be a safer method.

For those with a medial marginal fracture of the patella, also known as avulsion fracture of the MPFL, the MPFL should be reattached after excising the bony fragment because its midsubstance usually remains intact [10].

For patients with recurrent dislocation, we generally apply the MPFL reconstruction using a semitendinosus tendon [20].

In patients with open physes without osteochondral fracture, first-time dislocators are conservatively treated by orthosis, followed by later MPFL reconstruction after physeal closure. However, those with osteochondral fracture should be surgically treated. If conservative treatment fails, MPFL reconstruction should be performed without the use of any bone tunnels such as the one used to suture the graft to the periosteum. Medial reefing combined with lateral retinacular release might provide an alternative salvage procedure.

11.4.3 Our Strategy for Surgical Treatments

Our strategy at the time of surgical treatment is simply to stabilize patellar tracking by reconstructing the checkrein to prevent dislocation, without any aggressive correction of predisposing anatomies.

According to our previous morphological study, patients have more proximally extended articular cartilage of the femoral trochlea with increasing patellar height [25] (Figs. 11.8–11.10), and the articular cartilage of their femoral trochlea is more laterally distributed [25] (Fig. 11.11). These observations suggest that most individuals have their own morphological adaptation between the patella and the femur. Therefore, we consider it enough to simply restore their natural conformity rather than

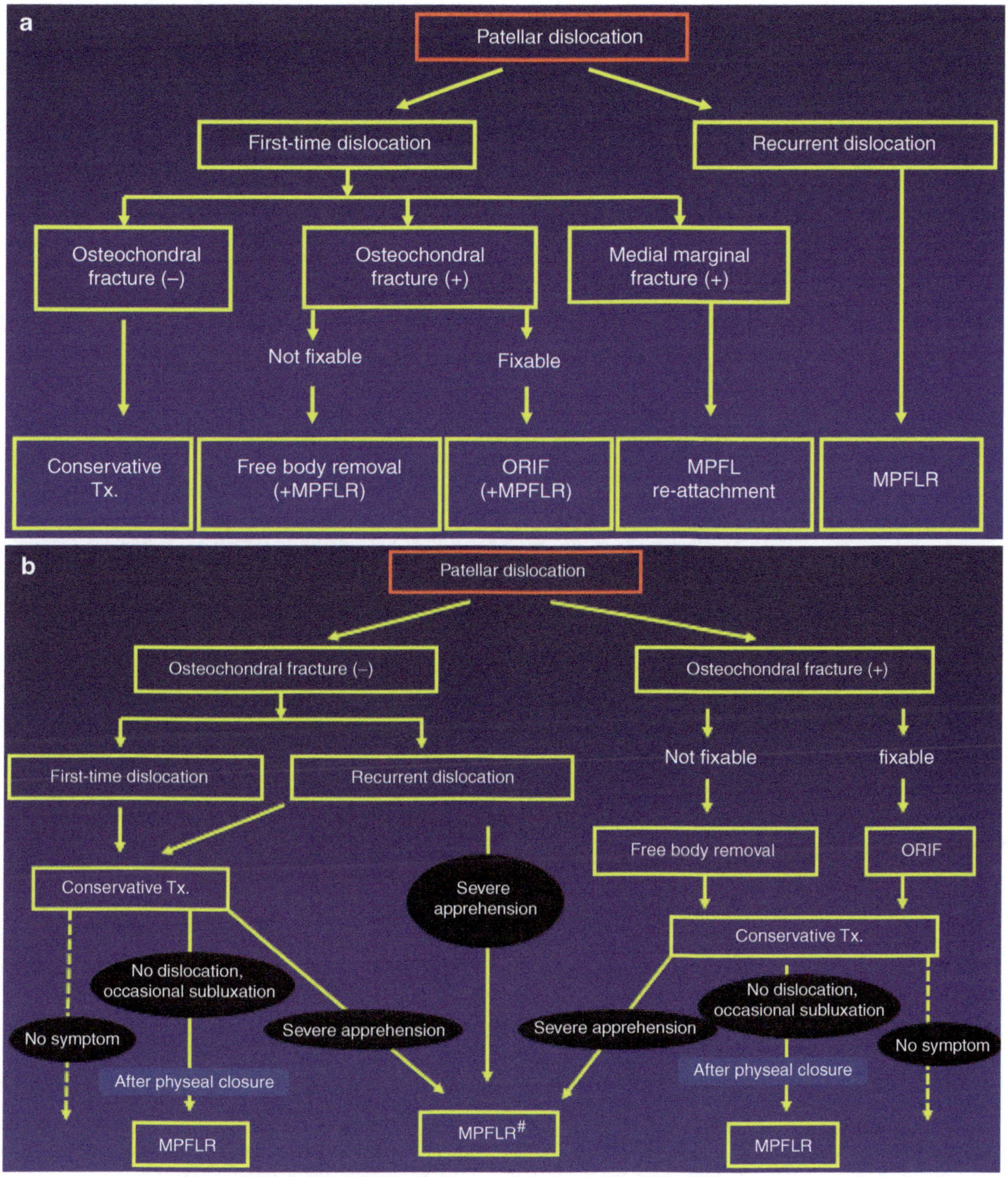

Fig. 11.7 The treatment algorithms at our institute. (**a**) For patients with closed physes, (**b**) for patients with open physes. *ORIF* open reduction and internal fixation, *Tx* treatment, *MPFLR* MPFL reconstruction, *MPFLR#* MPFL reconstruction without any bone tunnels

correcting malalignments based on measured angles or ratios from radiographs. An aggressive change of alignment or structure, we believe, under the guise of corrective surgery, might result in an unfavorable pressure distribution on the articular cartilage leading to secondary problems.

In addition, our previous 3-D motion analysis revealed that patellar tracking appeared unstable when the contact on the convex trochlea was

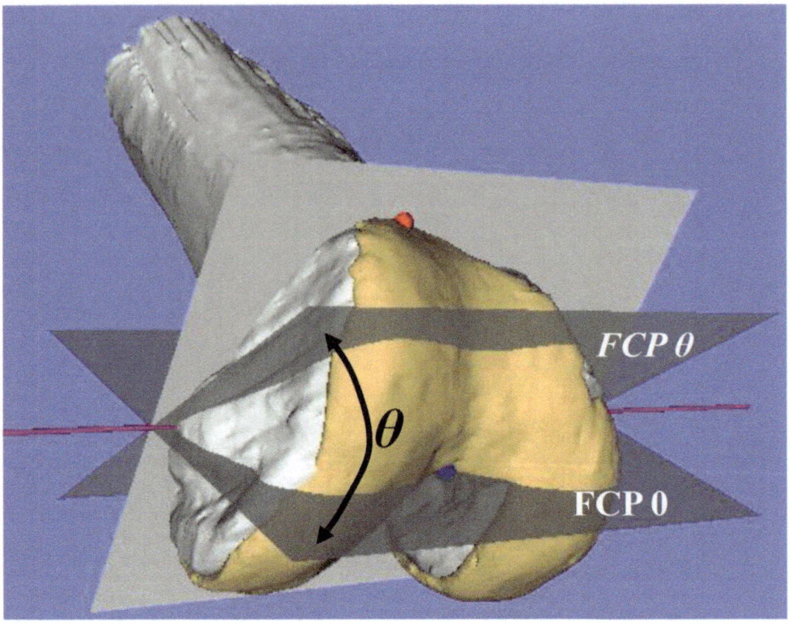

Fig. 11.8 Femoral condyle planes (*FCPs*). The Femoral Condyle Planes (*FCPs*) are the virtual cross sections including the transepicondylar axis (*TEA*). FCP 0 is defined as the reference plane including the top of the bone-articular cartilage border of the intercondylar notch (*blue spot*). FCP θ is defined as the plane with the angle θ from FCP 0 [25]

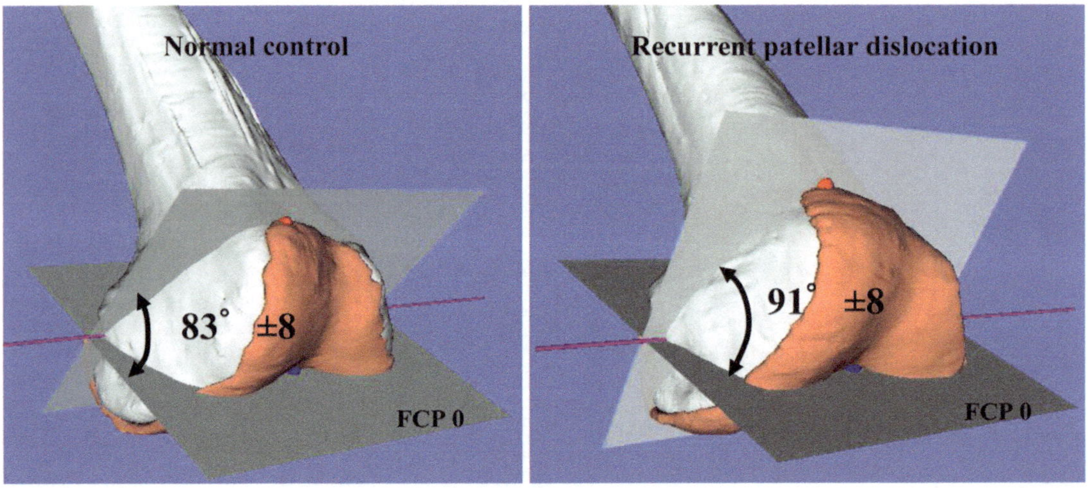

Fig. 11.9 Proximally-extended articular cartilage of the femoral trochlea. The proximal distribution of the articular cartilage is expressed by the angle between the FCP 0 and the FCP including the proximal edge of the articular cartilage. The mean value in normal controls and in patients with recurrent patellar dislocation is 83° ± 8 and 91° ± 8, respectively ($P = 0.03$) [25]

smaller than in normal controls but nevertheless appeared congruent at every flexion angle (Fig. 11.12) [24]. Therefore, we consider it sufficient to create a checkrein that acts only when laterally directed traction force is applied to the patella in order to stabilize the tracking while maintaining most of the patient's original congruity.

11.4.4 Future Directions

An understanding of the essential pathology of patellar dislocation is important when considering future treatment strategies, since unstable maltracking of the patella is generated by individual predisposing anatomies rather than by a single entity of malalignment. Recently, a new method of noninvasive in vivo 3-D motion analysis has been developed using computer technology. This allows creation of 3-D computer models and accurate analysis of kinematics of the patella using image-matching techniques for each patient [24], leading to individually customized treatment. Thus, as a first step in devising a customized strategy, we have tried to classify recurrent patellar dislocations according to patellar tracking. This classification could be applied to 95 % of patients, suggesting its usefulness [29].

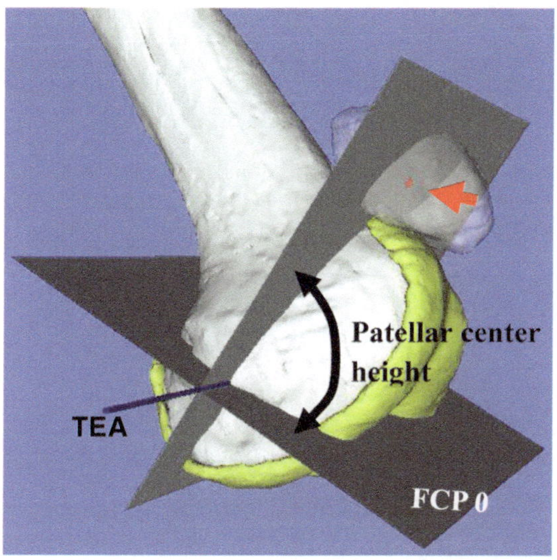

Fig. 11.10 Patellar center height. The patellar center height was expressed as the angle between FCP 0 and the FCP which included the patellar center (*arrow*). The patellar center height was strongly correlates with the proximal distribution of the articular cartilage ($r = 0.753$, $P<0.001$) [25]

Our findings can be summarized as follows:

1. Patellae moved medially with knee flexion in only half of the patients with recurrent patellar dislocation, although this tracking pattern is generally believed to be the most common in recurrent dislocation cases (Fig. 11.13a).
2. It was noteworthy that approximately 20% of the patellae moved laterally with knee flexion, which is regarded as the tracking pattern of habitual dislocation. In these cases, lateral release may be required in addition to MPFL reconstruction (Fig. 11.13b).
3. Approximately 20 % of the patients might be regarded as having a kind of mild type of recurrent patellar dislocation because the patellae showed smaller changes in mediolateral translation (Fig. 11.13c).
4. Less than 10 % of the patients showed an equivalent tracking pattern to that of the normal knee. These may be good candidates for conservative treatment in view of the patellar tracking, if they are first-time dislocators (Fig. 11.13d).

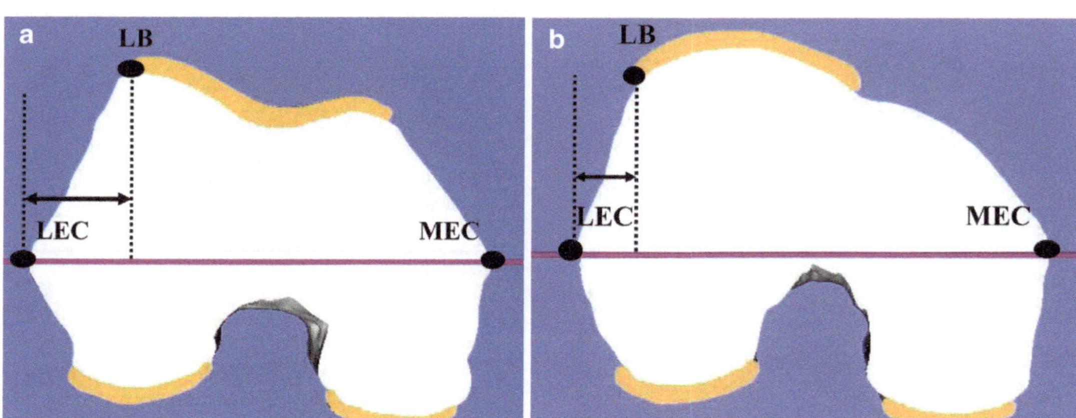

Fig. 11.11 Lateral distribution of articular cartilage of the femoral trochlea. These are representative images from subjects in each group shown in FCP 70. (**a**) A representative normal control. (**b**) A patient with recurrent patellar dislocation (RPD). The lateral edge of the articular cartilage of the femoral trochlea in patients with RPD is more laterally located [25]. *LEC* lateral epicondyle, *MEC* medial epicondyle, *LB* lateral border of the articular cartilage

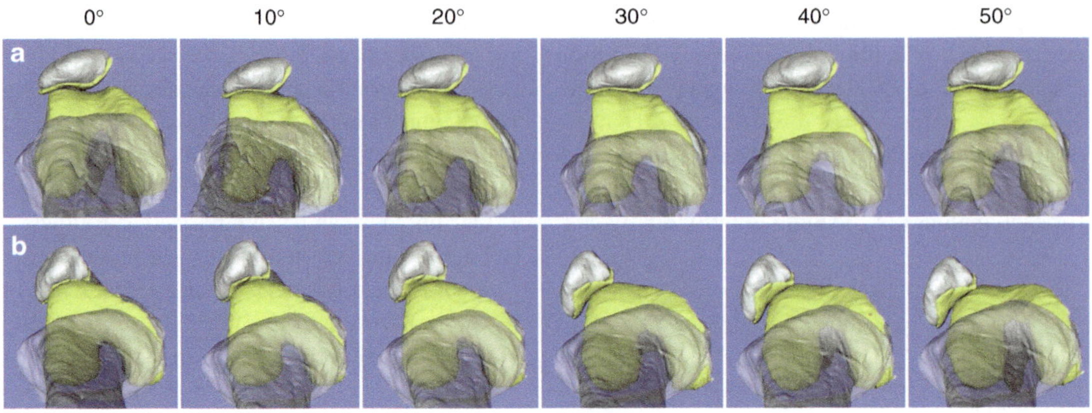

Fig. 11.12 Patellar tracking. (**a**) The *upper row* of images shows the tracking pattern of medial shift with knee flexion. (**b**) The *lower row* shows a pattern of lateral shift with knee flexion. These patterns of movement are totally different but each patella appears to have its own congruity during tracking

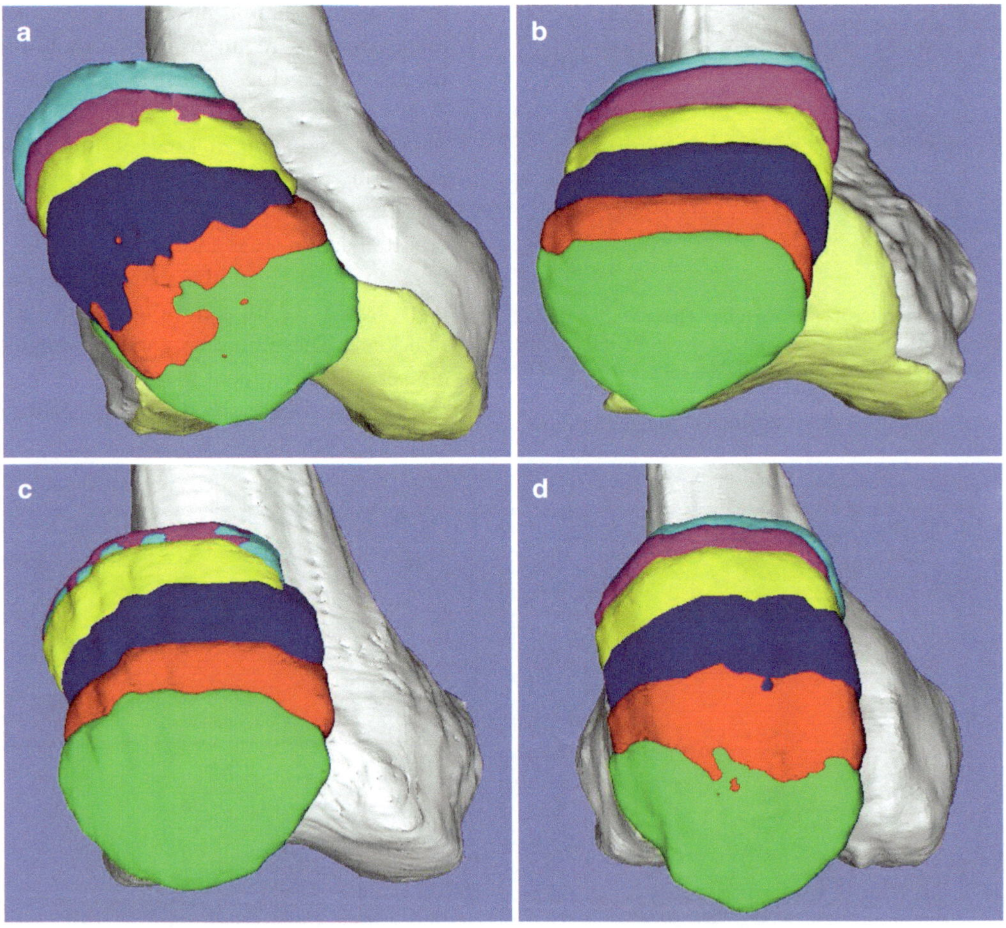

Fig. 11.13 Classification of recurrent patellar dislocation based on patellar tracking. Diagrams show representative four types of patellar tracking by 3-D computer models superimposed over images at 0–50° of knee flexion using voxel-based registration. The patella at 0° is shown in *light blue*, 10° in *pink*, 20° in *yellow*, 30° in *blue*, 40° in *red*, and 50° in *green*. (**a**) A severely subluxated-medial shift (SS-MS) type. (**b**) A severely subluxated-lateral shift (SS-LS) type. (**c**) A mildly subluxated-lateral path (MS-LP) type. (**d**) A mildly subluxated-central path (MS-CP) type

In conclusion, patients with recurrent patellar dislocation should not be managed as a single entity of patellar disorder, but instead they should be classified into several types at the time of treatment. This will become a first step to individually customized treatment leading to better clinical outcomes.

References

1. Halbrecht JL, Jackson DW. Acute dislocation of the patella. In: Fox JM, Del Pizzo W, editors. The patellofemoral joint. New York: McGraw-Hill, Inc; 1993. p. 123–34.
2. Atkin DM, Fithian DC, Marangi KS, Stone ML, Dobson BE, Mendelsohn C. Characteristics of patients with primary acute lateral patellar dislocation and their recovery within the first 6 months of injury. Am J Sports Med. 2000;28:472–9.
3. Sallay PI, Poggi J, Speer KP, Garrett WE. Acute dislocation of the patella: a correlative pathoanatomic study. Am J Sports Med. 1996;24:52–60.
4. Nomura E. Classification of lesions of the medial patello-femoral ligament in patellar dislocation. Int Orthop. 1999;23:260–3.
5. Gomes JLE. Medial patellofemoral ligament reconstruction for recurrent dislocation of the patella: a preliminary report. Arthroscopy. 1992;8:335–40.
6. Feller JA, Feagin Jr JA, Garrett Jr WE. The medial patellofemoral ligament revisited: an anatomical study. Knee Surg Sports Traumatol Arthrosc. 1993;1:184–6.
7. Nomura E, Inoue M, Osada N. Anatomical analysis of the medial patellofemoral ligament of the knee, especially the femoral attachment. Knee Surg Sports Traumatol Arthrosc. 2005;13:510–5.
8. Ahmad CS, Stein BE, Matuz D, Henry JH. Immediate surgical repair of the medial patellar stabilizers for acute patellar dislocation: a review of eight cases. Am J Sports Med. 2000;28:804–10.
9. Vainionpaa S, Laasonen E, Silvennoinen T, Vasenius J, Rokkanen P. Acute dislocation of the patella. A prospective review of operative treatment. J Bone Joint Surg Br. 1990;72:366–9.
10. Toritsuka Y, Horibe S, Hiro-Oka A, Mitsuoka T, Nakamura N. Medial marginal fracture of the patella following patellar dislocation. Knee. 2007;14:429–33.
11. Rosemberg NJ. Osteochondral fractures of the lateral femoral condyle. J Bone Joint Surg Am. 1964;46(5):1013–26.
12. Runow A. The dislocating patella. Etiology and prognosis in relation to generalized joint laxity and anatomy of the patellar articulation. Acta Orthop Scand. 1983;201(Suppl):1–5.
13. Huberti HH, Hayes WC. Patellofemoral contact pressures: the influence of Q-angle and tendofemoral contact. J Bone Joint Surg Am. 1984;66(5):715–24.
14. Steiner T, Parker RD. Subluxation and dislocation. In: DeLee JC, Dretz D, Miller MD, editors. DeLee & Drez's orthopaedic sports medicine: principles and practice. 3rd ed. Philadelphia: Saunders Elsevier Inc; 2010. p. 1548–72.
15. Nakagawa K, Wada Y, Minamide M, Tsuchiya A, Moriya H. Deterioration of long-term clinical results after the Elmslie-Trillat procedure for dislocation of the patella. J Bone Joint Surg [Br]. 2002;84B:861–4.
16. von Knoch F, Böhm T, Bürgi ML, von Knoch M, Bereiter H. Trochleaplasty for recurrent patellar dislocation in association with trochlear dysplasia: a 4- to 14-year follow-up study. J Bone Joint Surg Br. 2006;88:1331–5.
17. Nomura E, Inoue M, Kobayashi S. Long-term follow-up and knee osteoarthritis change after medial patellofemoral ligament reconstruction for recurrent patellar dislocation. Am J Sports Med. 2007;35:1851–8.
18. Ostermeier S, Holst M, Bohnsack M, Hurschler C, Stukenborg-Colsman C, Wirth CJ. Dynamic measurement of patellofemoral contact pressure following reconstruction of the medial patellofemoral ligament: an in vitro study. Clin Biomech (Bristol, Avon). 2007;22:327–35.
19. Ostermeier S, Holst M, Bohnsack M, Hurschler C, Stukenborg-Colsman C, Wirth CJ. In vitro measurement of patellar kinematics following reconstruction of the medial patellofemoral ligament. Knee Surg Sports Traumatol Arthrosc. 2007;15:276–85.
20. Toritsuka Y, Amano H, Mae T, Uchida R, Hamada M, Ohzono K, Shino K. Dual tunnel medial patellofemoral ligament reconstruction for patients with patellar dislocation using a semitendinosus tendon autograft. Knee. 2011;18:214–9.
21. Kolowich PA, Paulos LE, Rosenberg TD, Farnsworth S. Lateral release of the patella: indications and contraindications. Am J Sports Med. 1990;18:359–65.
22. Mountney J, Senavongse W, Amis AA, Thomas NP. Tensile strength of the medial patellofemoral ligament before and after repair or reconstruction. J Bone Joint Surg Br. 2005;87:36–40.
23. Noyes FR, Grood ES. The strength of the anterior cruciate ligament in humans and Rhesus monkeys. J Bone Joint Surg Am. 1976;58:1074–82.
24. Yamada Y, Toritsuka Y, Horibe S, Sugamoto K, Yoshikawa H, Shino K. In vivo movement analysis of the patella using a three-dimensional computer model. J Bone Joint Surg Br. 2007;89:752–60.
25. Yamada Y, Toritsuka Y, Yoshikawa H, Sugamoto K, Horibe S, Shino K. Morphological analysis of the femoral trochlea in patients with recurrent dislocation of the patella using three-dimensional computer models. J Bone Joint Surg Br. 2007;89:746–51.
26. Conlan T, Garth WP. Evaluation of the medial soft-tissue restraints of the extensor mechanism of the knee. J Bone Joint Surg Am. 1993;75:682–93.

27. Drez Jr D, Edwards TB. Results of medial patellofemoral ligament reconstruction in the treatment of patellar dislocation. Arthroscopy. 2001;17:298–300.
28. Nomura E, Horiuchi Y. A mid-term follow-up of medial patellofemoral ligament reconstruction using an artificial ligament for recurrent patellar dislocation. Knee. 2000;7:211–5.
29. Yamada Y, Toritsuka Y, Horibe S, Nakamura N, Nakata K, Shiozaki Y, Matsumoto N, Sugamoto K, Yoshikawa H, Shino K. Classification of patellar tracking in patients with recurrent dislocation of the patella using three-dimensional computer models. Presented in: 9th biennial ISAKOS congress. 2013.

Medial Patellar Instability: A Little Known Cause of Anterior Knee Pain

12

Vicente Sanchis-Alfonso, Cristina Ramírez-Fuentes, Francisco Martínez-Soriano, Joan C. Monllau, and Alan C. Merchant

Contents

V. Sanchis-Alfonso (✉)
Department of Orthopaedic Surgery, Hospital Arnau de Vilanova, C/San Clemente 12, 46015 Valencia, Spain

Department of Orthopaedic Surgery, Hospital 9 de Octubre, Valle de la Ballestera 59, 46015 Valencia, Spain
e-mail: vicente.sanchis.alfonso@gmail.com

C. Ramírez-Fuentes
Department of Radiology, Hospital Universitario y Politécnico La Fe, Valencia, Spain

F. Martínez-Soriano
Department of Human Anatomy and Embriology, University of Valencia, Valencia, Spain

J.C. Monllau
Department of Orthopaedic Surgery, Parc de Salut Mar (Hospital del Mar/Hospital de l'Esperança), ICATME-Hospital Universitari Quirón-Dexeus, Universitat Autònoma de Barcelona (UAB), Barcelona, Spain

A.C. Merchant
Department of Orthopedic Surgery (Emeritus), Stanford University School of Medicine, Stanford, California, USA

Department of Orthopedic Surgery (Emeritus), El Camino Hospital, Mountain View, California, USA

12.1 Background

Anterior knee pain (AKP) is one of the most common patient complaints heard by orthopedic surgeons. A wide variety of causes of AKP have been ascribed to it. Medial patellar instability (MPI) is an objective condition with its own personality provoking incapacitating AKP that should be included in the differential diagnosis of AKP patients, above all in disabling AKP patients after realignment surgery. However, this condition can be difficult to diagnose because it is not well described in the medical literature. Its clinical repercussion was identified at the end of the 1980s [1]. We can now state that it is an objective cause of AKP and that it is more frequent than we had thought, although it might be underdiagnosed because it is still not a well-known entity. The first author has seen many patients with MPI who have seen more than three physicians without a true diagnosis of their condition.

Lateral retinacular release (LRR), described by Merchant and Mercer in 1974 [2], has been one of the most frequently performed

A. Gobbi et al. (eds.), *The Patellofemoral Joint*,
DOI 10.1007/978-3-642-54965-6_12, © ISAKOS 2014

procedures in orthopedic surgery to treat AKP patients. This procedure was considered to be relatively benign and had a low associated complication rate. In 1988, Hughston and Deese [1] described MPI as a complication of LRR for the first time. Often, MPI is a late complication of both open and arthroscopic "extensive" LRR. Of the 166 cases reported in the literature, 152 (91 %) occurred in patients who had undergone a previous LRR, either isolated or associated with patellofemoral realignment surgery. In these cases, the surgery was followed, at varying intervals, by a paradoxical increase in their symptoms: incapacitating AKP, difficulty with stairs, intermittent swelling, patellar crepitus, and disabling giving way. In the remaining 14 reported cases (non-iatrogenic), 8 were the result of trauma, and 6 occurred spontaneously. We can conclude that MPI after LRR or realignment surgery is a rare entity, and non-iatrogenic MPI, either traumatic or spontaneous, is even more uncommon. However, the incidence of MPI after LRR is not known. Out of 60 knees referred to Hughston and Deese [1] due to LRR failure, 30 (50 %) were found to have painful MPI. Shellock et al. [3] evaluated 40 patients (43 knees) with persistent symptoms after LRR with kinematic magnetic resonance imaging (MRI) of the patellofemoral joint. They found that 27 (63 %) had MPI. Kolowich et al. [4] found that, after LRR, 16 of 57 knees (28 %) needed reconstruction of the lateral patellar anatomy because of painful MPI. More recently, Pagenstert et al. [5] have shown that 5 of 14 patients (36 %) in their LRR group showed MPI.

Recent evidence suggests that the indications for LRR must be more limited. Indeed, LRR and realignment surgery are currently performed less frequently [6]. Therefore, we might think that the cases of MPI would decrease. This, however, is not true. Nowadays, more and more medial patellofemoral ligament (MPFL) reconstructions are performed, and we must take note that an anterior and proximal placement of the MPFL femoral graft insertion may also lead to iatrogenic medial patellar subluxation [7]. Moreover, when a malpositioning of the MPFL femoral graft insertion is combined with an overtightened MPFL and an LRR previously performed or performed during MPFL surgery, an iatrogenic MPI may develop [7]. It is interesting to note that many of the patients feel as if the patella subluxes laterally when it actually subluxes medially. This fact makes the diagnosis more difficult.

12.2 The Problem

The patient with MPI is an unusual one who comes into our office with incapacitating chronic AKP and a disabling disability with very poor Kujala score, as well as important psychological problems. Thus, the MPI patient is different from the "typical" AKP patient. In all the cases of the series of the first author, there was a previous surgical procedure consisting of an "extensive" isolated LRR, sometimes associated with a proximal and/or distal realignment, to treat patellofemoral pain resistant to conservative treatment or to treat lateral patellar instability. The patient is a female in eight out of ten cases (80 %), which is consistent with epidemiological studies on AKP. She had been evaluated by several physicians who had told her that there was nothing that could be done to solve her problem, and then she comes to our office with a large folder full of radiological studies (MRI, CT scan, x-rays) that are identified as normal or as "lateral patellar subluxation" or "chondromalacia patellae" at the most. However, the patient feels a new pain and a new instability with day-to-day activities that are distinct and much worse than the ones prior to the LRR or realignment surgery. The question we should ask ourselves is: Are these radiologically "normal" tests enough to rule out an objective condition that justifies the severe pain the patients suffer? The answer is an emphatic no. We can understand this answer by analyzing case # 1 (Fig. 12.1).

12.3 Diagnosis: Functional Evaluation

The diagnosis is based upon the physical examination. The clinical findings are crucial for the diagnosis of MPI. The most important findings are (1) pain and tenderness localized at the site

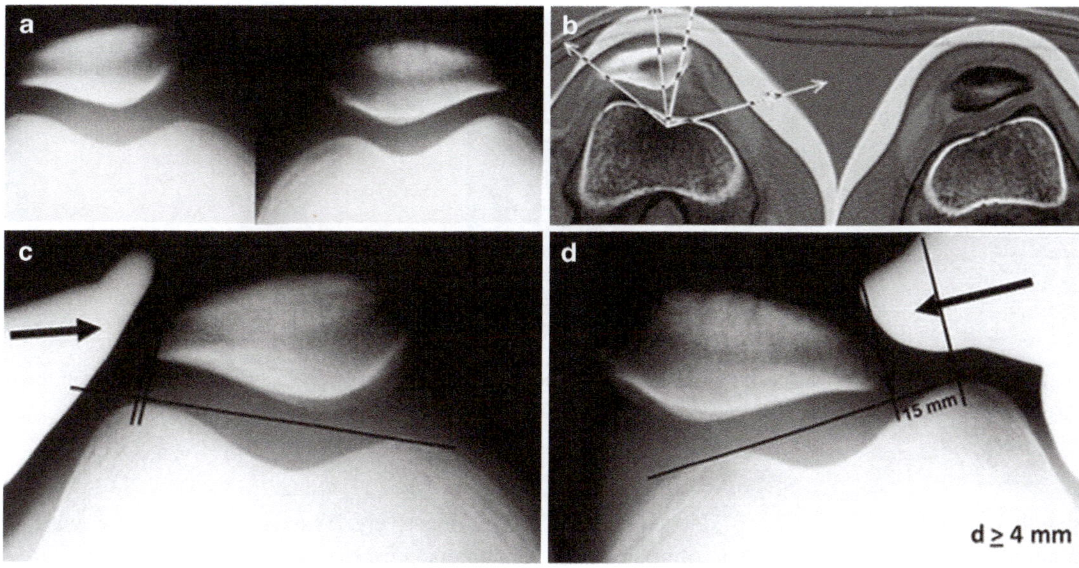

Fig. 12.1 Case # 1. A 24-year-old woman with severe AKP (10/10 – VAS), a disabling disability (4/100 – Kujala score), and patellar instability in the left knee distinct and much worse than the previous instability, anxiety, depression, catastrophizing ideas with pain and kinesiophobia. This knee was operated on 2 years ago, performing an Insall proximal realignment and LRR. She came to our office with conventional radiographs, which were normal (**a**), and a CT at 0° that showed correct radiological patellofemoral congruence (**b**). Nevertheless, the right knee was asymptomatic despite the patellar subluxation and patellar tilt (**b**). The Fulkerson relocation test for medial subluxation was positive. An axial stress radiograph of the left knee allowed us to detect an iatrogenic medial subluxation of the patella (medial displacement of 15 mm) (**d**). Axial stress radiograph of the asymptomatic right knee (**c**). The symptoms disappeared after an isolated surgical correction of the medial subluxation of the patella using iliotibial band and patellar tendon for repairing the lateral stabilizers of the patella (From Sanchis-Alfonso [8]. Reproduced with permission from Springer)

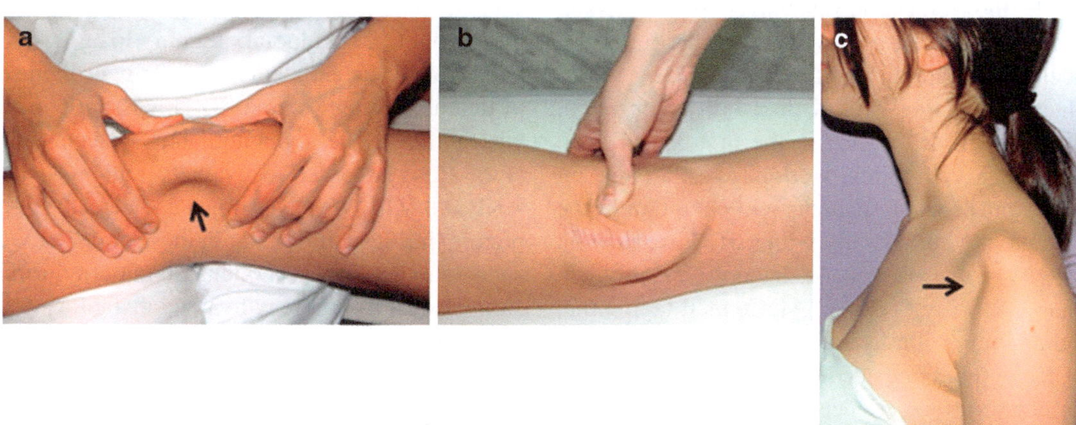

Fig. 12.2 Patellar glide test in a patient with multidirectional instability. Pathologic lateral displacement of the patella (**a**). Pathologic medial displacement of the patella (**b**). We have seen an image (**a**) similar to the sulcus sign (*black arrow*) observed in patients with multidirectional instability of the shoulder (**c**) (From Sanchis-Alfonso [8]. Reproduced with permission from Springer)

of the lateral retinacular defect, (2) increased passive medial patellar mobility (Fig. 12.2), (3) a positive medial patellar apprehension test, (4) a positive "gravity subluxation test" [9], and (5) a positive Fulkerson's relocation test [10]. As always with a knee exam, the contralateral

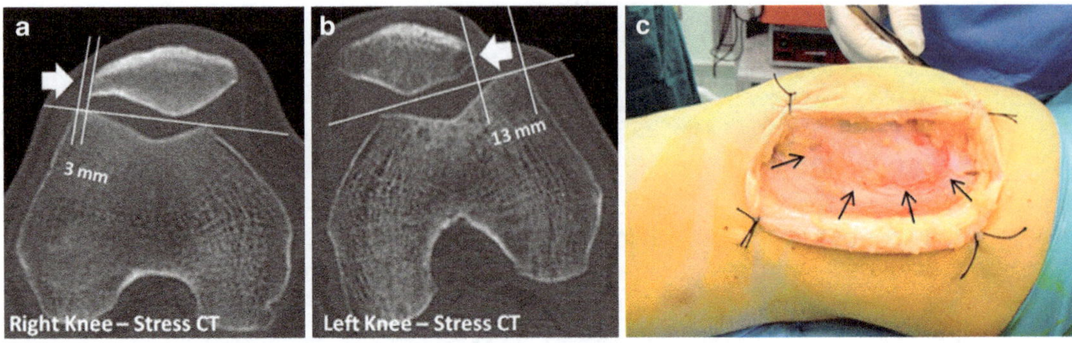

Fig. 12.3 Patellofemoral imbalance secondary to lateral retinaculum release (iatrogenic medial patellar instability of the left knee). (**a**, **b**) Stress CT. (**c**) Absence of lateral retinaculum of the same patient delimited by *arrows*

patella should be assessed to determine the amount of normal medial patellar subluxation in a particular patient. Our preference is the relocation test described by Fulkerson [10]. This is a very important diagnostic test, particularly in patients who are symptomatic after previous realignment surgery or LRR. Fulkerson's test consists of applying a medial translation force to the patella with the knee extended and then passively flexing the knee, which causes a sudden entry of the patella into the trochlea from medial to lateral. In patients with symptomatic MPI, the relocation maneuver of Fulkerson will usually cause considerable discomfort, a feeling of apprehension and a dramatic reproduction of the painful instability sensation. In general, MPI occurs in the first 30° of knee flexion. MPI is often overlooked as a cause of symptoms because patients will complain of the patella moving laterally with early knee flexion. One way that we help to diagnose the problem is to perform a "reverse" McConnell taping that holds the patella laterally to attempt to prevent it from subluxing medially. All of the patients in our series experienced significant relief from their pain with this kind of tape. In the same way, application of a patellar brace with the buttress pad or strap on the medial side will minimize or eliminate symptoms of MPI. Both taping and brace serve as a diagnostic tool.

Stress radiography, with the technique described by Teitge et al. [11], has proven to be a useful diagnostic tool to document and to quantify MPI objectively. An axial stress CT scan, with the technique described by Biedert [12], also confirms the diagnosis (Fig. 12.3). If possible, a comparison of the normal with the abnormal side is more important than the absolute amount of displacement.

Stair climbing is one of the most painful and challenging activities of daily living for subjects with AKP. Moreover, it is universally accepted that going downstairs is more challenging than going upstairs due to the level of eccentric control required during step descent. Therefore, stair descent can be demanding enough from a biomechanical standpoint, not only to aggravate pain in AKP patients but also to trigger the use of defense strategies as well. This is the reason why we use the dynamic stair descent test (Fig. 12.4) in order to evaluate patients with AKP and MPI. We have shown a knee extension pattern during stair descent in these patients. We speculate that the knee extension pattern during stair descent is a strategy to avoid instability and pain. With knee flexion, the patella migrates medially, and this displacement is greater in the absence of the lateral retinaculum in MPI patients. Moreover, with this knee extension pattern, there is a reduction in the knee extensor moment and therefore a reduction in pain. Finally, with this knee extension pattern, the posterior muscles work in a persistent elongated eccentric condition; this situation might be responsible at least in part for the posterior knee pain in MPI patients.

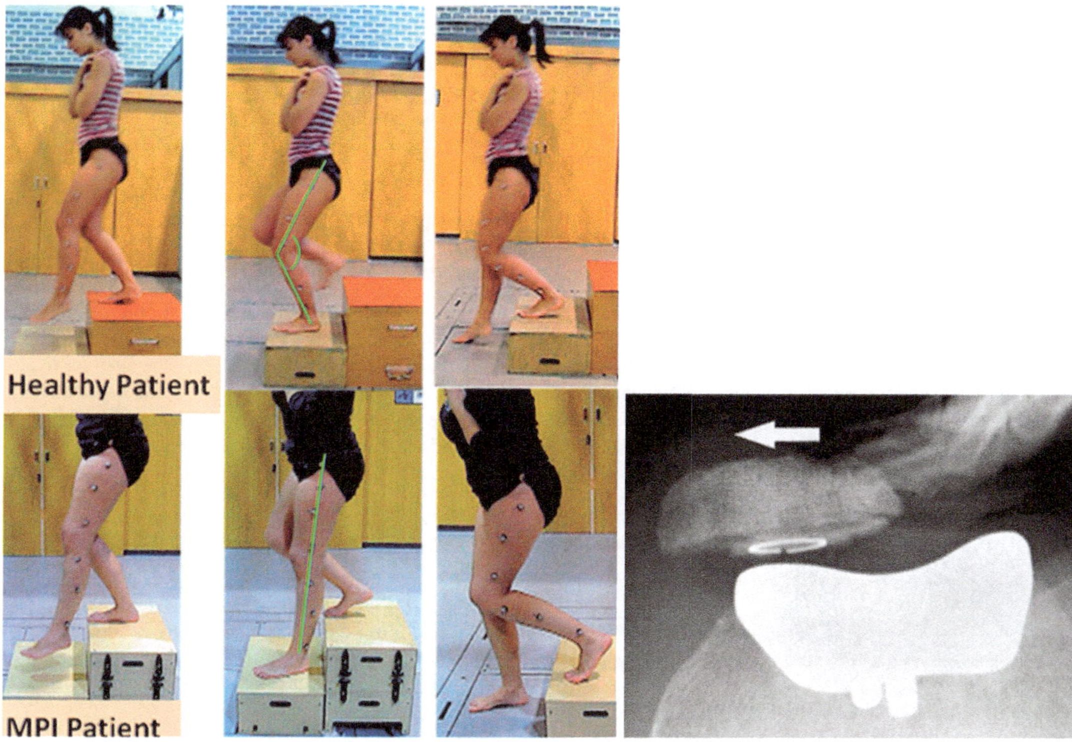

Fig. 12.4 Knee joint angle during stair descent. In this case the patient had five previous surgeries and presented with severe AKP in spite of the patellofemoral arthroplasty. The cause of pain was MPI secondary to a previous "extensive" LRR (From Sanchis-Alfonso [8]. Reproduced with permission from Springer)

12.4 Treatment: A Paradigm of "Salvage" Procedures

Nonoperative treatment of these patients, which includes increasing quadriceps strength and stretching of the medial retinaculum, usually fails to relieve their symptoms or improve their function. Thus, the majority of these patients require surgical treatment.

We advocate diagnostic arthroscopy before open surgery to address concomitant patellofemoral chondral lesions, which can be common in this patient population, and rule out other concomitant intra-articular knee conditions that can cause pain.

There are two surgical therapeutic approaches that are completely opposite: medial retinacular release [13] (functional surgery) and lateral reticular repair or reconstruction [14–19] (reconstructive surgery). Reconstructive surgery includes direct ligament repairs or reconstructions of the "lateral patellofemoral ligament." The reconstructive procedures described use (1) the iliotibial band, (2) the patellar tendon, and (3) the quadriceps tendon with a bone block. Hughston et al. [17] reviewed 65 cases of chronic medial patellar subluxation that were treated with either a direct repair (39 knees) or a reconstruction of the lateral patellofemoral ligament (26 knees). They reported that surgical revisions were required in 6 knees (9 %), complications occurred in 16 knees (25 %), and subsequent surgical procedures were necessary in 15 knees (23 %).

Given that the literature regarding surgical treatment offers only evidence levels of IV or V, the surgical therapeutic approach should be based on common sense, experience, and the surgeon's criteria. A logical approach should be based on three elements: (1) repair what is damaged, (2) restore native anatomy [20–22] (Fig. 12.5) and functionality, and (3) do not make the same mistakes again.

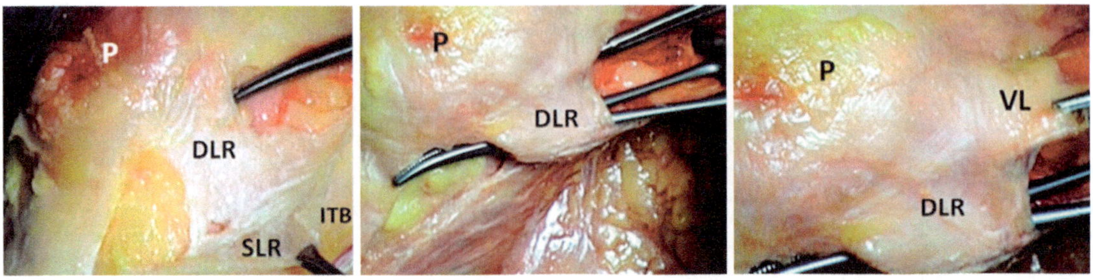

Fig. 12.5 Anatomy of the lateral retinaculum. Patella (*P*), deep lateral retinaculum (*DLR*), superficial lateral retinaculum (*SLR*), iliotibial band (*ITB*), and vastus lateralis (*VL*). The true lateral patellofemoral ligaments are thickenings of the lateral capsule. There is a lateral epicondylopatellar ligament described and present in some individuals, to varying degrees of frequency, but the superficial oblique and deep transverse retinacular layers are more consistent. The superficial oblique retinaculum is quite thin. The deep transverse retinaculum is stout, oriented in an optimal direction to restrain the patella and attached to the lateral boarder of the patella and the deep surface of the iliotibial band

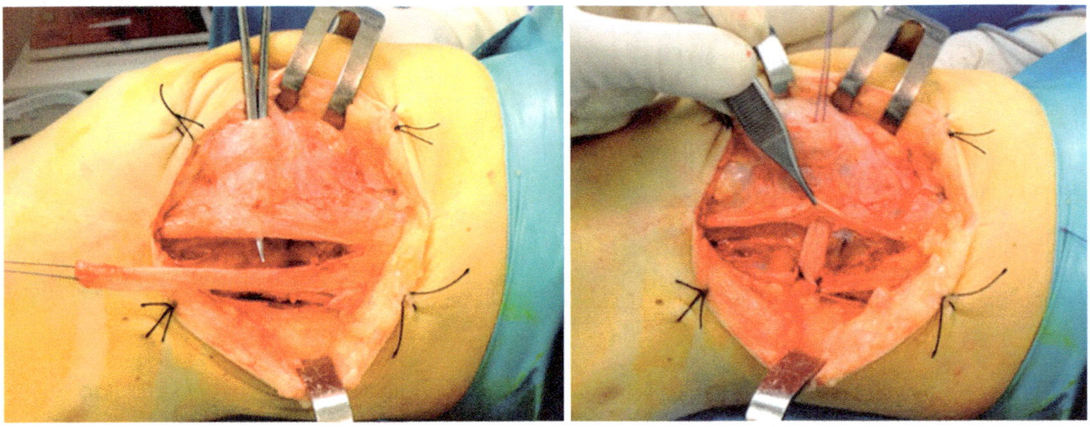

Fig. 12.6 Reconstruction of the deep lateral retinaculum following the technique described by Jack Andrish

In this way, medial retinacular release can lead to future problems similar than those caused by "extensive" LRR. We must note that a certain procedure is not better because it is less invasive. According to Teitge and Torga [19], MPI reappears after the first postoperative year after lateral retinacular repair and imbrication. For this reason, the most logical approach should be to reconstruct the lateral patellar retinaculum. The preference of the first author (V S-A) at the present time is the technique described by Jack Andrish [15] because it is very anatomic (Figs. 12.5 and 12.6) and allows us fine-tuning of the graft tension by adding sutures to further tighten the graft. As in reconstruction of the MPFL, the surgeon should tension the lateral reconstruction with the patella engaged within the trochlea with the knee flexed 30°, approximating the orientation of the native deep transverse retinacular ligament. The purpose of this technique is to reconstruct the deep transverse layer of the lateral retinaculum and not the lateral patellofemoral ligament. The deep layer of the lateral patellar retinaculum is reconstructed using a central strip of the iliotibial band leaving it attached proximally and attaching it to the midpoint of the patella (Fig. 12.6). We must note that it is a "salvage" procedure. It does not address the original source of complaint. Moreover, it cannot improve or reverse symptoms from osteoarthritis, bony malalignment, or lateral instability caused by a MPFL injury.

Patients with symptomatic iatrogenic MPI have chronic pain and the etiology of chronic pain is multifactorial with a different pathoneurophysiology than acute pain, including psychological factors like pain modulators [8]. Moreover, we must note that the mere fact that the patient can sublux and even dislocate their patella medially is no guarantee that her pain and disability are directly due to the instability. Normally, reconstructive surgery is associated with other surgical procedures. In 80 % of our cases, we accompanied the lateral patellar retinaculum reconstruction with a partial arthroscopic synovectomy of the inferior pole of the patella with its resultant denervation [23]. In 5 % of the cases, we added a trochleoplasty, a revision of previous anterior tibial tubercle osteotomy or a reconstruction of the MPFL. Therefore, we perform "a la carte" surgery in which we attempt to address each possible cause of pain. Thus, it is difficult to objectively determine how important the absence of the lateral patellar retinaculum is in the genesis of pain symptoms.

12.5 Prevention of Postoperative Medial Patellar Instability

It is clear that "extensive" LRR is a major cause of MPI [3, 9, 17, 18, 24, 25]. We believe that the role of LRR is limited to the patient with clear signs and symptoms of lateral patellar hypercompression syndrome. In these patients, lateral retinacular lengthening has been advocated to prevent postoperative MPI [26, 27] (Fig. 12.7). Indeed, many years ago in 1978, when all LRRs were performed open, Larson et al. [28] stated that it is better to lengthen the lateral retinaculum rather than release it. Recently, Pagenstert et al. [5] published a prospective double-blind study comparing open LRR with open lateral retinacular lengthening using the same end point in both groups to assure an "adequate" release. As an end point, they chose to use the 90° turnup test (rotational elevation of the lateral patella up to 90° in relation to the epicondylar axis) published by Henry et al. 26 years previously [29]. Half the patients underwent a repair of this release using a pants-over-vest lengthening technique, and the other half were left unrepaired. Not surprisingly, the release-only cohort experienced an unacceptably high incidence of severe quadriceps atrophy, worse outcome scores, and MPI. Not too long after Henry's publication of this 90° turnup test, most knowledgeable surgeons abandoned its use for the same reasons demonstrated in the Pagenstert study: an unacceptably high rates of quadriceps atrophy, worse clinical outcomes, and iatrogenic medial patellar subluxation, all the result of over-release of the lateral retinaculum [30].

Pagenstert et al. [5] did conclude that the turnup test be reduced to 70°, but it is unfortunate that the release-only cohort had to suffer such a high incidence of permanent damage in the process. It would be interesting to repeat this study using a standard lateral release technique compared with the lengthening technique. Such a study would be a more realistic comparison.

Unfortunately, many orthopedic surgeons attribute the severe adverse complications caused by over-release ("extensive" LRR) to all lateral release procedures. To our knowledge and in our experience, an isolated LRR performed properly

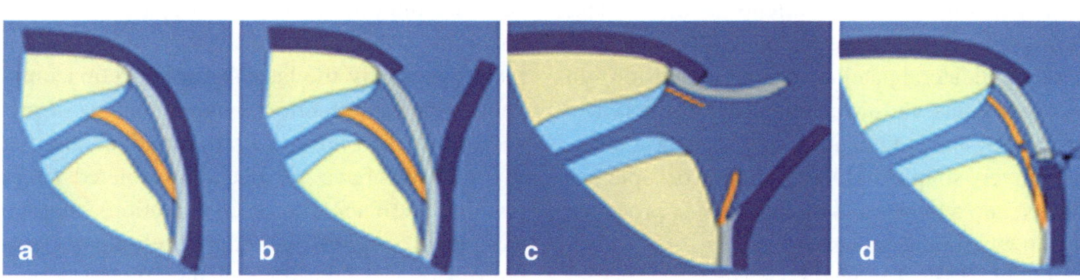

Fig. 12.7 Schematic diagram showing the lengthening of the lateral retinaculum (technical note according to Roland Biedert. MD). (**a**) The lateral retinaculum consists of a superficial oblique (*in blue*) and deep transverse part (*in gray*). Synovial layer (*in orange*). (**b**) Lengthening is started incising longitudinally the superficial layer from its attachment to the lateral border of the patella. Then it is separated from the deep transverse layer. (**c**) Then the deep transverse layer is incised longitudinally and the synovial layer is opened. (**d**) Finally, the two parts of the lateral retinaculum are sutured together in 90° of knee flexion. (From Sanchis-Alfonso [8]. Reproduced with permission from Springer)

has never caused an iatrogenic medial subluxation with severe quadriceps atrophy. "Excessive" LRR plus improper patient selection (cases with severe trochlear dysplasia, patella alta, or hyperelasticity), such as releasing a lateral retinaculum that is not tight, are the major reasons for poor results after LRR surgery. Because the purpose of a lateral release is to normalize the tight soft-tissue restraints, there is no reason to release the retinaculum beyond the goal of 1–2 patellar quadrants of medial patellar glide or a lateral tiltup of approximately 60° as advocated by Merchant [30, 31] and Ewing et al. [32]. A satisfactory LRR should not cut the muscle or tendon of the vastus lateralis [31]. However, it is safe to release the LR distally to the joint line if necessary to achieve these end points [31]. Marumoto et al. [33] in 1995 stated: "A lateral patellar retinacular release that transects the tendon of the vastus lateralis muscle may result in significant complications." Their summary states: "Complications of lateral releases include medial patellar subluxation, vastus lateralis muscle atrophy and persistent quadriceps muscle weakness. These are likely due to excessive superior extension through the tendon of the vastus lateralis muscle that eliminates its function as a dynamic lateral stabilizer of the patella, and a major extensor of the knee. Maximizing the inferior extent of a lateral release while preserving the tendon of the vastus lateralis muscle may allow an adequate release of the patella while maintaining the physiologic function of the vastus lateralis muscle."

Our goal as surgeons is to achieve the best results possible for our patients using the least invasive and safest techniques available. By treating all lateral patellar hypercompression syndrome patients who are candidates for surgery with open LRR and lengthening, surgeons have turned a relatively simple and low-risk arthroscopic procedure into a longer and more complex procedure for no measurable advantage [30].

Finally, MPI may also be secondary to MPFL reconstruction. An anterior and proximal placement of the MPFL femoral graft insertion may also lead to iatrogenic medial patellar subluxation. Therefore, as far as possible, one must perform anatomic MPFL reconstructions. Apart from the femoral attachment position, another key factor in MPFL reconstruction is to maintain the appropriate tension of the graft throughout knee range of motion. It is important to tension the graft with the patella centered on the trochlea. Securing fixation without maintaining correct patellar position can lead to symptoms of medial overload and possibly medial patellar subluxation. In the case of revision surgery for failed patellofemoral realignment or MPFL reconstruction, one should not forget to establish a competent lateral retinaculum, even if the only observable instability is lateral. Moreover, the first author has seen cases of lateral patellar instability where the only technique required to correct the problem was an isolated lateral retinaculum reconstruction. Indeed the lateral retinaculum is very important in the treatment of lateral patellar instability as well as medial. In fact, Jack Andrish (personal communication) uses lateral retinaculum reconstruction more often in revision surgery (combined with whatever else is needed) for recurrent lateral patellar dislocations (when the lateral retinaculum has been destroyed) than he does for medial instability.

12.6 Case Studies: A "Snapshot" of the Key Role of the Lateral Retinaculum in the Knee. Implications in the Etiology of Medial Patellar Instability

The key question would be: How can one measure the importance of the lateral patellar retinaculum? The obvious answer would be to reconstruct only the lateral patellar retinaculum, and not to pay attention to other sources of pain such as Hoffa's infrapatellar fat pad, chondropathy, etc. We have verified that isolated lateral patellar retinaculum reconstruction, without associated surgical procedures, eliminates pain and restores the knee's function from a kinetic and kinematic point of view (see Cases # 2 and 3 – Figs. 12.8, 12.9, and 12.10). This leads us to believe that the lateral patellar retinaculum is an important structure of the knee.

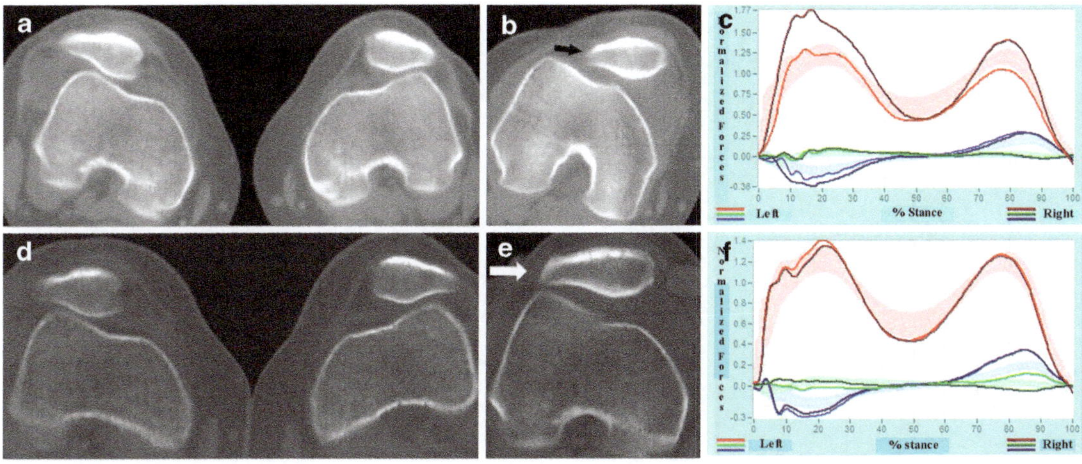

Fig. 12.8 Case # 2. Computed tomography (CT) examination in 0° extension and quadriceps contraction shows lateralization of the patella (**a**). Documentation of medial patellar instability (**b**). Preoperative gait analysis (**c**). Follow-up CT scan at 0° extension with quadriceps contraction demonstrates similar lateral displacement of the patella in both knees (**d**), and stress CT revealed medial patellar stability (**e**). Postoperative gait analysis (**f**) (From Sanchis-Alfonso et al. [25]. Reproduced with permission from Elsevier)

There are several possibilities that can explain a medial instability of the patella: (1) a dominating vastus medialis over the vastus lateralis, (2) a weak or severed vastus lateralis, and (3) the absence of the lateral patellar retinaculum (passive element). Nonweiler and De Lee [9] believe that the main reason for medial patellar subluxation following lateral release was vastus lateralis insufficiency secondary to the detachment of vastus lateralis from the patella. On the contrary, Kramers-de Quervain et al. [34] have shown that MPI is mostly caused by the insufficiency of the lateral retinaculum following its release. They found that subluxation was more prominent in the unloading phase of gait when the quadriceps was inactive. In the cases that occurred spontaneously, Akşahin et al. [35] thought that the reason for medial subluxation was not only the vastus medialis overdominance but also the insufficiency of lateral passive structures due to irreversible elongation of the ligaments as a result of the chronic imbalance of the active structures. But the question is: Why don't all the patients operated with an LRR operation develop an MPI? In this way, Shellock et al. [3] showed that 17 of 40 patients (43 %) with unilateral arthroscopic LRR had medially subluxated patellae on the unoperated joints. They concluded that medial subluxation of the patella may have been present before the LRR but was not recognized in these patients.

If before the reconstruction of the lateral patellar retinaculum in patients with MPI, the physical therapist had focused on regaining balance between vastus lateralis and vastus medialis and this had not improved the pain and it was not until the lateral patellar retinaculum reconstruction that the pain improved, we can surmise that the key element in the genesis of pain was a deficient lateral patellar retinaculum with subsequent patellofemoral imbalance and its physiopathologic consequences [8].

Case # 2 (See Fig. 12.8)

A 25-year-old female came to our institution with a history of chronic severe anterior right knee pain, severe disability, and patellofemoral instability refractory to conservative treatment, for about 5 years. The Kujala preoperative score was 36 points. The exercise rehabilitation program performed in our institution was unsuccessful in improving her symptoms.

The patient underwent an Insall's proximal realignment with LRR procedure at the age of 18 due to recurrent lateral patellar dislocation.

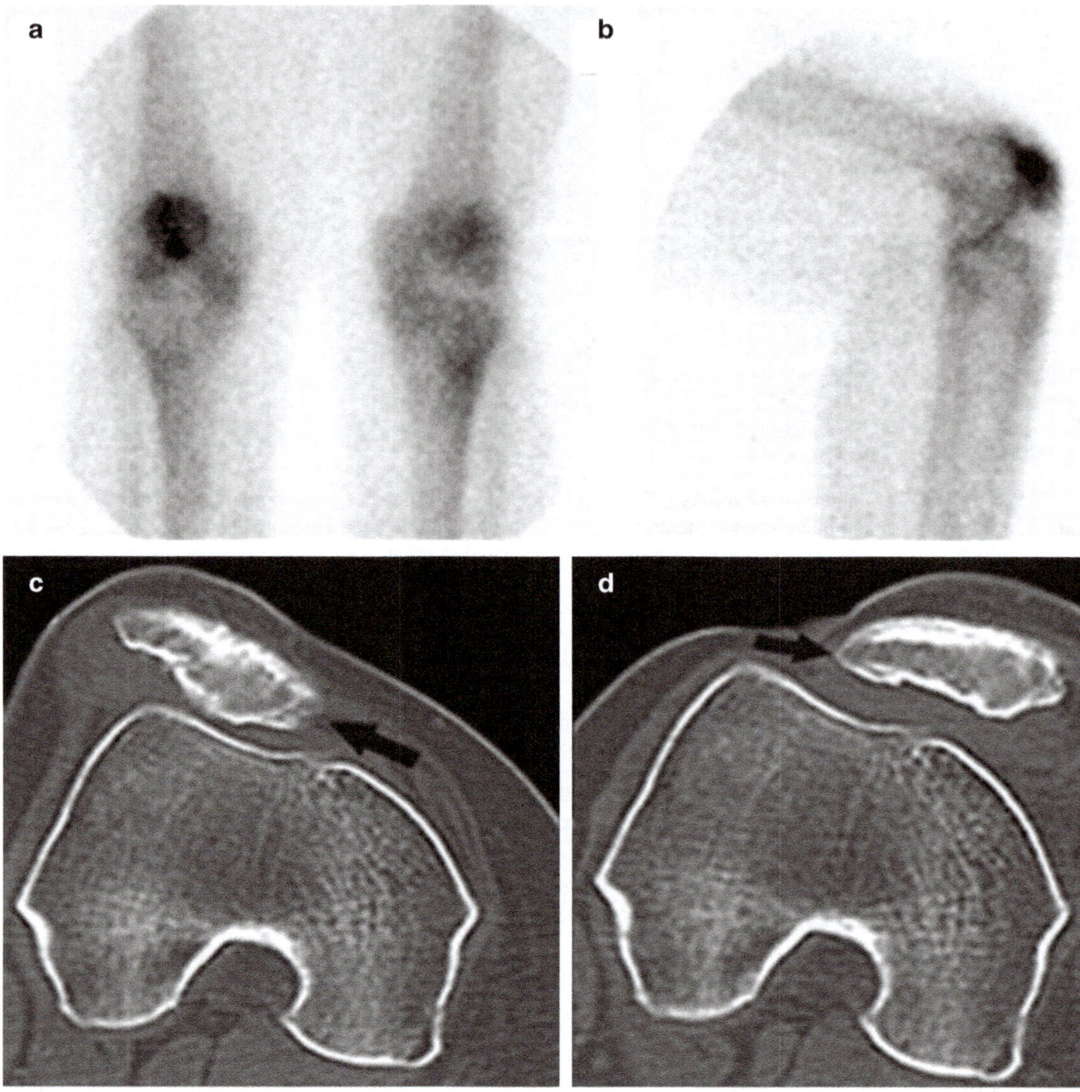

Fig. 12.9 Case # 3. Loss of vascular homeostasis, expressed by a "hot" patella (**a**, **b**), may also be associated with AKP, and could be secondary to patellofemoral imbalance, in this case a multidirectional instability as is seen in the stress CTs (**c**, **d**). Increased metabolic activity correlates to regions of increased mechanical stress or loading. Intermittent hypoxia is considered one of the most important stimuli in bone remodeling. In this case the patellofemoral osteoarthritis was asymptomatic at 5 years follow-up after the last surgery. *Arrows* indicate direction of the stress applied on the patella during stress CT (From Sanchis-Alfonso [8]. Reproduced with permission from Springer)

Following surgery, the previous symptoms disappeared, and she could play basketball again, but a different and worse type of knee pain around the patella developed with time and 1 year later she gave up playing basketball.

Physical examination of the knee revealed peripatellar and posteromedial pain, joint effusion, a positive apprehension sign upon pressing the patella medially, and a positive Fulkerson's relocation test. Conventional radiography, including skyline views, revealed no abnormalities. MRI examination showed lateral subluxation of the patella and joint effusion. A computed tomography (CT) examination in 0° extension with the quadriceps relaxed showed mild lateralization of the patella that increased with quadriceps

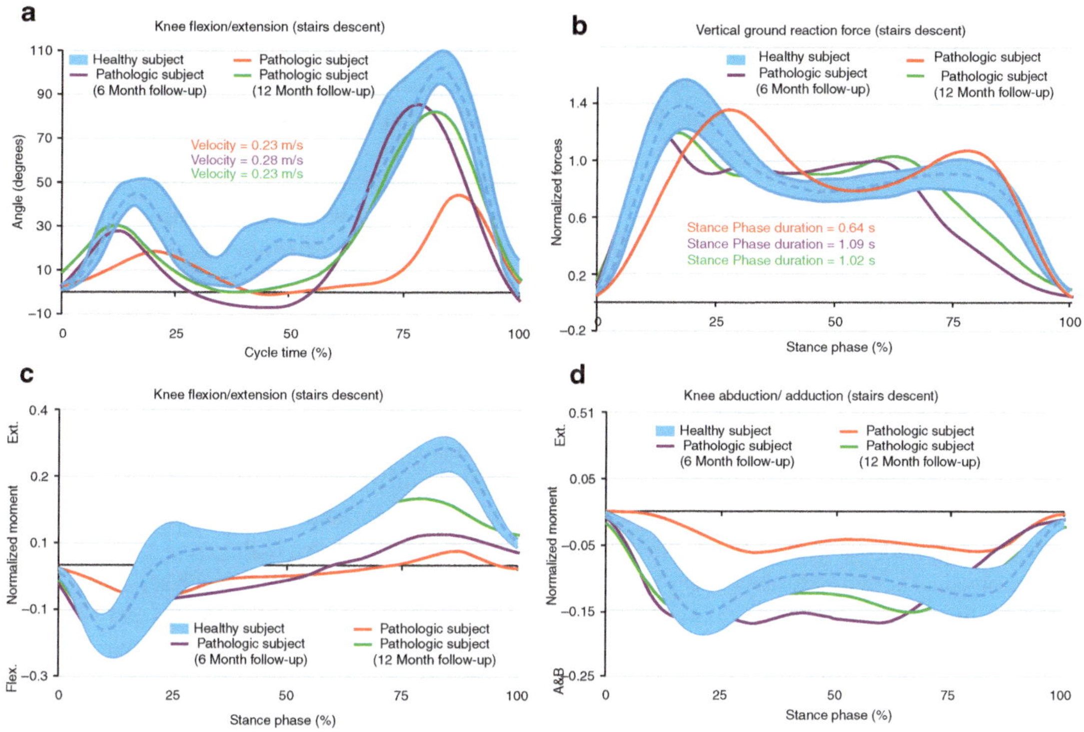

Fig. 12.10 Case # 3. Knee kinetics and kinematics during stair descent in case #3. (**a**) Knee joint angle during stair descent. (**b**) Ground force reactions during stair descent. (**c**) Flexion-extension knee moments during stair descent. (**d**) Abduction-adduction knee moments during stair descent (From Sanchis-Alfonso [8]. Reproduced with permission from Springer)

contraction being greater on the right side than on the left side. CT of the patellofemoral joint in extension with manual patellar pressure (stress CT) revealed medial patellar instability. Gait analysis was performed for documentation purposes prior to a subsequent reconstructive surgical procedure. Gait analysis revealed a significant increment of the vertical heel contact peak force as a result of a knee extension gait pattern.

At the time of surgery, an arthroscopy of the right knee was performed. All the intra-articular structures were intact, except for a patellar chondropathy grade III, according to the Outerbridge classification, located medially and a peripatellar synovitis. We did not perform chondroplasty or peripatellar synovectomy. After the arthroscopy, we performed an open reconstruction of the lateral patellotibial ligament according to the technique described by Hughston using the iliotibial band and the patellar tendon.

Four months after surgery, she was symptom-free. Gait analysis was performed at this time to evaluate the effects of surgical reconstruction of the lateral retinaculum on gait parameters. No significant differences were seen when compared to the contralateral limb, the gait pattern being normal. One year after surgery, she was symptom-free. The follow-up Kujala score was 91 points. The follow-up CT scan at 0° extension with quadriceps contraction demonstrated a similar lateral displacement of the patella, and stress CT revealed medial patellar stability. Eight years after surgery, the clinical result is excellent.

If in a patient with severe AKP, significant patellofemoral osteoarthritis, and a patellofemoral imbalance, due to a previous realignment surgery, one only reconstructs the lateral patellar retinaculum, making the pain completely disappear; one can conclude that the element responsible for the pain is the imbalance and not the osteoarthritis.

This means that the lateral retinaculum plays a role in stabilizing the patella medially and laterally and is the key element in the genesis of pain, and therefore its reconstruction is the key of the surgical treatment.

Case # 3 (See Figs. 12.9 and 12.10)
A 41-year-old woman came to our institution complaining mainly of right patellofemoral instability and also of severe right AKP that had not improved with physical therapy. The preoperative Kujala score was 24 points. The contralateral knee was completely asymptomatic.

This patient had been operated on 3 years before, with an Insall proximal realignment and LRR for a lateral patellar instability that was the main symptom, along with mild occasional pain during physical activity as a secondary symptom. After the first surgery, as time went by, the patient mentioned that the patellar instability had increased, but it was different and more incapacitating than the one she had before surgery. She also had more AKP. A year and a half after her realignment surgery along with worsening of the symptoms, another surgeon suggested a knee arthroscopy to which the patient agreed. With this second procedure (partial synovectomy and denervation), the pain worsened considerably as did the patellar instability that the patient had with day-to-day activities.

A physical examination of the knee showed AKP with a positive apprehension sign when pressing the patella medially and a positive Fulkerson's relocation test. Moreover, there was an apprehension sign when pressing the patella laterally. The rest of the physical examination was completely normal. Conventional radiography showed patellofemoral osteoarthritis. The radiographs prior to the first surgery had shown no degenerative changes. An MRI examination showed a lateral subluxation of the patella and a severe patellar chondropathy. A CT examination at 0° extension and with a relaxed quadriceps showed mild lateralization of the patella. The TT-TG index was 10 mm. The stress CT of the patellofemoral joint in extension revealed medial patellar instability. Moreover the lateral stress CT of the patellofemoral joint in extension showed a lateral patellar displacement significantly greater in the right knee compared to the left knee. A bone scan with Tc 99 m showed an increased pathologic uptake only in the patella. Kinetic and kinematic analyses were performed during stair descent, which showed that the patient had a stair descent pattern with knee extension, a decrease in the stance phase duration on the platform, reduced values of the extensor moment, and reduced values of the abduction moment.

Before reconstruction of the lateral retinaculum, arthroscopy was performed. A severe patellofemoral osteoarthritis was noted but not treated. The rest of the findings were normal. A reconstruction of the lateral retinaculum using fascia lata was performed according to the technique described by Jack Andrish.

Kinetic and kinematic analyses during stair descent were performed at 6 months and 12 months after surgery and showed a progressive recovery of the kinetic and kinematic parameters. Twelve months after surgery, the patient was asymptomatic and could go down the stairs in a natural way without any problem. The postoperative Kujala score was 94 points. Five years after surgery, the clinical result is excellent.

12.7 Personal Experience: Clinical Results

We have analyzed 17 cases from the personal series of the first author (V S-A) of patients with MPI causing disabling AKP with a mean follow-up of 55 months (range, 24–96 months). In all cases, the relocation test of Fulkerson was positive, and palpation on the distal pole of the patella and the proximal patellar tendon was very painful. All were operated on using the technique described by Jack Andrish combined with a partial arthroscopic synovectomy of the inferior pole of the patella. We excluded any patients with workman compensation issues in this series, and all belonged to a medium-high socioeconomic level. Moreover, we excluded those cases associated with trochleoplasty, anterior tibial tubercle re-osteotomy, and MPFL reconstruction. The mean preoperative VAS was 7.58 ± 1.54 (range,

5–9), compared with the mean postoperative score of 1.91 ± 2.42 (range, 0–8). Mean preoperative Lysholm score was 36.35 ± 14.04 (range, 20–55), and the knee was described as bad in 100 % of cases. Mean postoperative Lysholm score was 86.11 ± 7.77 (range, 70–94), with 35 % of the knees described as fair, 17 % as good, and 47 % as excellent. There was an improvement in the postoperative Lysholm score compared with the preoperative value in all the cases. Thus, even in the patients with a non-satisfactory result, there was improvement. Twenty-three percent of the patients had clinical criteria of depression before surgery, 58 % of anxiety, 41 % had catastrophizing ideas with pain, and 100 % suffered kinesiophobia. After surgery, none of the patients had the criteria for depression or anxiety. None of the patients had catastrophizing ideas, and only 53 % had kinesiophobia. Thirteen percent had sued the physician who had operated on them previously, not because of the failed surgery but because of the psychological disdain they had been subject to. This indicates how important psychological factors are in these patients, more so than in other cases of chronic pain. We must also note that pain had an anatomic objective basis and that psychological factors are only modulating factors that, however, are still very important in this condition. Relative to satisfaction with the treatment received, we must keep in mind that 100 % of the patients would decide to have the same type of procedure again. This opinion demonstrates the benefit of this technique as well as how disabling MPI can be.

Conclusion

MPI is an objective condition with its own personality causing incapacitating AKP. We must suspect an MPI in a patient with AKP and previous realignment surgery that has made this pain more disabling. Unfortunately, many orthopedic surgeons attribute the severe adverse complications caused by over-release of the lateral retinaculum to all lateral release procedures. To our knowledge and in our experience, an isolated LRR performed properly has never caused an iatrogenic medial subluxation with severe quadriceps atrophy.

Acknowledgments The authors wish to thank the members of the International Patellofemoral Study Group (IPSG), whose collaboration inspires and informs their work.

References

1. Hughston JC, Deese M. Medial subluxation of the patella as a complication of lateral retinacular release. Am J Sports Med. 1988;16:383–8.
2. Merchant AC, Mercer RL. Lateral release of the patella. A preliminary report. Clin Orthop Relat Res. 1974;103:40–5.
3. Shellock FG, Mink JH, Deutsch A, Fox JM, Ferkel RD. Evaluation of patients with persistent symptoms after lateral retinacular release by kinematic magnetic resonance imaging of the patellofemoral joint. Arthroscopy. 1990;6:226–34.
4. Kolowich PA, Paulos LE, Rosenberg TD, Farnsworth S. Lateral release of the patella: indications and contraindications. Am J Sports Med. 1990;18:359–65.
5. Pagenstert G, Wolf N, Bachmann M, Gravius S, Barg A, Hintermann B, Wirtz DC, Valderrabano V, Leumann AG. Open lateral patellar retinacular lengthening versus open retinacular release in lateral patellar hypercompression syndrome: a prospective double-blinded comparative study on complications and outcome. Arthroscopy. 2012;28:788–97.
6. Fithian DC, Paxton EW, Post WR, Panni AS, International Patellofemoral Study Group. Lateral retinacular release: a survey of the International Patellofemoral Study Group. Arthroscopy. 2004;20:463–8.
7. Bollier M, Fulkerson J, Cosgarea A, Tanaka M. Technical failure of medial patellofemoral ligament reconstruction. Arthroscopy. 2011;27:1153–9.
8. Sanchis-Alfonso V. Anterior knee pain and patellar instability. 2nd ed. London: Springer; 2011.
9. Nonweiler DE, DeLee JC. The diagnosis and treatment of medial subluxation of the patella after lateral retinacular release. Am J Sports Med. 1994;22:680–6.
10. Fulkerson JP. A clinical test for medial patella tracking. Tech Orthop. 1997;12:165–9.
11. Teitge RA, Faerber WW, Des MP. Stress radiographs of the patellofemoral joint. J Bone Joint Surg. 1996;78-A:193–203.
12. Biedert RM. Computed tomography examination. In: Biedert RM, editor. Patellofemoral disorders. Diagnosis and treatment. West Sussex: Wiley; 2004. p. 101–11.
13. Shannon BD, Keene JS. Results of arthroscopic medial retinacular release for treatment of medial subluxation of the patella. Am J Sports Med. 2007;35:1180–7.
14. Abhaykumar S, Craig DM. Fascia lata sling reconstruction for recurrent medial dislocation of the patella. Knee. 1999;6:55–7.

15. Andrish J. Reconstruction of the lateral patellofemoral ligament: how i do it. In: Sanchis-Alfonso V, editor. Anterior knee pain and patellar instability. 2nd ed. London: Springer; 2011.
16. Heyworth BE, Carroll KM, Dawson CK, Gill TJ. Open lateral retinacular closure surgery for treatment of anterolateral knee pain and disability after arthroscopic lateral retinacular release. Am J Sports Med. 2012;40:376–82.
17. Hughston JC, Flandry F, Brinker MR. Surgical correction of medial subluxation of the patella. Am J Sports Med. 1996;24:486–91.
18. Johnson DP, Wakeley C. Reconstruction of the lateral patellar retinaculum following lateral release: a case report. Knee Surg Sports Traumatol Arthrosc. 2002; 10:361–3.
19. Teitge RA, Torga SR. Lateral patellofemoral ligament reconstruction. Arthroscopy. 2004;20:998–1002.
20. Amis A. Current concepts on anatomy and biomechanics of patellar stability. Sports Med Arthrosc. 2007;15:48–56.
21. Fulkerson J, Gossling H. Anatomy of the knee joint lateral retinaculum. Clin Orthop Relat Res. 1980;153: 183–8.
22. Starok M, Lenchik L, Trudell D. Normal patellar retinaculum: MR and sonographic imaging with cadaveric correlation. AJR Am J Roentgenol. 1997;168:1493–9.
23. Sanchis-Alfonso V, Roselló-Sastre E, Subías-Lopez A. Neuroanatomic basis for pain in patellar tendinosis ("jumper's knee"): a neuroimmunohistochemical study. Am J Knee Surg. 2001;14:174–7.
24. Biedert RM, Friederich NF. Failed lateral retinacular release: clinical outcome. J Sports Traumatol. 1994; 16:162–73.
25. Sanchis-Alfonso V, Torga-Spak R, Cortes A. Gait pattern normalization after lateral retinaculum reconstruction for iatrogenic medial patellar instability. Knee. 2007;14:484–8.
26. O'Neill DB. Open lateral retinacular lengthening compared with arthroscopic release. A prospective, randomized outcome study. J Bone Joint Surg Am. 1997;79:1759–69.
27. Biedert RM. Is there an indication for lateral release and how I do it. In: Proceedings of the International Patellofemoral Study Group, Garmisch-Partenkirchen; 2000.
28. Larson RL, Cabaud HE, Slocum DB. The patellar compression syndrome: surgical treatment by lateral retinacular release. Clin Orthop Relat Res. 1978;134: 158–67.
29. Henry JH, Goletz TH, Williamson B. Lateral retinacular release in patellofemoral subluxation. Indications, results, and comparison to open patellofemoral reconstruction. Am J Sports Med. 1986;14:121–9.
30. Merchant AC. Is it lateral retinacular lengthening versus lateral retinacular release or over-release? Arthroscopy. 2013;29(3):403. doi:10.1016/j.arthro.2013.01.002.
31. Merchant AC. Diagnosis and treatment of patellofemoral disorders must be individualized. 2013. http://www.healio.com/orthopedics/blogs/patellofemoral-update/diagnosis-and-treatment-ofpatellofemoral-disorders-must-be-individualized
32. Ewing JW. Arthroscopic patellar shaving and lateral retinacular release. In: McGinty JB, editor. Operative arthroscopy. New York: Raven; 1991. p. 282.
33. Marumoto JM. A biomechanical comparison of lateral retinacular releases. Am J Sports Med. 1995;23:151–5.
34. Kramers-de Quervain IA, Biedert R, Stussi E. Quantitative gait analysis in patients with medial patellar instability following lateral retinacular release. Knee Surg Sports Traumatol Arthrosc. 1997; 5:95–101.
35. Akşahin E. Role of pathophysiology of patellofemoral instability in the treatment of spontaneous medial patellofemoral subluxation: a case report. J Med Case Reports. 2010;4:148.

Nonoperative Treatment of Patellofemoral Joint

13

Lorenzo Boldrini, Furio Danelon, Francesco Della Villa, and Stefano Della Villa

Contents

L. Boldrini (✉) • F. Danelon
Isokinetic Medical Group – FIFA Medical Centre of Excellence, Milano, Italy
e-mail: l.boldrini@isokinetic.com

F. Della Villa • S. Della Villa
Isokinetic Medical Group – FIFA Medical Centre of Excellence, Bologna, Italy

A. Gobbi et al. (eds.), *The Patellofemoral Joint*,
DOI 10.1007/978-3-642-54965-6_13, © ISAKOS 2014

Patellofemoral disorders are a complex argument still debated in the scientific community; many classifications have been proposed, and treatment cannot be considered by a single algorithm.

We refer to the classification of the French school [1] that considers three groups of pathologies: objective patellar instability, potential patellar instability, and patellofemoral pain syndrome (PFPS).

Nonoperative treatment can be considered the primary approach for patellofemoral disorders in particular for PFPS, an extremely common problem especially in the young population, characterized by anterior knee pain, without episodes of dislocation and without the four major factors of patellar instability (trochlear dysplasia, patella alta, increased TA GT distance, patellar tilt).

To underline the role of conservative treatment in patellofemoral pathologies, it is noteworthy a sentence from Grelsamer in a paper of 1998 [2]: "there should be no distinction between the terms conservative and surgical when it comes to treatment of the patella. Treatment should always be conservative, be it surgical or not."

It is generally assumed that an adequate and prolonged rehabilitative treatment must be indicated up to 6–8 months before considering unsuccessful the conservative approach.

The rationale of nonoperative treatment derives from the multifactorial etiology of PFPS. Besides overuse and biomechanical and muscle imbalance causes that are traditionally reported, we have to consider other more recently described factors such as alterations in vascular, nervous and functional characteristics of peripatellar tissue [3]. Moreover we have to consider the psychological aspects that are usually involved in patients affected by anterior knee pain.

Different theories can be proposed in the etiology of PFPS [3]:

- Biomechanical theory: anatomical and biomechanical alterations, such as malalignments and postural defects, can cause the release of cartilage fragments inducing an inflammatory synovial reaction that produces cytokines and pain.
- Functional theory: excessive mechanical load on the extensor mechanism can induce an alteration of joint homeostasis. Activities must be kept in a frequency and load applied within the available "envelope of function" [4].
- Biological theory: it is the case of idiopathic chondromalacia and synovial plicae.
- Neurogenic theory: a neuromuscular alteration or a malalignment can induce, through repetitive microtrauma, an ischemic damage of the retinaculum and a consequent degenerative neuropathy [5]. This alteration produces an increase of substance P, NGF, and other hyperalgesic factors, as described in the "fall of Sanchis," leading to pain [6, 7].

The anamnesis and the clinical examination are fundamental aspects in order to define the proper conservative treatment. Patients may refer a history of traumatic episodes, previous surgery, overuse, overload, and modification in training habits, including sudden decrease in physical activity. Psychological aspects can be related to previous unsuccessful experience of treatment with poor clinical results; therefore, patients are often demotivated or affected by character fragility, depression, and anxiety that can reduce compliance to a prolonged rehabilitation, leading the patient to a detrimental avoidance attitude.

On clinical examination, it is possible to reveal several alterations that must be kept in mind for treatment. Common conditions are malalignments of lower limbs, hyper-pronation of the foot (investigate the subtalar joint!), different length of the limbs, reduction in muscle flexibility and tone, hypotrophy, poor dynamic control during squat test, patellar laxity, positive provocative pain test, skin alterations (hyperemia, scars, etc.), and emotional aspects.

Evaluation of radiological findings is also necessary to define anatomical, biomechanical, and biological factors that must be considered in the rehabilitation protocol. Radiographs, CT scan, and MRI can be helpful in diagnosing patellofemoral disorders, depending on clinical evaluation and according to specific protocols.

Based on all these factors, rehabilitation protocols must be customized and adapted to the single patient. An important criterion is the progression of the rehabilitation loads depending on adaptation of the joint and patient's reactions to applied stimuli (specific adaptation to imposed demand). Pain and swelling reaction must always be avoided because they can worsen clinical conditions inducing a delay in recovery and a diminished trust in the rehabilitation program by the patient. The evaluation of pain is very important, and a good experienced physician has to distinguish among pain, soreness, and fear of pain through an accurate series of questions to the patient.

The key points of the nonoperative treatment of PFPS can be resumed as follows [8]:

- Therapeutic exercise
- Manual therapy
- Treatment modalities
- Pharmacological therapy
- Taping, braces, and plantar foot orthotics
- Psychological approach

13.1 Therapeutic Exercise

A proper exercise program is the "core" of the treatment of patellofemoral pathologies. The recovery of muscle balance, in terms of flexibility and strength, together with the recovery of normal function with a correct neuromuscular and dynamic control during movement, can be alone the key of the resolution of symptoms and disabilities in these patients.

Therefore, a fine evaluation must be done by the clinician to investigate all the anatomical, postural, biomechanical, and functional factors we have mentioned above in order to define the proper rehabilitation program for each patient.

In our opinion the common prescription of simple "muscle quadriceps isometric strengthening and swimming" can be often useful but is too generalized and inadequate compared to the potential benefits of a complete and well-designed exercise rehabilitation program.

When possible, it is suggested to investigate knee muscle strength through isokinetic testing. This analysis can provide several information about pain-free range of motion, muscle recruitment, and muscle group balance and strength. It will be surprising that in many cases, we will find a muscle strength deficit in the knee flexor muscles and not in the extensors as normally supposed.

In general, strengthening exercises in patellofemoral problems must be performed on pain-free range of motion, starting with submaximal and isometric exercises improving to concentric and eccentric modalities with different resistances and speeds of movement. At the beginning, closed kinetic exercises between 0° and 50° and open kinetic exercises between 0° and 30° are normally better tolerated.

Vastus medialis obliquus (VMO) is the main medial dynamic stabilizer and must be trained especially in patellofemoral disorders characterized by hyper-pressure of the lateral facet of the patella and lateral malalignments. It is notable that, during strengthening exercises of the quadriceps muscle, rotation of the hip (modifying the position of the foot) and concomitant co-contraction of flexor muscles don't activate VMO in a preferential way as normally believed [9].

To enhance muscle mass and tone, it is recommended the use of electrical stimulation, in particular during the first phases of rehabilitation, when pain can limit normal execution of therapeutic exercises. Quadriceps and VMO are the preferential muscle groups treated with electrical stimulation.

A key point in the control of dynamic movement of the knee is represented by hip muscles, in particular abductor and external rotator muscles. Tensor fasciae latae, gluteus medius, gluteus maximus, and other muscles involved in the external rotation and extension-abduction must be trained to control internal rotation and adduction of the femur during functional movement that can lead to an increased valgus of the knee. This characteristic is more frequent in female patients with increased femoral internal rotation and increased Q angle of the knee.

Many papers have underlined the positive effects of hip external rotation and abductor muscles in patellofemoral syndromes. It is reported a reduction of pain and an increase of function with exercise programs that involve strengthening of hip muscles, stretching, and neuromuscular control [10–13].

Patients can exercise at the beginning lying down against manual, elastic, or weight resistance moving to standing exercises (Fig. 13.1) and gradually increasing the modalities and loads of exercises. The goal is to exercise in dynamic conditions like step up-down exercises and jumping or plyometric exercises involving the control of all the kinetic chain from the foot to the upper body (Fig. 13.2). Thus, it is necessary during the program to train trunk and pelvis muscles stabilizers, both in static and dynamic conditions. Exercises of "core strengthening" and "core stability" are therefore recommended to increase tone of gluteus, abdominal, and extensor trunk muscles (Fig. 13.3).

During dynamic exercises, the physical therapist must correct erroneous movements in particular, the tendency at knee valgus and intra-rotation of the hip.

Proprioceptive and neuromuscular exercises are fundamental in the rehabilitation process of these patients in the last phase of rehabilitation. Many unstable devices and balance training paths are available to train coordination and neuromuscular characteristics.

In case of alterations of plantar foot anatomy and biomechanics, especially in subjects with static and dynamic hyper-pronation of the foot, it is recommended to train anti-pronator muscles and muscles that stabilize the ankle, both with strengthening and proprioceptive exercises without shoes.

Another key point of treatment is the need to improve physical fitness status, frequently decreased in these patients as a consequence of their movement avoidance. Consequently, lowered physical fitness can contribute to enhance physical disability and psychological distress.

Fig. 13.1 Hip muscle strengthening with elastic resistance

Fig. 13.2 Step down eccentric exercise

To increase muscle endurance, it is suggested aerobic exercises on the cyclergometer with high seat and/or exercises on a cross-trainer before running on flat surface. Every activity must be done in a pain-free manner, gradually increasing resistance and duration of exercise.

Stretching exercises are important weapons too, necessary to help to resolve muscle flexibility deficits. Often patients affected by PFPS present a reduced flexibility of the posterior muscle chain (hamstring and calf muscles) and lateral tight muscles (tensor fasciae latae and vastus lateralis). This situation can increase the conflict at the patellofemoral joint even in the activity of daily living. Therefore, it is important to restore a balance between all the muscles that act on the knee. Stretching of quadriceps, hamstrings, tensor fascia latae, gluteus, iliopsoas, and gastrocnemius can be recommended depending the individual situation and postural characteristics. The position of each exercise must be kept for at least 30" and repeated daily for five to six times for every muscle group without discomfort or pain.

Fig. 13.3 Core stability exercises

Another important chapter is the hydrokinesis therapy that is particularly useful in the first phase of the rehabilitation to exercise in a gravity-reduced environment that offers many possibilities to improve range of motion, flexibility, strengthening, and neuromuscular control. Furthermore the pool offers the chance to experience advanced motor skills like running and jumping in a soft environment helping the patient both psychologically and physically towards the full recovery.

The last phase of functional recovery is the rehabilitation of specific gesture, including sport activities. To train running ability, stop and go, change of directions, skill pathways, jumping, and other exercises typical of every sport activity, we use to perform a part of the rehabilitation process on the field, always under the supervision of a qualified rehabilitation trainer. This work can allow the patient to come back to complete functional activity which is a delicate transition for patients affected by patellofemoral problems even by a psychological point of view.

13.2 Manual Therapy

Many advantages can be obtained from proper use of manual treatment in PFPS. The rationale is to treat stiff, retracted, or fibrotic tissues and to restore a balance between the structures acting on the patella. Often manual therapy is associated to stretching and mobilization techniques to improve the results of treatment.

Mobilization of the patella and massage of peripatellar tissue can help the recovery of a better function of the patellofemoral joint during movements, leading to a reduction of pain.

Patella is more often mobilized in a lateral to medial direction in order to stretch the lateral retinaculum, contrasting the frequent lateral subluxation.

The treatment of trigger points and the massage therapy of muscles and tendons help to reduce contractures and painful areas. Quadriceps, hamstrings, gastrocnemius, and adductor and abductor muscles (including fascia latae) are the main muscle groups to be treated with these techniques.

A chiropractic evaluation can be indicated to investigate and resolve alterations in the postural habits involving lower limbs, pelvis, and spine. A typical finding is the reduced mobility of the sacroiliac joint that can induce postural changes and asymmetric overload on the spine and lower limb joints.

13.3 Treatment Modalities

There is no consensus in the scientific community on the efficacy of treatment modalities in patellofemoral pain syndrome. Many treatments can be address for the joint, peripatellar tissue, muscle, and tendon depending on the individual clinical situation and according to the physiatrist.

Pulsed electromagnetic fields (magneto therapy), ionophoresis with anti-inflammatory drugs, and ultrasounds can be indicated for cartilage and joint inflammation.

High power laser therapy is recommended for peripatellar tissues.

The application of warmth, like hyperthermia, can be useful to reduce stiffness and contractures of muscle and tendons.

Ice is indicated in every inflammatory condition and at the end of the rehabilitation sessions. Transcutaneous electrical therapy (TENS) is generally used to reduce pain in association to other modalities or when a pain killer effect is desired to facilitate manual mobilization and functional training in a pain-free situation.

13.4 Pharmacological Therapy

The use of drugs is a valid resource in the hands of the physician. However, no medicine can substitute the rehabilitation treatment. Anti-inflammatory and analgesic medicines can be used to control the acute phase of inflammation and pain, together with treatment modalities (especially ice). In case of cartilage damage and incoming osteoarthritis, glucosamine and chondroitin supplements may be suggested [14]. Hyaluronic acid is often intra-articularly injected with the intent to feed the cartilage tissue, to create a protection between cartilage surfaces, and to stimulate autogenous hyaluronate self-synthesis [15, 16].

13.5 Taping, Braces, and Plantar Foot Orthotics

The use of taping and braces has a long history in PFPS. The goal of these tools is to restore a better patellofemoral tracking during active movements and exercises. The stabilization obtained with taping and braces may decrease pain, reduce malalignments, and improve the ability of performing exercises [17, 18]. It is important to underline that the use of these devices doesn't substitute the recovery of a proper muscle tone and function. Therefore, it is suggested to limit the use of taping and braces to selected cases and to remove it as soon as possible.

Many techniques are available. More than 25 years ago, McConnell [19] proposed the use of taping. Since then, many other techniques have been proposed, and different materials have been produced to improve stabilization and proprioceptive effects of bracing or taping.

The patella can be pushed medially or laterally and superiorly or inferiorly, depending on clinical individual condition. Commonly the patella is taped to the medial compartment to correct excessive lateralization.

Plantar foot orthotics are a useful tool to correct postural and anatomical alterations, in particular conditions of foot hyper-pronation that can lead to an increased valgus of the knees. It is suggested a static and dynamic evaluation of the foot and a study of walking and running patterns. Based on a computerized and video analysis, functional compensations can be identified and corrected by using plantar supports.

13.6 Psychological Approach

Patients affected by PFPS are usually involved in a long time clinical history. Often the patient refers many previous treatments without clinical success. These situations associated to a prolonged rehabilitation process (up to 6–8 months) may explain how these patients are normally "energy and time consuming." Therefore, the psychological and motivational aspects are really important and often underestimated. The case manager medical doctor and the physiotherapist must investigate psychological aspect and strictly follow the patient during the rehabilitation, keeping high his/her motivation and giving him or her the proper stimuli. Also for these reasons, pain must be avoided during the exercises as it may have negative effects on the patient.

In selected cases, psychological consultation can be proposed.

References

1. Dejour H, Walch G, Neyret P, et al. Dysplasia of the femoral trochlea. Rev Chir Orthop Reparatrice Appar Mot 1990;76:45–54.
2. Grelsamer RP, Mc Connell J. The patella: a team approach. Aspen Publisher, Gaithersburg, MD, 1998.
3. Gigante A, Enea D, Greco F. Dolore anteriore di ginocchio ed instabilità rotulea. Tecniche chirurgiche in Ortopedia e Traumatologia. 2007;5(1–2):25–30.
4. Dye S. Patellofemoral pain: current concepts. Sports Med Arthrosc Rev. 2001;9(4):264–272.
5. Mori Y, et al. Clinical and histological study of patellar chondropathy in adolescents. Arthroscopy. 1991;7(2):182–97.
6. Witonski D, et al. Distribution of substance-P fibers in the knee joint in patient with anterior knee pain syndrome. A preliminary report. Knee Surg Sports Traumatol Arthrosc. 1999;7:177–83.
7. Sanchis Alfonso V, et al. Quantitative analysis of nerve changes in the lateral retinaculum in patients with isolated symptomatic patellofemoral malalignment. A preliminary study. Am J Sports Med. 1998;26(5):703–9.
8. Severini G. La rieducazione funzionale nel dolore e nell'instabilità femoro-rotulea. Tecniche Chirurgiche in ortopedia e Traumatologia. 2007;5(1–2):63–6.
9. Smith TO, et al. Can vastus medialis oblique be preferentially activated? A systematic review of electromyographic studies. Physiother Theory Pract. 2009;25(2):69–98.
10. Boling MC, et al. Outcomes of a weight-bearing rehabilitation program for patients diagnosed with patellofemoral pain syndrome. Arch Phys Med Rehabil. 2006;87(11):1428–35.
11. Nakagawa TH, et al. The effect of additional strengthening of hip abductor and lateral rotator muscles in patellofemoral pain syndrome: a randomized controlled pilot study. Clin Rehabil. 2008;22(12):1051–60.
12. Tyler TF, et al. The role of hip muscle function in the treatment of patellofemoral pain syndrome. Am J Sports Med. 2006;34(4):630–6.
13. Earl JE, et al. A proximal strengthening program improves pain, function, and biomechanics in women with patellofemoral pain syndrome. Am J Sports Med. 2011;39(1):154–63.
14. Sawitzke AD, et al. Clinical efficacy and safety of glucosamine, chondroitin sulphate, their combination, celecoxib or placebo taken to treat osteoarthritis of the knee: 2-year results from GAIT. Ann Rheum Dis. 2010;69(8):1459–64.
15. McArthur BA, et al. Long term safety, efficacy, and patient acceptability of hyaluronic acid injection in patients with painful osteoarthritis of the knee. Patient Prefer Adherence. 2012;6:905–10.
16. Colen S, et al. Hyaluronic acid in the treatment of knee osteoarthritis: a systematic review and meta-analysis with emphasis on the efficacy of different products. BioDrugs. 2012;26(4):257–68.
17. Crossley KM, et al. Can patellar tape reduce the patellar malalignment and pain associated with patellofemoral osteoarthritis? Arthritis Rheum. 2009; 61(12):1719–25.
18. Paoloni M, et al. Long-term efficacy of a short period of taping followed by an exercise program in a cohort of patients with patellofemoral pain syndrome. Clin Rheumatol. 2012;31(3):535–9.
19. Mc Connell J. The management of chondromalacia patellae: a long term solution. Aust J Physiother. 1986;32:215–23.

Treatment of Acute Patellar Dislocation: Current Concepts

14

André Costa, André Sarmento, Ana Leal, Nuno Sevivas, Pedro L. Ripoll, Ricardo Sampaio, Hélder Pereira, and João Espregueira-Mendes

Contents

A. Costa • A. Sarmento
Clínica do Dragão – Espregueira Mendes Sports Centre – Porto FIFA Medical Centre of Excellence, F.C. Porto Dragão Stadium, 4350-415 Porto, Portugal

Orthopedic Department, Centro Hospitalar de Vila Nova de Gaia, Vila Nova de Gaia, Portugal

A. Leal
Clínica Espregueira-Mendes F.C. Porto Stadium – FIFA Medical Centre of Excellence, F.C. Porto Dragão Stadium, 4350-415 Porto, Portugal

Mechanical Engineering Department, University of Minho, Minho, Portugal

CT2M – Centre for Mechanical and Materials Technologies, University of Minho, Campus Azurém, Azurém, Portugal

Orthopedic Department, Centro Hospitalar Póvoa de Varzim, Vila do Conde, Portugal

N. Sevivas
Clínica do Dragão – Espregueira Mendes Sports Centre – Porto FIFA Medical Centre of Excellence, F.C. Porto Dragão Stadium, 4350-415 Porto, Portugal

Orthopedic Department, Hospital de Braga, Braga, Portugal

ICVS/3B's – PT Government Associated Laboratory, Braga, Portugal

P.L. Ripoll
Orthopedic Department, Rypoll y De Prado Sports Clinic FIFA Medical Centre of Excellence, Murcia-Madrid, Spain

R. Sampaio
Imaging Department, Aveiro University, Aveiro, Portugal

H. Pereira
Clínica do Dragão – Espregueira Mendes Sports Centre – Porto FIFA Medical Centre of Excellence, F.C. Porto Dragão Stadium, 4350-415 Porto, Portugal

Orthopedic Department, Centro Hospitalar Póvoa de Varzim, Vila do Conde, Portugal

ICVS/3B's – PT Government Associated Laboratory, Braga, Portugal

3B's Research Group – Biomaterials, Biodegradables and Biomimetics, Univ. Minho, Headquarters of the European Institute of Excellence on Tissue Engineering and Regenerative Medicine, Caldas das Taipas, Guimarães, Portugal

J. Espregueira-Mendes (✉)
Clínica do Dragão – Espregueira-Mendes Sports Centre – FIFA Medical Centre of Excellence, F.C. Porto Stadium, 4350-415 Porto, Portugal

ICVS/3B's Research Group – Biomaterials, Biodegradables and Biomimetics, University of Minho, Headquarters of the European Institute of Excellence on Tissue Engineering and Regenerative Medicine, Caldas das Taipas, Guimarães, Portugal

Department of Orthopaedics, Minho University, Braga, Portugal
e-mail: jem@espregueira.com

A. Gobbi et al. (eds.), *The Patellofemoral Joint*, DOI 10.1007/978-3-642-54965-6_14, © ISAKOS 2014

14.1 Introduction

The incidence of primary patellar dislocation is 5.8 per 100,000 and this increases to 29.0 per 100,000 in the 10–17-year-old age group [21, 26].

The recurrence rate ranges from 15 to 44 % after nonoperative treatment of an acute injury [26]; 58 % of patients continue to experience pain and mechanical symptoms after the initial dislocation episode [2] and 55 % fail to return to full sports activity [2].

Instability of the patellofemoral joint is a multifactorial problem related with limb alignment, osseous architecture of the patella and trochlea, the integrity of the soft-tissue constraints and the interplay of the surrounding muscles [64].

Treatment of patellar instability requires an understanding of the aforementioned relationships and how to evaluate them. Conservative treatment for acute patellofemoral dislocation has been the classical approach for many years. Maenpaa and Lehto [34] presented their results with conservative treatment reporting a recurrence of 44 % of cases.

Surgical treatment was first described by Boring and O'Donoghue who repaired the medial capsule in 18 patients with no recurrences in the follow-up period [8]. Kaplan was the first to describe the medial patellofemoral ligament (MPFL) in 1957 [30]. However, it was only in 1996 that its repair was described for acute patellofemoral dislocation with no recurrent dislocation [47].

Other surgical procedures have been described to treat acute patellofemoral dislocation; however, it is still a matter of discussion whether surgical treatment is eligible and which is the adequate procedure to achieve optimal results.

The outcomes of nonoperative and operative treatment for acute patellofemoral dislocation are very variable and there are no straightforward guidelines accepted. Only nowadays, papers are being published with comparable randomized series with identical populations and similar treatment.

This review is intended to address the following questions: (1) How should a first-time acute patellar dislocation be evaluated (which are the most important factors to consider)? (2) Which is the relevance of combined osteochondral fractures? (3) When should the initial management be surgical versus nonoperative treatment?

14.2 Pathophysiology of Acute Patellar Dislocation (Major Factors to Consider)

The incidence of primary patellar dislocation is in average 5.8 per 100,000 per year in the general population [21]. The highest incidence occurs between the age of 10 and 17 (29 per 100,000) [29]. Currently, it is considered that most of these patients will not suffer subsequent instability episodes; however recurrence rates of 15–44 % after conservative treatment have been reported [29]. In a recent study focused on pediatric and adolescent populations, patients with acute patellar dislocation combining immature physes and trochlear dysplasia had a recurrence rate of 69 % [32]. Patellofemoral dislocation might occur after a traumatic event causing disruption of normal patella position in the trochlear groove. Two common activities that have frequently been associated with episodes of patellar dislocation are sports activities (61 %) and dance (9 %) [21].

The required force to dislocate the patella probably varies according to individual patellofemoral characteristics. When the femur rotates internally while the tibia suffers external rotation, with a foot fixed on the ground, the patella may dislocate without presence of preexistent pathological patellofemoral characteristics [53]. However, more frequently, patellar dislocation will occur in knees presenting risk factors for patellar instability (Fig. 14.1).

Risk factors for patellofemoral instability include patella alta, trochlear and patellar dysplasia, lateral patellar tilt, increased Q angle, *vastus medialis obliquus* (VMO) insufficiency, excessive TT-TG, patellar tendon length [38], genu valgum, medial patellofemoral ligament (MPFL) hyperlaxity, increased femoral anteversion, and increased external tibial external torsion [42] (16).

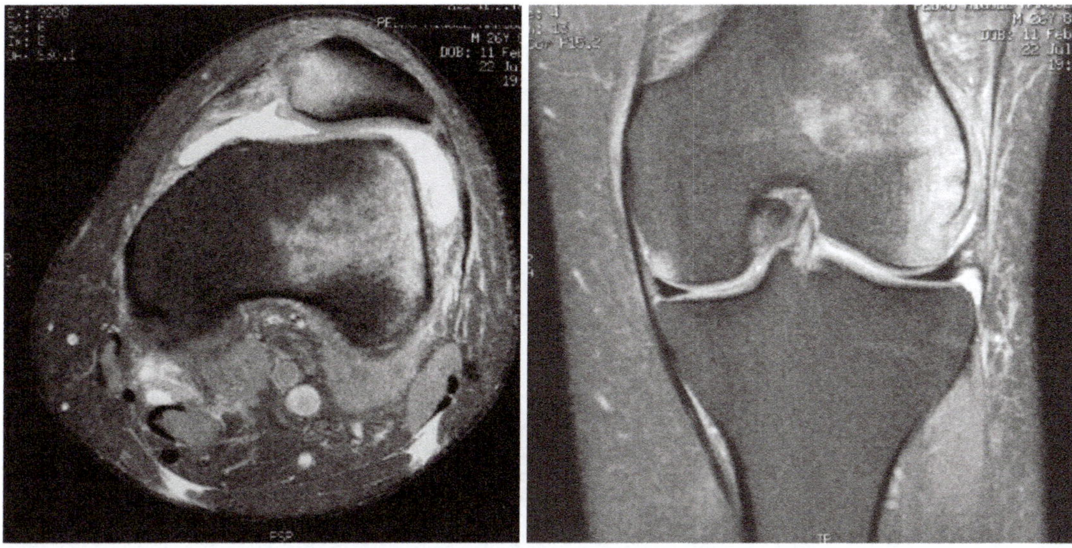

Fig. 14.1 Typical MRI aspect after acute patella dislocation, presenting effusion, MPFL injury, and bone edema in medial patella and lateral femur

These factors are described with further detail elsewhere within this publication.

In recent years, many researchers, like Kuroda [37] and Dejour, Arendt, and Zaffagnini [63], have focused on the importance of MPFL in patellar dislocation.

Anatomically, the MPFL is a thin band of retinacular tissue transversally connecting from the medial condyle to the medial aspect of patella, attaching to the undersurface of the VMO proximal to its patellar insertion. There is a region of common meshing fibers of approximately 20.3 mm between MPFL and VMO [31, 52]. The MPFL seems to be the most important dynamic stabilizer of the patella in early flexion [43]. Biomechanically, MPFL is the primary ligamentous restraint, providing about 50–60 % of the restraining force against lateral patellar displacement [6, 45]. Clinically, up to 94–100 % of patients suffer MPFL rupture after acute patellar dislocation. Some authors suggest that lateral patellar dislocation is frankly impossible without, at least a partial damage to the MPFL [18, 41].

MPFL injuries are located most frequently at the femoral attachment [41, 47] but are also located in the patellar attachment (Fig. 14.2) or in the mid-substance region [18, 54].

The origin of the medial patellofemoral ligament (MPFL) at the femur and its insertion at the patella are characterized by high individual variations. The origin on the medial femoral condyle is created by an arc of fibers originating from the anterior edge of the superficial medial collateral ligament near the medial epicondyle and fibers originating from the medial epicondyle or the adductor tubercle. This thin but wide ligament is located in layer 2 of the medial soft-tissue structures [5, 31, 42].

Senavongse et al. [51] reported that lateral patellar displacement occurred at the lowest restraining force (74 N) at 20° of knee flexion.

Currently, MPFL injury patterns have been identified and categorized into four types based on MRI findings: injuries in the patellar insertion, mid-substance, femoral origin, and combined injuries [3, 4, 61].

However, it is difficult to differentiate the mid-substance injury from patellar or femoral injury patterns when the injury is located at the mid-substance-patellar insertion or at the mid-substance-femoral insertion junction zones [64]. Additionally, the same mid-substance injury may result in different clinical outcomes, with or without the presence of VMO attachment [5].

Therefore, the VMO-based three-part classification, according to the injury location with or without VMO attachment, for acute MPFL injury was introduced: the overlap-region injury, non-overlap-region injury, and combined injures. Clinical nonsurgical outcomes for acute patellar dislocation were analyzed taking into account different injury types according to the latter classification system [5]. The hypothesis was that nonsurgical treatment would achieve better clinical results in stability and subjective patellofemoral function for the overlap-region injury than that for the non-overlap-region injury.

Nonsurgical treatment yielded satisfactory clinical outcomes when the injury was at the overlap region and therefore might be the treatment of choice for such type of injury. The optimal choice for the non-overlap-region injury still requires further researches.

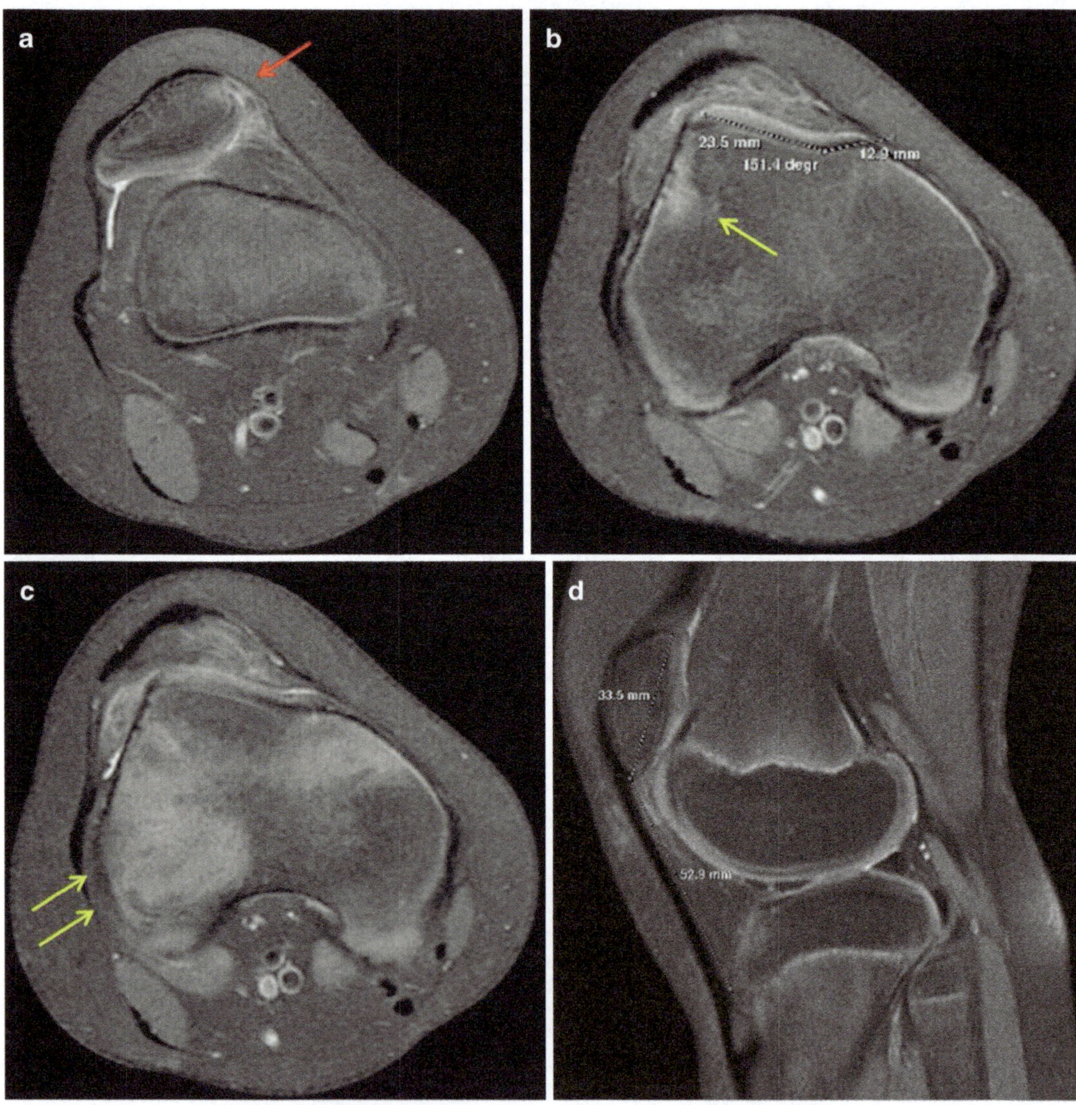

Fig. 14.2 MRI study specific for patellofemoral joint on a 10-year-old patient presenting trochlea dysplasia: rupture of MPFL with small avulsion (*red arrow* – **a**); sulcus angle 151° (**b**); bone edema (*yellow arrows* – **b** and **c**); patella *alta*, Insall-Salvati index 1.6 (**d**); TTGT = 21 (**e**)

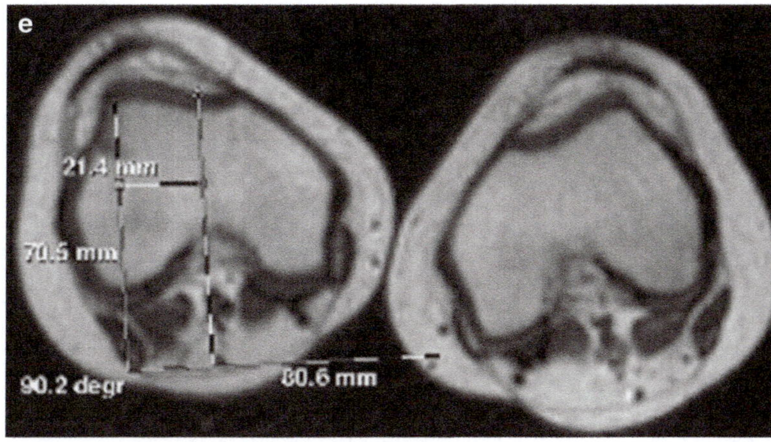

Fig. 14.2 (continued)

14.2.1 Evaluation of First-Time Acute Patellar Dislocation

The initial evaluation of a first-time traumatic patellar dislocation should include an appropriate patient history, family history of patellar dislocation, and hyperlaxity.

Aspiration of the knee joint might be performed as a diagnostic and therapeutic gesture in patients with moderate to severe effusions [46, 58]. Fatty globules visible on aspirate are suggestive of an osteochondral fracture. It increases patient comfort and diminishes pressure within the joint. Clinical and radiographic evaluation (particularly Merchant's view at 45° flexion, schuss view, and 30° lateral view, which might be difficult with concomitant severe hemarthrosis) are important. Following ACL ruptures, acute patellar dislocations are the second most frequent etiology of acute knee hemarthrosis [24]. A hemarthrosis around 50 mL volume or higher has been associated with a lower recurrence rate [58]. The proposed *rationale* is that bigger hemarthrosis might represent a more traumatic event versus a patient with dislocation after lower-energy mechanism which might have previous risk factors, thus requiring a less amount of energy to make the patella dislocate.

Physical examination (Fig. 14.3) is mandatory to rule out other injuries, such as anterior cruciate and/or medial collateral tears that involve similar mechanisms and might also occur concomitantly [29].

Alignment of lower extremities must be checked and presence of global laxity ruled out (hypermobility of the opposite knee, elbows, thumbs; small fingers) (Fig. 14.4).

Patellar mobility and apprehension might be inspected. However it can be difficult to assess on the acute setting. Global stability of the knee joint should be tested to check other structures. Palpation is important aiming to find specific areas of tenderness. Palpable defects in the VMO, adductor mechanism, medial patellofemoral ligament (MPFL), and an easily dislocatable patella have been considered prognostic factors for poor nonoperative outcomes [28].

Radiographic assessment should include an AP extended knee weight-bearing view, a Merchant's view with comparison of the opposite knee, Schuss view, and a 30° flexion lateral view. Osteochondral fractures have been reported to be missed in 30–40 % of initial radiographs (Fig. 14.5) based on both surgical and MRI studies [57].

Intra-articular loose bodies have been reported to be a major factor for poor outcome for nonoperative or late surgical treatment [29, 58]. In such cases, arthroscopic or open surgical approach should be performed in acute setting.

CT scan has played a relevant role in evaluating patellofemoral joint in last decades [16]. It is a less expensive method (comparing to MRI) for assessment of patellofemoral alignment and risk factors for instability or to detect osteochondral fractures or loose bodies. CT scanning is useful in

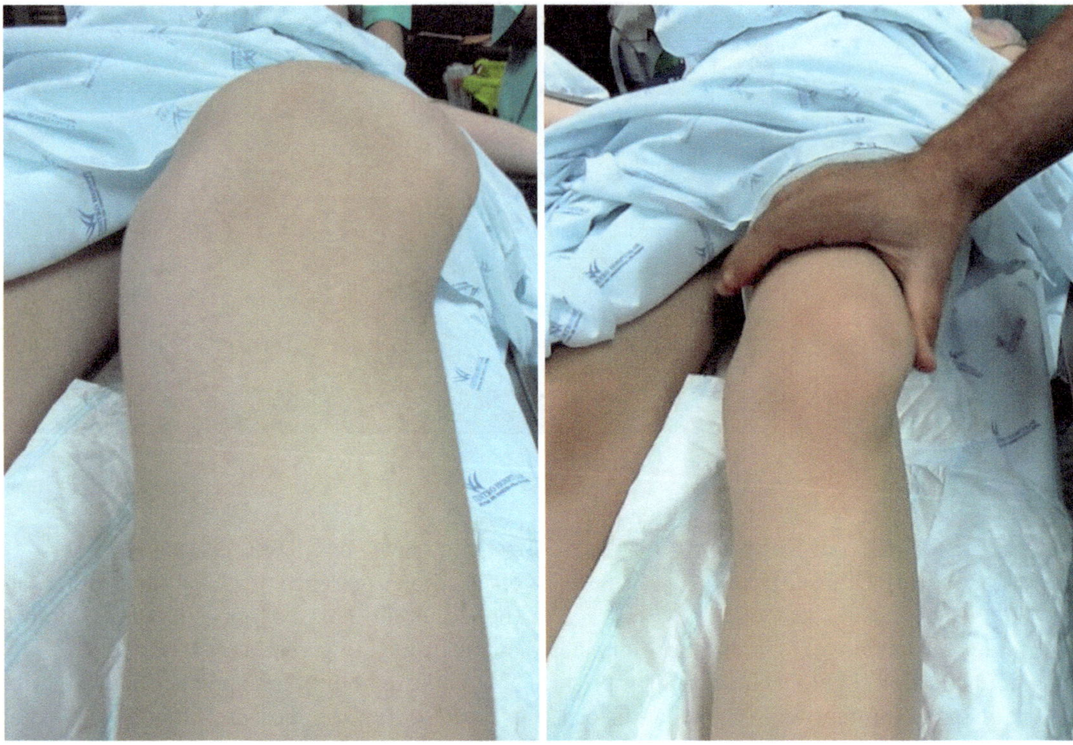

Fig. 14.3 Deformity and effusion after patellar dislocation

Fig. 14.4 Thumb-forearm apposition maneuver in a patient with hyperlaxity

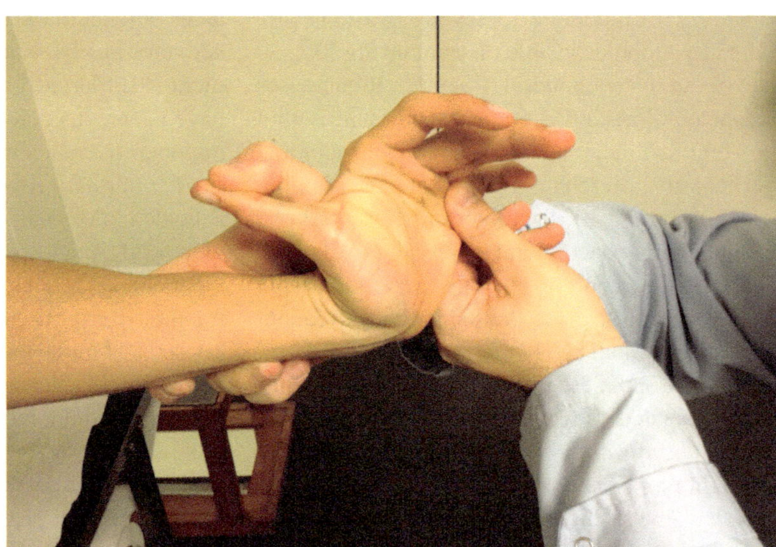

measuring patellar tilt, translation, tibial tuberosity trochlear groove (TTTG) distance, and trochlear dysplasia [27]. It is also useful in evaluating lower-limb alignment (torsional deformities) and determining the rotational relationship between the tibial tuberosity and femoral sulcus in different degrees of flexion. However, it has limited capacity to assess soft tissue. Furthermore, in

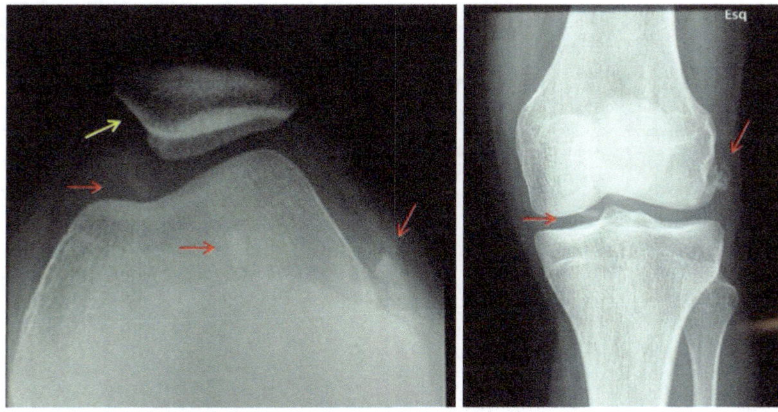

Fig. 14.5 X-ray presenting loose bodies (*red arrows*) and sit of patellar osteochondral defect (*yellow arrow*)

skeletally immature patients, the cartilaginous femoral sulcus contour is shallower than the underlying bone. For this reason, measurement of the bony femoral sulcus angle in these circumstances is less accurate than evaluating the cartilaginous femoral sulcus angle by ultrasound or MRI [39].

MRI assessment is important to assess cartilage status of patellofemoral joint and to evaluate the site and extent of soft-tissue damage to the medial patellar stabilizers (mainly MPFL). Moreover it has higher capacity to evaluate different soft tissues including meniscus and ligaments which might be concomitantly damaged or constitute differential diagnosis.

With the development of magnetic resonance sequencing, MRI is becoming more specific in assisting the surgeon in deciding on nonoperative versus operative management and also combines the possibility to measure "classical instability factors initially described on CT" [48] (Fig. 14.6).

Fithian et al. [21] described that, if evidences of acute injury in the MPFL or VMO are visible on MRI, it might represent a tendency for lower risk of subsequent patellar instability [21]. However, no statistical significance was achieved. Injury on the femoral side of MPFL might be predictive of higher-risk subsequent patellar instability [56]. Once more, it remains unclear if MPFL reconstruction in this setting leads to improved long-term clinical outcomes.

Dynamic MRI evaluation of patellofemoral joint might bring, in near future, further knowledge concerning guidelines for operative versus conservative treatment decision in acute setting (Fig. 14.7).

14.3 Current Treatment Options

The treatment of patellar dislocation involves the resolution of the acute situation (deformity, pain, functional impairment) but also aims to minimize squeals such as recurrent instability, painful subluxation, or osteoarthritis. However, controversial management of first-time patellar dislocation is the mainstay found in the literature, with little exceptions.

14.3.1 Nonoperative Treatment

Nonsurgical, "atraumatic" reduction of the patella should be performed as fast as possible. It delivers pain relief and reduces the risk of further osteochondral injury to the articular surface of either the patella or the lateral femoral trochlea. The prereduction and postreduction radiographs have to be analyzed and compared for evidence of intra-articular loose bodies.

From this step on, little evidence or consensus exists concerning conservative treatment after acute patellar dislocation [59].

Currently, treatment programs vary from immediate mobilization without orthoses or bracing to cast immobilization in extension for 6 weeks.

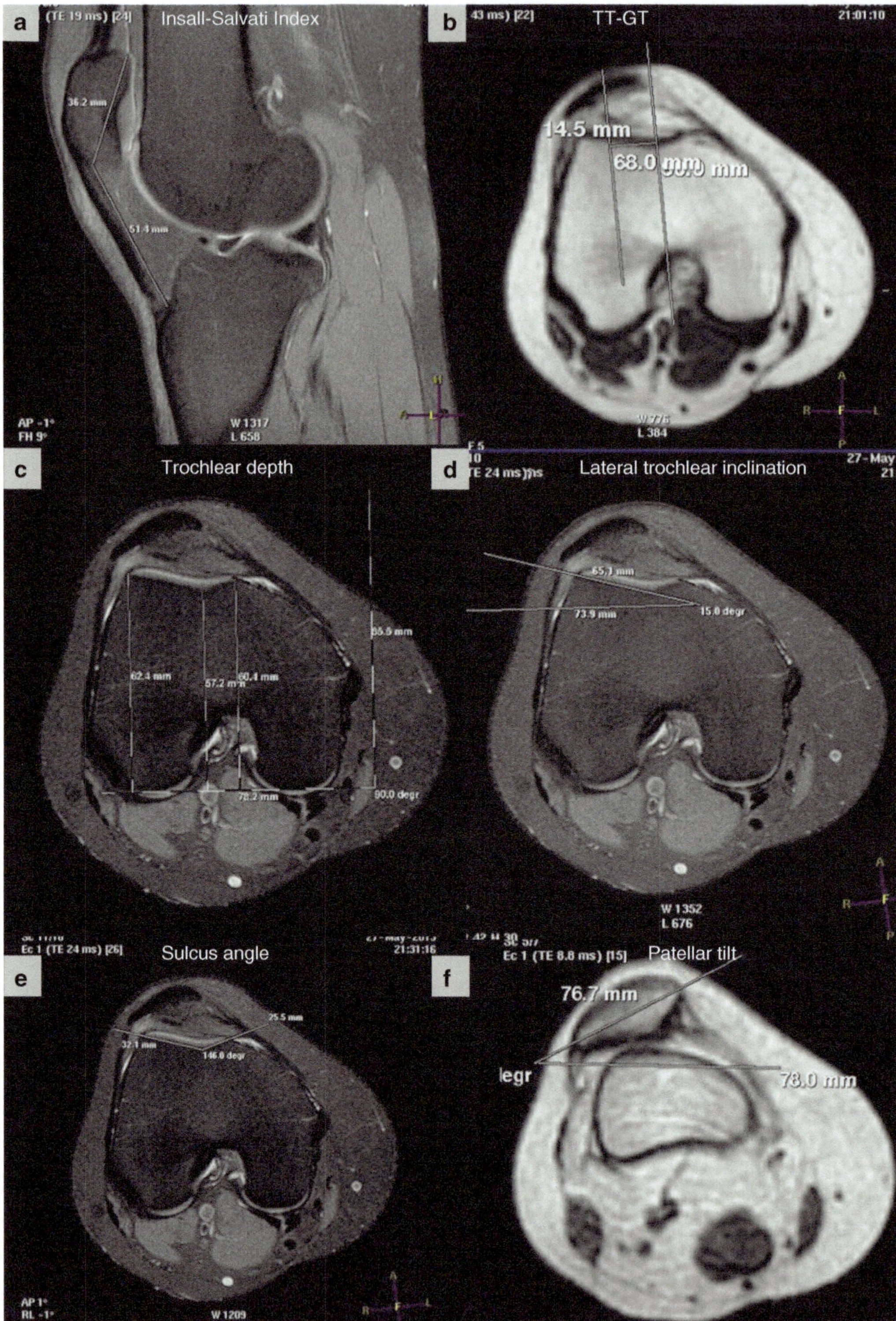

Fig. 14.6 MRI basic protocol for patellofemoral evaluation: Insall-Salvati Index (**a**); TT-GT (**b**); trochlear depth (**c**); lateral trochlear inclination (**d**); sulcus angle (**e**); patellar tilt (**f**)

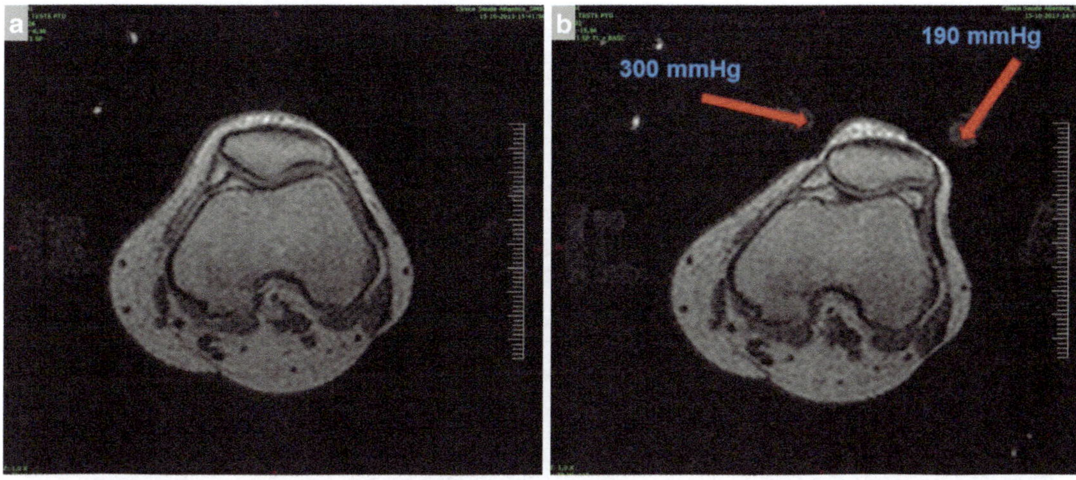

Fig. 14.7 Dynamic MRI evaluation: check patellofemoral joint at rest (**a**) opposing to dynamic evaluation (**b**) with external lateral translation and tilt forces (*arrows*)

Immobilizing the knee in extension might proportionate a better environment for healing to the medial structures. However, the risk of stiffness, muscle weakness, and loss of proprioception must be considered [29]. Patient's own will and compliance with treatment regimen might also be a factor implicated in decision for conservative treatment.

The role of patellar braces and straps on the outcome of acute primary patellar dislocation has not been determined to date.

In a controlled study enrolling 100 primary patellar dislocations [34], patellar bandage or brace, posterior splint, or plaster cast was compared. The immobilization in the splint and cast groups was performed for 6 weeks. The immediate mobilization group had a risk of redislocation three times higher. Stiffness was more frequent in the cast group.

Most clinicians propose a short period of immobilization, early weight bearing (as tolerated) with crutches, followed by rehabilitation of the knee, with or without bracing [59].

14.3.2 Surgical Treatment

The most consensual indication for operative intervention after acute first-time patellar dislocation is a large displaced osteochondral fracture with a loose body that may be possible to fix in place (Fig. 14.8).

Arthroscopic procedure can be performed for diagnostic purposes and the removal of intra-articular loose bodies such as large blood clots and osteochondral fragments (Fig. 14.9) or medial retinacular repair [50].

The recent literature does not support any use of an isolated lateral release for the treatment of patellar instability.

A number of different procedures have been advocated to reconstruct the medial structures [14] or repair the MPFL [11, 13–15, 17, 22, 23, 49].

The surgical objective is, in most cases, to stabilize the patella by an "anatomical" MPFL reconstruction at the patella and femur using a mini-open technique (Fig. 14.10).

A gracilis tendon graft is widely chosen for this purpose because the load to failure of the native MPFL is lower (208 N) and implicates lower surgical damage (leaving the stronger semitendinosus tendon available). Fixation at the patella is usually performed using resorbable anchors or bone tunnels. Care should be taken to avoid long tunnels in the frontal plane. Transversal drilling of the patella from side to side dictates significant risk of fracture [44]. Short- to long-term results showed a low redislocation rate and significant patient satisfaction. Extreme caution to avoid graft malpositioning, this may lead to a change in the patellofemoral joint forces and may lead to pain, restricted range of motion, redislocation, and articular cartilage

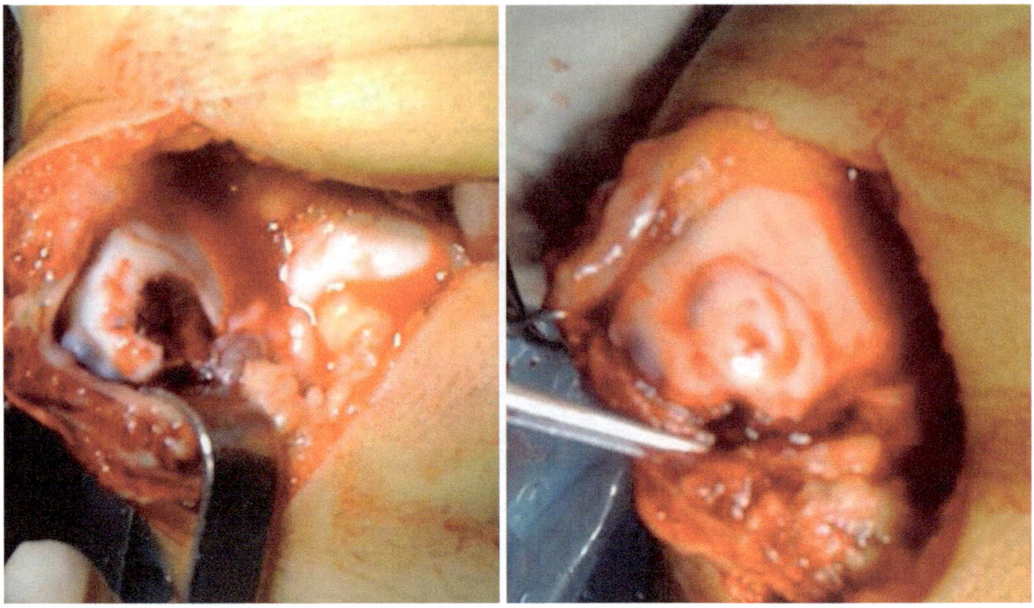

Fig. 14.8 Osteochondral defect of the patella fixed "in situ"

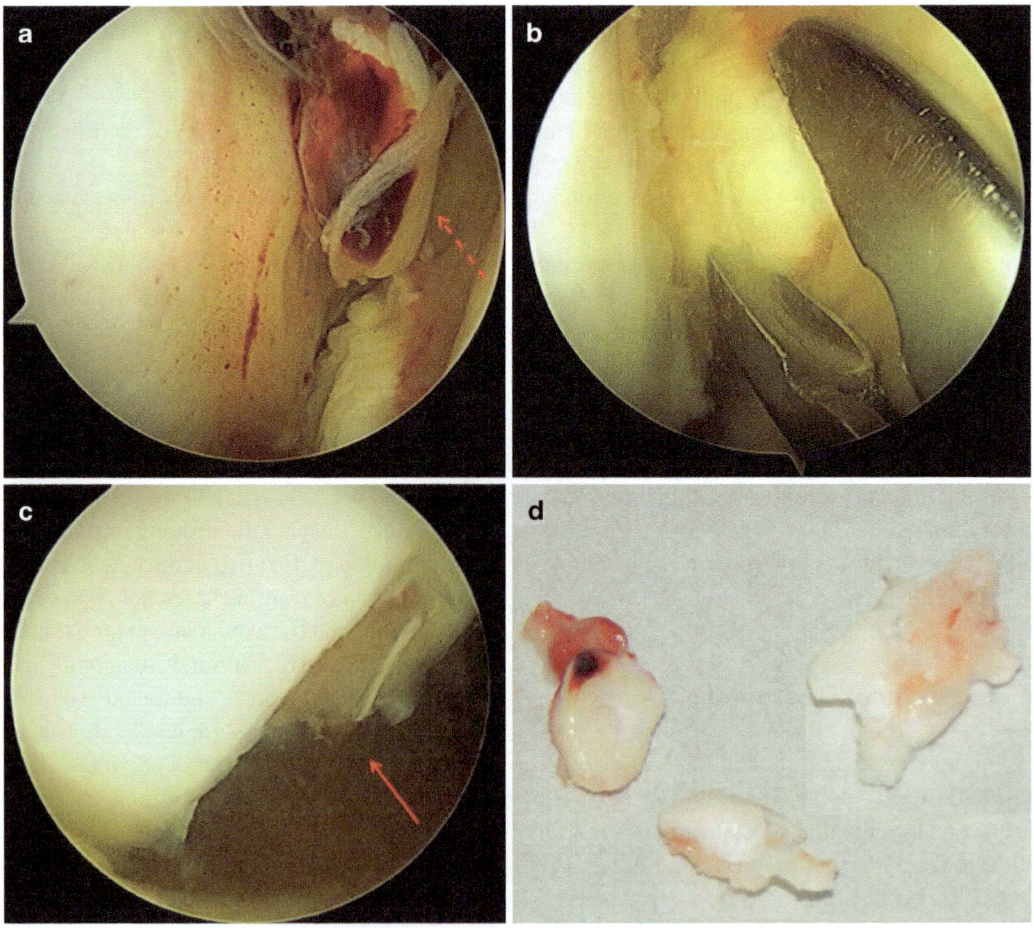

Fig. 14.9 Arthroscopic view of loose bodies after acute patellar dislocation (**a**, **b**); zone of the patellar defect (**c**); loose fragments removed (**d**)

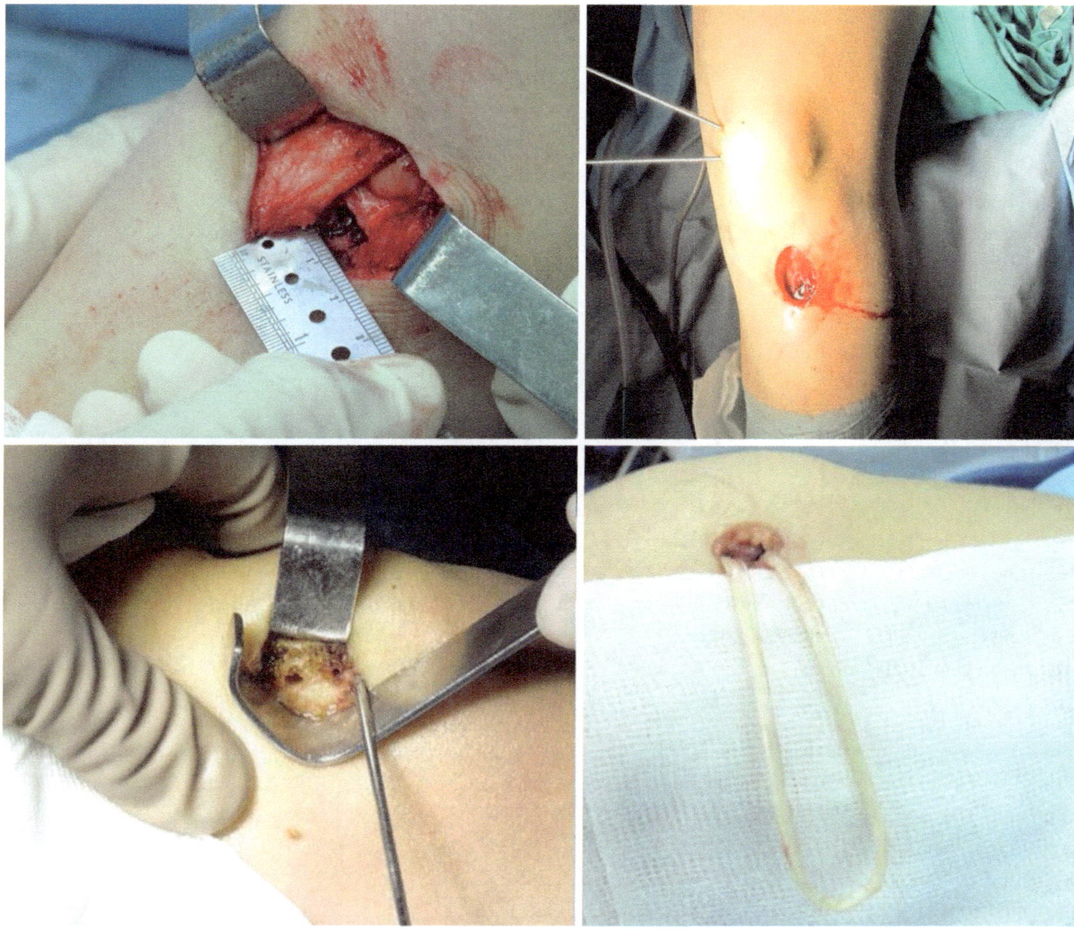

Fig. 14.10 Combined Elmslie-Trillat and MPFL repair procedures. Small anterior approach enables both harvesting of gracilis tendon and medialization of tibial tuberosity

deterioration over time. Therefore, correct tunnel placement is required [44].

Surgical approach might combine distal realignment procedures and might be considered in patients with predisposing risk factors (e.g., increased TT-TG or patella *alta*; Figs. 14.2 and 14.6). These approaches are described elsewhere within this publication.

14.4 Systematic Literature Review

Most studies in the literature are retrospective and nonrandomized level IV studies. We selected our review based on the following criteria: (1) English language, (2) level I–IV studies, (3) a minimum of ten patients in the series at baseline who underwent surgical or conservative treatment for acute patellofemoral dislocation, and (4) a minimum of 6 months of follow-up. Review articles, case reports, and technique articles without reported patient data, and studies which did not state inclusion criteria were excluded. Tables 14.1 and 14.2 summarize the results of papers matching inclusion criteria.

14.5 Discussion

Most of the studies published are not prospective and/or blinded randomized studies. Nonetheless, several conclusions can be depicted.

First, understanding the biomechanics of the patellofemoral joint is necessary to understand

Table 14.1 Demographics of studies

Study	Study type	Patients number	Age (years) (st/ct)	Treatment N=st/N=ct	Follow-up (years)
Bitar et al. [7]	RC	41	12-/38	21st/20ct	Minimum 2
Apostolovic et al. [1]	RNC	37	12–16	14 st/23 ct	6.1
Mariani et al. . [36]	CS	17	NR	All st	2.2
Camanho et al. [10]	RNC	33	24.6/26.8	17st/16 ct	3.4
Nietosvaara et al. [40]	RNC	71 (74knees)	<16	36st/28 ct	2
Sillanpaa et al. [56]	RC	40	20	18st/22 ct	7
Sillanpaa et al. [56]	RC	40	20	18st/22 ct	7
Silanpaa et al. [52]	RNC	76	19–22	30st/46 ct	7
Christiansen et al. [13]	RCT	80 (3 lost to f-up; final n=77)	20.0/19.9 (13–39)	42 st/35ct	2
Buchner et al. [9]	CC	126	NR	63st/63 ct	8.1
Atkin et al. [2]	CS	74	19.9	All ct	19.9
Maenpaa et al. [33]	CS	100	NR	All ct	13
Maenpaa et al. [33]	CS	270	Women: 23.5 (range 9–56) Men: 22.6 (range 12–42)	All st	4.1
Harilainen et al. [25]	CS	53	29.1 (range 17–57)	All st	6.5
Vainiopaa et al. [60]	CS	55	21 .5 years (range 14–54)	All st	2
Cash and Hughston [12]	CC	399	9–72	All ct	8
Yamamoto et al. [62]	CS	30	NR	All st	1–7
	CS	75	37 (median age 19) had a recurrence; 38 (median age 28) no recurrence	All ct	6–24
Lewallen et al. [32]	CC	222	14.9	All ct	12
Bitar et al. [7]	RC	41	12–38	21st/20ct	Minimum 2
Apostolovic et al. [1]	RNC	37	12–16	14 st/23 ct	6.1
Mariani et al. [36]	CS	17	NR	All st	2.2
Camanho et al. [10]	RNC	33	24.6/26.8	17st/16 ct	3.4
Nietosvaara et al. [40]	RNC	71 (74 knees)	<16	36st/28 ct	2
Sillanpaa et al. [56]	RC	40	20	18st/22 ct	7
Silanpaa et al. [52]	RNC	76	19–22	30st/46 ct	7
Buchner et al. [9]	CC	126	NR	63st/63 ct	8.1
Atkin et al. [2]	CS	74	19.9	All ct	19.9
Maenpaa et al. [33]	CS	100	NR	All ct	13
Maenpaa et al. [33]	CS	270	Women: 23.5 (range 9–56) Men: 22.6 (range 12–42)	All st	4.1
Harilainen et al. [25]	CS	53	29.1 (range 17–57)	All st	6.5
Vainiopaa et al. [60]	CS	55	21.5 years (range 14–54)	All st	2
Cash and Hughston [12]	CC	399	9–72	All ct	8
Yamamoto et al. [62]	CS	30	NR	All st	1–7
	CS	75	37 (median age 19) had a recurrence; 38 (median age 28) no recurrence	All ct	6–24
Lewallen et al. [32]	CC	222	14.9	All ct	12

RC randomized controlled, *RNC* randomized noncontrolled, *CC* case control, *CS* case series, *NR* not referred, *st* surgical treatment, *ct* conservative treatment

Table 14.2 Summary of outcomes according to treatment

Treatment	Study	Type of surgical procedure	Results
Conservative	Lewallen et al. [32]		62 % success rate for conservative treatment after first-time patellar dislocation 10.8 % required operation
	Atkin et al. [2]		58 % presented limitation in strenuous activities after 6 months
	Maenpaa and Lehto [34]		0.17 dislocations per year of follow-up
	Cash and Hughston [12]		75 % good/excellent results if no risk factors were found
Surgical	Mariani et al. [36]	Arthroscopic repair of MPFL	No redislocations; Lysholm 90 (72–100) 14/17 returned to sports at the same level
	Maenpaa and Lehto [35]	Medial capsular reefing (all) + lateral retinacular release (n=243)+Elmslie-Roux-Trillat procedure (n=2)	Excellent/good subjective results: Traumatic group: 76.8 % Non-traumatic group: 60.4 % Redislocation rate: Traumatic group: 2.4 % Non-traumatic group: 38.6 % No significant difference was found between methods of postoperative treatment: immobilization or mobilization. "The subjective result of operative treatment was better and the re-dislocation rate was lower if the injury mechanism was traumatic rather than non-traumatic and if there was no history for family occurrence of patellar dislocation"
	Harilainen and Sandelin [25]	Medial retinacular suturing (n=7) or reefing (n=46) and lateral capsular discission (within 1 week of injury)	17 % recurrence More recurrence with greater patellofemoral incongruence
	Vainiopaa et al. [60]	Medial capsular suture and lateral release when lateral retinaculum was tight	9 % recurrence;
		Removed/refixed displaced osteochondral fragments	Most returned to previous sports activities
	Yamamoto et al. [62]	Arthroscopic medial capsular repair/lateral release	Successful stabilization of the acute dislocation Early accurate diagnosis Accurate restoration of normal anatomy
Conservative *vs* surgical	Bitar et al. [7]	MPFL reconstruction	Mean Kujala operated 88.9/nonoperated 70.8 0 recurrence operated group; 35 % recurrences nonoperated group
	Apostolovic et al. [1]	Arthroscopic surgery: Medial retinacular and capsular repair and lateral retinacular release	No statistical difference between groups
	Camanho et al. [10]	MPFL repair vs conservative group	0 relapse operated; 50 % relapses nonoperated
	Nietosvaara et al. [40]	Medial repair/lateral release alone	66 % good/excellent operated; 75 % good/excellent nonoperated No difference in redislocation between treatment groups; Predisposing factor – family history
	Silanpaa et al. [56]	MPFL repair vs conservative group	0 redislocation operated 6/21 redislocation nonoperated
	Silanpaa et al. [55]	Arthroscopic medial capsular repair	19 %redislocation operated; 81 % pre-injury level 23 % redislocation nonoperated 56 % pre-injury level
	Christiansen et al. [13]	Reinsertion MPFL to the adductor tubercle vs conservative treatment	Redislocation rates were 17 and 20 % in the operative and conservative treatment groups, respectively (not significant)
	Buchner et al. [9]	MFPL reconstruction	No difference in redislocation between the two groups

the pathology of patellar dislocation. Recent studies have focused on the medial patellofemoral ligament (MPFL) and have shown that the MPFL is the most significant passive stabilizer of the patella. It is accepted that primary patellar dislocation leads to MPFL injury [18].

Because of the insufficient evidence in literature, there is currently no universally accepted, optimal strategy approach for acute primary patellar dislocation.

The complexity of patellar instability leads to challenges in decision making between different treatment modalities.

Most cases seem to be suitable for initial nonsurgical management in the first episode, although recurrent instability might occur [29].

Osteochondral fragments amenable for surgical fixation are an indication for surgery [58].

In the setting of surgical treatment, MPFL reconstruction might be a more reliable method of stabilizing the patella than repair, which has inherent limitations related to the MPFL injury location [53].

The MPFL injury location can be assessed by MRI with increasing feasibility. Despite the current thought that considers the femoral attachment as the most frequent site of lesion after acute dislocation, this still remains debatable and further prospective studies will be needed.

Several recent studies advocate MRI after acute patellar dislocation, as the acutely injured knee usually shows hemarthrosis as a sign of tissue damage and clinical diagnosis can sometimes be difficult [21, 54]. MRI can be used to diagnose the signs of acute patellar dislocation and associated injuries, such as osteochondral fractures and meniscal or ligament injuries [19].

Yamamoto [62] studied arthroscopic repair of the MPFL, showing overall good or excellent results, only 1 redislocation of 30 operated knees. Sillanpaa et al. [55, 56] reported limited efficacy of an arthroscopic MPFL repair compared with conservative treatment. The redislocation rate was similar in both treatments, and the authors stated that the explanation for these unsatisfying results was that the MPFL injury has different patterns and locations. Sillampaa [56] followed 44 patients after first-time patellar dislocation and subjected to conservative treatment for 7 years. They found femoral avulsions of the MPFL to be a significant predictor for subsequent patellar instability, and these were less likely to return to prior activity level, but without significant differences in the Kujala score.

A prospective trial from Camanho et al. [10] comparing conservative and operative treatment in 33 patients with acute patellar dislocations revealed a significantly better Kujala score (92/100) and no recurrence in the operative group compared to a Kujala score of 69/100 and 8 relapses in the conservative group.

On another study, no redislocation was observed in the operated group comparing to six cases (over 21) among those treated conservatively [56].

The median Kujala scores were 91 points for the surgically treated patients and 90 points for the nonoperatively treated patients. Thirteen (over 17) patients in the operated group and 15 (over 21) in the nonoperatively treated group returned to their pre-injury physical activity level. The authors concluded that the rate of redislocation for those treated with surgical stabilization was significantly lower than the rate for those treated without surgical stabilization. However, no clear patient-referred clinical benefits were seen at long-term follow-up from initial surgery [56].

It seems useful to define the exact location of the MPFL tear after a primary episode of dislocation; however there is no agreement on this issue. Considering this fact, dynamic MRI might be a significant step to diagnose, classify the grade of severity, and permit more effective guidelines for treatment. The Porto-knee testing device (PKTD) [20] is under development for patellofemoral application to assist on more accurate diagnosis capacity and understanding of patellofemoral anatomy and kinematics (Fig. 14.7).

It is common experience that there are different types of dislocations with different tear patterns. From a clinical point of view, this issue has clear surgical implications since the surgeon must be ready to change the treatment on the basis of pathoanatomical findings. In fact, the femoral attachment of MPFL is located in the second tissue layer, below the superficial fascia and above the capsule, and if torn the hemorrhage is extraarticular. When a patellar avulsion is suspected, then an arthroscopic examination can be carried out in order to define the exact location of tear and to plan the surgical repair. In cases of ligament avulsion from its patellar insertion, a direct repair could be applied [53]. However, there is no consensus on the matter and currently systematic arthroscopy is not a rule for first-time dislocation.

It seems that younger patients more often sustain patellar-based ruptures, while older patients more often sustain femoral-based ruptures of the MPFL. Incomplete MPFL ruptures are correlated with lower Insall-Salvati indices than complete ruptures, and trochlear dysplasia is correlated with higher rates of redislocation [53].

Osseous surgery is usually not needed if surgery is planned after primary dislocation, but corrections are needed in cases with severe bony abnormalities [21].

Previous randomized studies of primary patellar dislocations concluded that surgery is not superior to nonsurgical treatment if all the patients with different types of MPFL injuries are treated similarly [59].

In contrast, a recent prospective randomized study in comparing nonsurgical treatment with surgical reinsertion, surgery resulted in better stability than nonoperative treatment. This was true for either femoral or patellar surgical attachment [7].

Because of the high (44–70 %) redislocation rate after primary dislocation, some cases might benefit from initial surgery, and surgery should definitely be considered for cases with a high risk of failure after nonsurgical treatment [7].

Patients with patellar MPFL avulsion fracture and MPFL disruption at the femoral attachment seem to be at greater risk of subsequent dislocation. In these patients, restoring the integrity of the MPFL might be necessary to ensure better stability [53].

Ruptures at the MPFL mid-substance or patellar insertion regions are generally not related to significant subsequent patellar instability.

The clinical outcomes of various MPFL injuries, however, remain highly uncertain with regard to the well-known factors that predispose a patient to patellar instability, such as trochlear dysplasia, axial and torsional lower-limb alignment abnormalities, and MPFL injuries. Most likely, the more dysplastic the trochlear shape, the more devastating the injury to the MPFL is to patellar stability. Whenever required, surgical treatment needs to be tailored individually, based on the diagnosis (MRI findings of the MPFL injury and osseous anatomy).

The precise risk factors for redislocation could not be adequately calculated in this review due to lack of consistent and quality reporting in several articles. To date, there is no evidence that the natural history of a person suffering primary patellar dislocation is improved by surgical intervention in the acute setting. Considering the previous, surgical stabilization of the patella cannot be firmly recommended after the first event of patella dislocation. However, after the second episode, the risk of redislocation is known to be much higher (49 %), and surgical intervention should be considered [29].

Conclusion

It seems that the predominating factors for patellar dislocation are heterogenetic morphology in combination with individual predisposition. The only consensual indication for surgical treatment after acute primary patellar dislocation is the presence of a concomitant osteochondral fracture still suitable for "in situ" fixation or the presence of loose bodies inside the joint. When surgical treatment is required, it should specifically address the correction of the implicated pathomorphology in each case. The type and site of MPFL injury seem to be a relevant factor to be considered as prognostic for redislocation. Low-energy events causing dislocation usually occur in persons with risk factors of patellofemoral instability, thus having higher risk for recurrence.

Conservative treatment is currently performed in most cases of a first-episode patella dislocation.

The optimal treatment has not yet been established and further prospective randomized studies are required.

References

1. Apostolovic M, Vukomanovic B, Slavkovic N, Vuckovic V, Vukcevic M, Djuricic G, et al. Acute patellar dislocation in adolescents: operative versus nonoperative treatment. Int Orthop. 2011;35:1483–7.
2. Atkin DM, Fithian DC, Marangi KS, Stone ML, Dobson BE, Mendelsohn C. Characteristics of patients with primary acute lateral patellar dislocation and their recovery within the first 6 months of injury. Am J Sports Med. 2000;28:472–9.

3. Balcarek P, Ammon J, Frosch S, Walde TA, Schuttrumpf JP, Ferlemann KG, et al. Magnetic resonance imaging characteristics of the medial patellofemoral ligament lesion in acute lateral patellar dislocations considering trochlear dysplasia, patella alta, and tibial tuberosity-trochlear groove distance. Arthroscopy. 2010;26:926–35.
4. Balcarek P, Walde TA, Frosch S, Schuttrumpf JP, Wachowski MM, Sturmer KM, et al. Patellar dislocations in children, adolescents and adults: a comparative MRI study of medial patellofemoral ligament injury patterns and trochlear groove anatomy. Eur J Radiol. 2011;79:415–20.
5. Baldwin JL. The anatomy of the medial patellofemoral ligament. Am J Sports Med. 2009;37:2355–61.
6. Bedi H, Marzo J. The biomechanics of medial patellofemoral ligament repair followed by lateral retinacular release. Am J Sports Med. 2010;38:1462–7.
7. Bitar AC, Demange MK, D'Elia CO, Camanho GL. Traumatic patellar dislocation: nonoperative treatment compared with MPFL reconstruction using patellar tendon. Am J Sports Med. 2012;40: 114–22.
8. Boring TH,O'Donoghue DH. Acute patellar dislocation: results of immediate surgical repair. Clin Orthop Relat Res. 1978;(136):182–5.
9. Buchner M, Baudendistel B, Sabo D, Schmitt H. Acute traumatic primary patellar dislocation: long-term results comparing conservative and surgical treatment. Clin J Sport Med. 2005;15:62–6.
10. Camanho GL, Viegas Ade C, Bitar AC, Demange MK, Hernandez AJ. Conservative versus surgical treatment for repair of the medial patellofemoral ligament in acute dislocations of the patella. Arthroscopy. 2009;25:620–5.
11. Carmont MR, Maffulli N. Medial patellofemoral ligament reconstruction: a new technique. BMC Musculoskelet Disord. 2007;8:22.
12. Cash JD, Hughston JC. Treatment of acute patellar dislocation. Am J Sports Med. 1988;16:244–9.
13. Christiansen SE, Jacobsen BW, Lund B, Lind M. Reconstruction of the medial patellofemoral ligament with gracilis tendon autograft in transverse patellar drill holes. Arthroscopy. 2008;24:82–7.
14. Davis DK, Fithian DC. Techniques of medial retinacular repair and reconstruction. Clin Orthop Relat Res. 2002;(402):38–52.
15. Deie M, Ochi M, Adachi N, Shibuya H, Nakamae A. Medial patellofemoral ligament reconstruction fixed with a cylindrical bone plug and a grafted semitendinosus tendon at the original femoral site for recurrent patellar dislocation. Am J Sports Med. 2011;39:140–5.
16. Dejour H, Walch G, Nove-Josserand L, Guier C. Factors of patellar instability: an anatomic radiographic study. Knee Surg Sports Traumatol Arthrosc. 1994;2:19–26.
17. Dodson CC, Shindle MK, Dines JS, Altchek DW. Arthroscopic suture anchor repair for lateral patellar instability. Knee Surg Sports Traumatol Arthrosc. 2010;18:143–6.
18. Elias DA, White LM, Fithian DC. Acute lateral patellar dislocation at MR imaging: injury patterns of medial patellar soft-tissue restraints and osteochondral injuries of the inferomedial patella. Radiology. 2002;225:736–43.
19. Engelhardt LV, Raddatz M, Bouillon B, Spahn G, Dàvid A, Haage P, et al. How reliable is MRI in diagnosing cartilaginous lesions in patients with first and recurrent lateral patellar dislocations? BMC Musculoskelet Disord. 2010;11:1–8.
20. Espregueira-Mendes J, Pereira H, Sevivas N, Passos C, Vasconcelos JC, Monteiro A, et al. Assessment of rotatory laxity in anterior cruciate ligament-deficient knees using magnetic resonance imaging with Porto-knee testing device. Knee Surg Sports Traumatol Arthrosc. 2012;20:671–8.
21. Fithian DC, Paxton EW, Stone ML, Silva P, Davis DK, Elias DA, et al. Epidemiology and natural history of acute patellar dislocation. Am J Sports Med. 2004;32:1114–21.
22. Giordano M, Falciglia F, Aulisa AG, Guzzanti V. Patellar dislocation in skeletally immature patients: semitendinosus and gracilis augmentation for combined medial patellofemoral and medial patellotibial ligament reconstruction. Knee Surg Sports Traumatol Arthrosc. 2012;20:1594–8.
23. Han H, Xia Y, Yun X, Wu M. Anatomical transverse patella double tunnel reconstruction of medial patellofemoral ligament with a hamstring tendon autograft for recurrent patellar dislocation. Arch Orthop Trauma Surg. 2011;131:343–51.
24. Harilainen A, Myllynen P, Antila H, Seitsalo S. The significance of arthroscopy and examination under anaesthesia in the diagnosis of fresh injury haemarthrosis of the knee joint. Injury. 1988;19:21–4.
25. Harilainen A, Sandelin J. Prospective long-term results of operative treatment in primary dislocation of the patella. Knee Surg Sports Traumatol Arthrosc. 1993;1:100–3.
26. Hawkins RJ, Bell RH, Anisette G. Acute patellar dislocations. The natural history. Am J Sports Med. 1986;14:117–20.
27. Hing CB, Shepstone L, Marshall T, Donell ST. A laterally positioned concave trochlear groove prevents patellar dislocation. Clin Orthop Relat Res. 2006;447:187–94.
28. Hinton RY, Sharma KM. Acute and recurrent patellar instability in the young athlete. Orthop Clin North Am. 2003;34:385–96.
29. Jain NP, Khan N, Fithian DC. A treatment algorithm for primary patellar dislocations. Sports Health. 2011;3:170–4.
30. Kaplan EB. Factors responsible for the stability of the knee joint. Bull Hosp Joint Dis. 1957;18:51–9.
31. LaPrade RF, Engebretsen AH, Ly TV, Johansen S, Wentorf FA, Engebretsen L. The anatomy of the

medial part of the knee. J Bone Joint Surg Am. 2007;89:2000–10.
32. Lewallen LW, McIntosh AL, Dahm DL. Predictors of recurrent instability after acute patellofemoral dislocation in pediatric and adolescent patients. Am J Sports Med. 2013;41:575–81.
33. Maenpaa H, Huhtala H, Lehto MU. Recurrence after patellar dislocation. Redislocation in 37/75 patients followed for 6–24 years. Acta Orthop Scand. 1997;68:424–6.
34. Maenpaa H, Lehto MU. Patellar dislocation. The long-term results of nonoperative management in 100 patients. Am J Sports Med. 1997;25:213–7.
35. Maenpaa H, Lehto MU. Surgery in acute patellar dislocation – evaluation of the effect of injury mechanism and family occurrence on the outcome of treatment. Br J Sports Med. 1995;29:239–41.
36. Mariani PP, Liguori L, Cerullo G, Iannella G, Floris L. Arthroscopic patellar reinsertion of the MPFL in acute patellar dislocations. Knee Surg Sports Traumatol Arthrosc. 2011;19:628–33.
37. Matsushita T, Kuroda R, Araki D, Kubo S, Matsumoto T, Kurosaka M. Medial patellofemoral ligament reconstruction with lateral soft tissue release in adult patients with habitual patellar dislocation. Knee Surg Sports Traumatol Arthrosc. 2013; 21:726–30.
38. Neyret P, Robinson AH, Le Coultre B, Lapra C, Chambat P. Patellar tendon length – the factor in patellar instability? Knee. 2002;9:3–6.
39. Nietosvaara Y, Aalto K. The cartilaginous femoral sulcus in children with patellar dislocation: an ultrasonographic study. J Pediatr Orthop. 1997;17:50–3.
40. Nietosvaara Y, Paukku R, Palmu S, Donell ST. Acute patellar dislocation in children and adolescents. Surgical technique. J Bone Joint Surg Am. 2009; 91(Suppl 2 Pt 1):139–45.
41. Nomura E. Classification of lesions of the medial patello-femoral ligament in patellar dislocation. Int Orthop. 1999;23:260–3.
42. Nomura E, Inoue M, Osada N. Anatomical analysis of the medial patellofemoral ligament of the knee, especially the femoral attachment. Knee Surg Sports Traumatol Arthrosc. 2005;13:510–5.
43. Panagiotopoulos E, Strzelczyk P, Herrmann M, Scuderi G. Cadaveric study on static medial patellar stabilizers: the dynamizing role of the vastus medialis obliquus on medial patellofemoral ligament. Knee Surg Sports Traumatol Arthrosc. 2006; 14:7–12.
44. Parikh SN, Nathan ST, Wall EJ, Eismann EA. Complications of medial patellofemoral ligament reconstruction in young patients. Am J Sports Med. 2013;41:1030–8.
45. Philippot R, Boyer B, Testa R, Farizon F, Moyen B. The role of the medial ligamentous structures on patellar tracking during knee flexion. Knee Surg Sports Traumatol Arthrosc. 2012;20:331–6.
46. Piper SL, Kramer JD, Kim HT, Feeley BT. Effects of local anesthetics on articular cartilage. Am J Sports Med. 2011;39:2245–53.
47. Sallay PI, Poggi J, Speer KP, Garrett WE. Acute dislocation of the patella. A correlative pathoanatomic study. Am J Sports Med. 1996;24:52–60.
48. Sanders TG, Morrison WB, Singleton BA, Miller MD, Cornum KG. Medial patellofemoral ligament injury following acute transient dislocation of the patella: MR findings with surgical correlation in 14 patients. J Comput Assist Tomogr. 2001;25: 957–62.
49. Schottle PB, Romero J, Schmeling A, Weiler A. Technical note: anatomical reconstruction of the medial patellofemoral ligament using a free gracilis autograft. Arch Orthop Trauma Surg. 2008;128:479–84.
50. Schottle PB, Scheffler SU, Schwarck A, Weiler A. Arthroscopic medial retinacular repair after patellar dislocation with and without underlying trochlear dysplasia: a preliminary report. Arthroscopy. 2006;22: 1192–8.
51. Senavongse W, Farahmand F, Jones J, Andersen H, Bull AM, Amis AA. Quantitative measurement of patellofemoral joint stability: force-displacement behavior of the human patella in vitro. J Orthop Res. 2003;21:780–6.
52. Sillanpaa P, Mattila VM, Iivonen T, Visuri T, Pihlajamaki H. Incidence and risk factors of acute traumatic primary patellar dislocation. Med Sci Sports Exerc. 2008;40:606–11.
53. Sillanpaa PJ, Maenpaa HM. First-time patellar dislocation: surgery or conservative treatment? Sports Med Arthrosc. 2012;20:128–35.
54. Sillanpaa PJ, Maenpaa HM. Repair of the medial patellofemoral ligament injury? Arthroscopy. 2009; 25:702–3; author reply 3.
55. Sillanpaa PJ, Maenpaa HM, Mattila VM, Visuri T, Pihlajamaki H. Arthroscopic surgery for primary traumatic patellar dislocation: a prospective, nonrandomized study comparing patients treated with and without acute arthroscopic stabilization with a median 7-year follow-up. Am J Sports Med. 2008;36: 2301–9.
56. Sillanpaa PJ, Mattila VM, Maenpaa H, Kiuru M, Visuri T, Pihlajamaki H. Treatment with and without initial stabilizing surgery for primary traumatic patellar dislocation. A prospective randomized study. J Bone Joint Surg Am. 2009;91:263–73.
57. Stanitski CL, Paletta Jr GA. Articular cartilage injury with acute patellar dislocation in adolescents. Arthroscopic and radiographic correlation. Am J Sports Med. 1998;26:52–5.
58. Stefancin JJ, Parker RD. First-time traumatic patellar dislocation: a systematic review. Clin Orthop Relat Res. 2007;455:93–101.
59. Tsai CH, Hsu CJ, Hung CH, Hsu HC. Primary traumatic patellar dislocation. J Orthop Surg Res. 2012;7:21.
60. Vainionpaa S, Laasonen E, Silvennoinen T, Vasenius J, Rokkanen P. Acute dislocation of the patella.

A prospective review of operative treatment. J Bone Joint Surg Br. 1990;72:366–9.
61. Weber-Spickschen TS, Spang J, Kohn L, Imhoff AB, Schottle PB. The relationship between trochlear dysplasia and medial patellofemoral ligament rupture location after patellar dislocation: an MRI evaluation. Knee. 2011;18:185–8.
62. Yamamoto RK. Arthroscopic repair of the medial retinaculum and capsule in acute patellar dislocations. Arthroscopy. 1986;2:125–31.
63. Zaffagnini S, Dejour D, Arendt EA (editors). Patellofemoral pain, instability, and arthritis: clinical presentation, imaging, and treatment. Berlin/London: Springer; 2010.
64. Zaffagnini S, Giordano G, Bruni D, Muccioli GM, Marcacci M. Pathophysiology of lateral patellar dislocation. In: Zaffagnini S, Dejour D, Arendt EA, editors. Patellofemoral pain, instability, and arthritis. Berlin/Heidelberg: Springer; 2010. p. 17–27.

15 Surgical Treatment of the Patellofemoral Joint: Lateral Release

Onur Bilge, Koen Lagae, Francis Bonte, Rene Verdonk, and Peter Verdonk

Contents

O. Bilge
Department of Orthopaedic Surgery, Antwerp Orthopedic Center, Monica Hospitals, Antwerp, Belgium

Department of Orthopaedic Surgery, Konya N.E. University Hospital, Konya, Turkey

K. Lagae
Department of Orthopaedic Surgery, Antwerp Orthopedic Center, Monica Hospitals, Antwerp, Belgium

F. Bonte • R. Verdonk
Department of Orthopaedic Surgery, Ghent University Hospital, Ghent, Belgium

P. Verdonk (✉)
Department of Orthopaedic Surgery, Antwerp Orthopedic Center, Monica Hospitals, Antwerp, Belgium

Faculty of Medicine, Ghent University, Ghent, Belgium
e-mail: pverdonk@yahoo.com

15.1 Introduction

Recent articles have ranked Lateral Release (LR) 47th among all procedures by orthopaedic surgeons [9, 14]. Despite its frequency the indications and results of an LR remain controversial. The acceptable results (good and excellent) have a wide range from 14 to 99 % [1, 4, 12, 14].

LR can be subdivided as both an isolated and an associated procedure (as part of a proximal or distal realignment). The procedure can also be performed as an open, mini-open or arthroscopic procedure [14]. O'Neill et al. proved that there is no significant difference in outcome between arthroscopic and open LR [18]. In arthroscopic LR, however, the risk for postoperative hemarthrosis and swelling is considered higher. The use of electrocautery is advised to prevent this major complication. Until today, this study is the only prospective randomized clinical trial concerning LR.

15.1.1 Anatomy of the Lateral Retinaculum

The lateral retinaculum consists of two separate layers (Fig. 15.1):

1. The superficial oblique layer: This originates from the iliotibial band and interdigitates with the longitudinal fibres of vastus lateralis.
2. The deep layer: This consists of the deep transverse retinaculum with the epicondylopatellar ligament proximally and the patellotibial ligament distally.

A. Gobbi et al. (eds.), *The Patellofemoral Joint*, DOI 10.1007/978-3-642-54965-6_15, © ISAKOS 2014

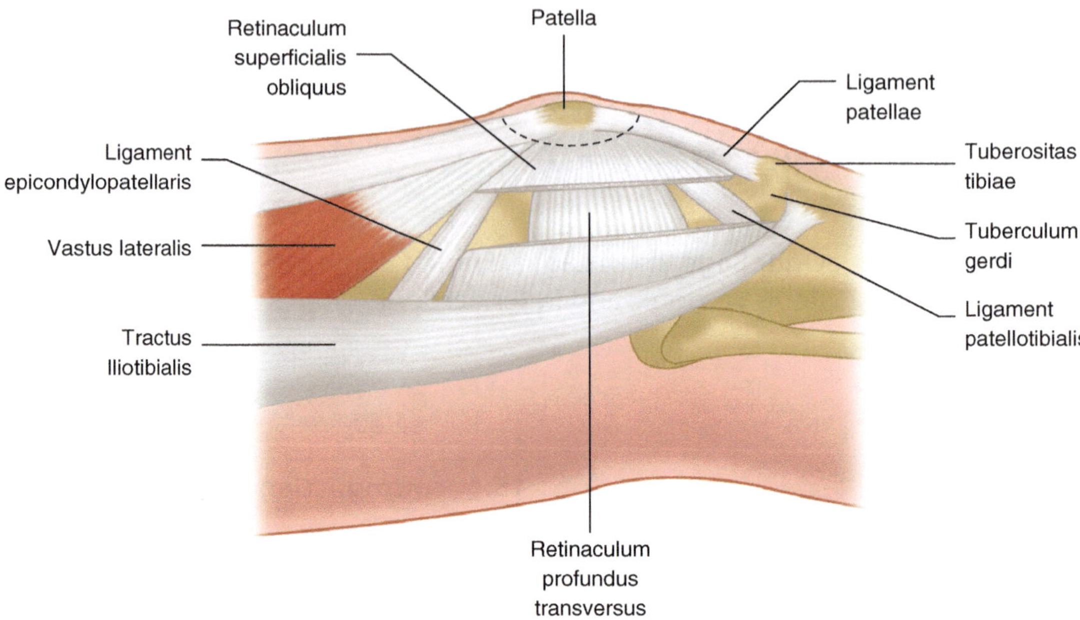

Fig. 15.1 Anatomy of the lateral retinaculum

Beneath the deep transverse retinaculum is the thin capsulosynovial layer that gives little retinacular support to the lateral side of the knee. Immediately posterior to the oblique and transverse retinacular ligaments lies the fascia lata. It is fixed proximal and distal to the lateral joint line, lending static as well as dynamic support to the lateral knee [11].

Important structures are well described by Merican and Amis [15] (Fig. 15.1):

- Deep fascia: not attached to the patella. It thickens laterally to become the iliotibial band.
- Quadriceps aponeurosis and iliotibial band: the bulk of the fibres of the iliotibial band run in a longitudinal direction to Gerdy's tubercle. The anterior fibres curve anteriorly to meet the descending fibres of the quadriceps aponeurosis. The fibres on the superficial surface proceed obliquely. They fuse with the aponeurotic layer of the quadriceps.
- Vastus lateralis obliquus.
- Deeper transverse fibres of the iliotibial band: connect to the patella and vastus lateralis obliquus. There is no attachment to the lateral epicondyle of the femur.
- Lateral patellofemoral and patellomeniscal ligaments: these capsular ligaments vary considerably and are not always found.
- Patellotibial ligament: the same as the quadriceps aponeurosis layer.
- Lateral superior genicular artery.

Merican and Amis also describe the lateral retinaculum as a complex structure which is difficult to delineate because of converging and interdigitating structures. The lateral retinacular complex of the knee is subdivided into three layers:

1. Superficial: deep fascia
2. Intermediate: quadriceps aponeurosis and iliotibial band
3. Deep: joint capsule

The deeper, more transverse fibres from the iliotibial band may be termed the iliotibial band-patella fibres; they are not lateral patellofemoral fibres.

15.1.2 Biomechanical Studies

The two biomechanical studies done by Merican AM et al. were standing out recently. In his first study, the lateral soft tissues of eight fresh-frozen cadaveric knees were dissected and tested to distinguish the different tensile loads for failure [17]. They identified three distinct structures as lateral soft tissues: ITB-patella band, patellofemoral ligament and patellomeniscal ligament. In conclusion, the transverse fibres of lateral retinaculum originating from ITB were found to be the strongest load transmitters to patella.

In his second study, nine cadaveric knees were used to determine the reduction in patellofemoral stability with progressively more extensive LR [16]. As the result of this study, they found that the main lateral restraint was the joint capsule in extension, whereas the transverse fibres at mid-patellar level were the main contributors to lateral restraint at 30° of flexion.

15.2 Materials and Methods

Using Pubmed, more than 30 relevant articles were found. Relevance was based on subject (abstract), language (English) and cited index.

Keywords: Lateral release, Knee, Patellofemoral pain, Lateral tightness, Patellar dislocation, Chondromalacia patellae

15.3 Physical Examination

Physical examination includes preoperative passive patellar tilt (also postoperative), medial and lateral patellar glides, measurement of the tubercle–sulcus angle, the lateral pull sign and lower extremity alignment [12]:

- Standing position: weightbearing alignment, rotational deformities, foot position
- Seated position, with the knees flexed 90°: effusion, patellar position (alta, baja, lateralization), tibial torsion, tubercle–sulcus angle.
- Supine position with the knee extended and the quadriceps relaxed: passive patellar tilt test to diagnose an excessively tight lateral restraint, subjective estimate of patellar crepitation, medial or lateral facet tenderness, retinacular tenderness.
- Supine position with the knees flexed 2030° and the quadriceps relaxed: patellar glide test to diagnose medial or lateral retinacular tightness and/or integrity (Fig. 15.2).

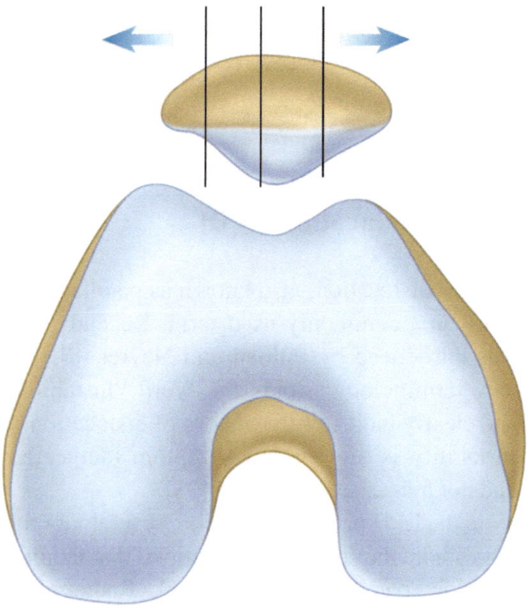

Fig. 15.2 Patellar quadrants and patellar glide test (*arrows*)

  - Lateral patellar glide of three quadrants. Incompetent medial restraint.
  - Lateral glide of four quadrants. Deficient medial restraint.
  - Medial glide of one quadrant. Tight lateral restraint.
  - Medial glide of three or four quadrants. Hypermobile patella.
- Supine position with the knees flexed 90°: Q-angle for evaluation of the distal restraint vector. A normal tubercle–sulcus angle is 0°, while greater than 10° is definitely abnormal.
- Supine position with the knee in extension: the lateral pull sign is useful to determine the vector of an active quadriceps contraction. The patella should be pulled in a straight superior direction or superior and lateral in equal proportions.

15.4 Surgical Options

Before considering surgery of course, an aggressive, nonoperative approach with rehabilitation, of at least 6 months, should be attempted.

Proximal realignments, distal realignments or combinations of these two are the surgical possibilities:

15.4.1 Episodic Patellar Dislocation (EPD)

Patellar dislocation, also known as patellar instability, is a commonly used term. Recently EPD was introduced by Fithian and Neyret [8]. This new terminology avoids the word "instability" and clearly indicates the history of dislocation(s). Instability is moreover a symptom (subjective) and not a disease (objective) [25].

In the EPD patient population, several morphologic anomalies have been identified that facilitate or allow patellar dislocation [6]. Radiographic examination will detect, in more than 96 % of cases, at least one of the four following features in EPD group: (1) trochlear dysplasia, (2) patella alta, (3) tibial tubercle–trochlear groove distance (TT-TG) > 20 mm and (4) patellar tilt >20° [6].

Studies Concerning EPD

To this date, there are no published randomized controlled clinical trials (level 1 evidence) assessing the effect of an isolated lateral retinacular release on the outcome of patellar instability. All currently available material is at best level 4 evidence (retrospective case series or review articles) [14].

Lateral Release (Isolated)

Panni et al. [21] set up a long-term retrospective clinical follow-up study, with two groups. The outcomes of lateral release were evaluated after 5 and 12 years. Each group contained 50 patients. Group I contained patients with patellofemoral pain, group II, patients with patellofemoral instability.

Compared with the 5 year follow-up evaluation, the percentage of satisfactory Lysholm scores after 12 years in group I remained stable, 71 % vs. 70 % ($P = 1.0$), whereas the percentage of satisfactory scores in group II decreased, 72 % vs. 50 % ($P < 0.5$).

Conclusion: Isolated LR is a procedure offering a good percentage of success in the management of a stable patella with excessive lateral pressure and elective location of pain on the lateral retinaculum. In patellar instability, the results are less favourable in long-term follow-up evaluation. The presence of high-grade joint surface injury is a poor prognostic indicator for lateral release.

From a mechanical perspective, isolated LR cannot correct the actual causes of patellar instability whether the cause is deficient trochlea, deficient ligamentous tethers or deficient of abnormal vastus medialis.

In his review, Lattermann et al. [14] evaluated several published case series. While some authors initially reported acceptable success of isolated LR for patella instability, most studies showed disappointing mid- and long-term results. The average percentage of satisfaction of patients with more than 4 years follow-up is only 63.5 %, whereas the short-term (<4 years) satisfaction is 80 %.

Conclusion: Isolated LR has little or no role in the treatment of acute or recurrent patella instability. LR may be added as an adjunct procedure to a proximal or distal realignment of the extensor mechanism. Isolated LR can be a successful procedure in patients with isolated lateral patellar tightness (Excessive Lateral Hyperpression Syndrome).

Lateral Release as Adjunct to Patellofemoral Alignment Procedures for Patellar Instability (Associated)

Scuderi et al. [24] compared two groups of 52 patients (60 knees). Group I consisted of 21 patients (26 knees) who had had one or more patellar dislocations. Group II comprised 31 patients (34 knees) who had knee pain (anterior, anterolateral, anteromedial or occasional popliteal).

All patients had an operation consisting of a lateral release and proximal realignment of the patella.

In group I, postoperative results were excellent in 18.6 % and good in 62.5 % on short term (<5 years). On long term (>5 years), results were excellent in 40 % and good in 40 %. Group II showed similar results on short term but worse results on long term: <5 years – 36 % excellent, 48 % good results; >5 years – 0 % excellent, 66.7 % good results.

A poor outcome was always associated with progression to patellofemoral osteoarthritis in both groups.

Ricchetti et al. [22] reported a systematic review of level III and IV studies to compare surgical success of lateral release or lateral release with medial soft-tissue realignment (MR) for recurrent lateral patellar instability. In total, there were 467 knees in 14 studies: 247 knees with a minimum of 2-year follow-up after LR and 220 after LR with MR. The frequency-weighted mean success with respect to instability in the LR studies was 77.3 % compared with 93.6 % in the LR with MR studies.

Conclusion: Isolated LR yields significantly inferior long-term results with respect to symptoms of recurrent lateral patellar instability compared with LR with MR.

In a controlled laboratory study performed by Bedi and Marzo, they hypothesized that LR would reduce the force required to displace patella laterally, when performed after medial patellofemoral ligament (MPFL) repair [3]. They measured the amount of force required to displace the patella 1 cm laterally at different degrees of flexion in eight fresh-frozen human cadaveric knees. After the cut of MPFL, the force measured was reduced by 14–22 % compared with native knee. The repair of MPFL was found to restore the ability to resist lateral force. The addition of LR to the repair reduced the force by 7–11 % compared with the repaired knee.

Conclusion: After repair of the MPFL, adding an LR lowered the ability of the patella to resist lateral displacement. LR may not be routinely appropriate as a part of the stabilizing procedure to address acute patellar dislocation when the MPFL is avulsed from the patella.

Conclusion: Isolated LR does not restore normal orientation of the malalignment extensor mechanism and thus results in long-term inferior results compared to a combination of LR and proximal realignment of the patella [14].

Distal or Combined (Proximal + Distal) Realignment Often Added by LR (Associated)

LR after tubercle transposition (Elmslie-Trillat/ Fulkerson) to allow a free passage of the patellar tendon throughout the entire range of motion [19].

To this date, there are no published studies comparing patient groups with tubercle transposition combined with LR and the patient groups with only tubercle transposition.

15.4.2 Isolated Patellofemoral Osteoarthritis (PF OA)

Studies Concerning LR for Isolated PF OA

Osborne and Fulford [19] compared two groups of patients: 70 patients in Outerbridge grades I and II were placed in group A. Five patients with more severe changes of grades III and IV were placed in group B. Seventy-four patients were reviewed at 1 year and again at 3 years after lateral release for established chondromalacia patellae.

At 1 year, 61 of 70 patients in group A had a good result, giving an initial success rate of 87 %. Only one of the five patients in group B gained relief. Review at 3 years showed that only 26 of the 70 patients in group A (37 %) continued to have a good result from the operation. Twenty-eight patients (40 %) had poor results at 3 years. In group B, all five patients had undergone patellectomy and are therefore considered to have had poor results from lateral release.

Conclusion: In the early stages of chondromalacia, release of the lateral retinaculum was successful in relieving the symptoms for a year or more; review at 3 years showed a significant number of relapses.

Christensen et al. [5] published a study comparing two groups with 58 patients in total, treated with isolated lateral release for symptoms of patellofemoral pain. All knees exhibited signs and symptoms of chondromalacia patellae (grades I–IV). Patients in group I had recurrent subluxation of the patella, whereas those in group II had no symptoms of instability.

In group I (30 knees), the initial response after 1 year was good in 36.7 %, but 0 % after 4.5 years. The number of poor results increased from 27 % after 1 year to 70 % after 4.5 years. In group 2, the number of poor results increased from 21 % after 1 year to 24 % after 4.5 years. Thus, at follow-up evaluation, the results of lateral release were significantly better in group 2 ($P<0.01$).

Conclusion: Lateral release is an acceptable short-term treatment of chondromalacia patellae without patellar subluxation when the disease does not respond to conservative treatment with isometric quadriceps exercises. In cases with patellar subluxation, the release is unable to correct the basic instability.

Aderinto and Cobb [1] set up 1 group of 50 patients that all underwent lateral release for symptomatic patellofemoral osteoarthritis. Lateral release was only performed in those for whom the anterior knee pain of patellofemoral arthritis appeared to predominate. None of the patients had patellar malalignment, patellar instability or tight lateral retinaculum. Despite 80 % of patients reporting an improvement in pain (VAS, OKS), 42 % were dissatisfied, which may be due to high expectations or reflect an initial improvement followed by deterioration with time.

Conclusion: LR provides temporary benefit, delaying the need for alternative surgical intervention such as patellofemoral resurfacing or total knee replacement.

Alemdaroglu et al. [2] examined the 35 patients in their fifth and seventh decades with grades 2–4 chondral lesions of the patellofemoral joint without patellar instability and malalignment. The patient underwent arthroscopic joint debridement and LR by bipolar radiofrequency. They aimed to define the postoperative course of patients at 3 months and at 24 months with VAS and WOMAC scores. They found that the greatest decrease of pain levels was observed at 3 months and that the pain level continued at about the same level over the next 24 months without significant changes.

Conclusion: Although this was not a randomized controlled study, arthroscopic thermal LR was shown to improve subjective pain scores significantly, however only within the postoperative 3 months.

Conclusion: Isolated LR does not result in a significant long-lasting improvement for the treatment of frank isolated PFOA. A high number of failures and relapses of pain have been observed in most studies.

15.4.3 Excessive Lateral Hyperpression Syndrome (ELHS)

Excessive Lateral Hyperpression Syndrome is a condition of lateral tightness of the lateral patellar retinaculum and decreased lateral patellar tilt because of hypertrophy of the lateral retinaculum [7]. The main symptom is lateral retinacular pain. It is important to distinguish ELHS from any other cause of patellofemoral pain. There are six major anatomic structural sources of patellofemoral pain: subchondral bone, synovium, retinaculum, skin, muscle and nerve [10].

Studies Concerning ELHS

Ceder and Larson [4] performed an isolated lateral release in 52 patients (64 knees) with Excessive Lateral Hyperpression syndrome. Results were rated as excellent, good, fair or poor, regarding subjective relief of pain, grating, giving way and swelling and ability to return to desired activities. Preoperatively, there was moderate pain in 77 %, moderate swelling in 15 %, moderate giving way in 6 % and moderate grating in 31 %. Postoperatively, most complaints were reduced: moderate pain in 10 %, moderate swelling in 1.5 %, only 1 case of severe giving way and grating in 9 %.

Conclusion: Isolated LR provides relief of symptoms in EHLS. It is a satisfactory initial procedure in cases not responding to conservative therapy.

Lattermann et al. [13] published a review of nine studies concerning lateral release for anterior knee pain. A total of 450 patients were included in the review.

Outcome parameters were improvement of pain postoperatively, incidence of excessive postoperative bleeding, incidence of postoperative infections and number of subsequent operative procedures on the involved leg.

Overall 76 % of all treated patients reported less pain after procedure. Postoperative bleeding appeared in 2 %. In 0.9 % of the cases, infection was reported. Revision surgery was required in 12 % of the cases.

Conclusion: If done in an appropriate population, an isolated lateral release has good chance for success. The overall number of patients that qualify for this procedure, however, is low. Less than 15 % of all patients that are being seen in the office for anterior knee pain require surgical treatment.

In a recent level II prospective, double-blinded comparative study published by Pagenstert and Wolf et al. [20], 28 patients (mean age of 48 years) with the diagnosis of ELHS were divided into two groups: lateral retinacular release and lateral retinacular lengthening. They excluded strictly other causes of anterior knee pain including malalignment and patella alta. The aim was to compare the complication rates and clinical outcomes of open LR and open lateral retinacular lengthening in the above-specified group of patients.

Outcome parameters: complications, muscle atrophy, Kujala patellofemoral outcome scores (preoperative, at 3, 6, 12 and 24 months).

Conclusion: Lateral retinacular lengthening group was shown to have less medial instability, less quadriceps atrophy and a better clinical outcome at 2 years compared with LR group significantly. The controlled preservation of the lateral patellar muscle–capsuloligamentous continuity after retinacular lengthening was shown to play an important role clinically in patients with isolated ELHS.

Combination: EPD–OA–ELHS

Including 70 patients with mild lateral tracking and lateral compression of the patella (group I), recurrent patellofemoral dislocation (group II) and intact or defective cartilage of the patellofemoral joint (group III), Schneider et al. [23] presents the results for the lateral release and medial imbrication of the vastus medialis obliquus. In group I (ELHS), patients complaining of a retropatellar pain syndrome were satisfied postoperatively in 77 %, using VAS. The results of the patients in group II (EPD) were also good (68 %). Distinctly worse results were attained in the patients in group III (PF OA); only 45 % of these patients were satisfied with their postoperative outcome ($P < 0.05$).

Conclusion: ELHS without PF OA and patients with EPD are good candidates for LR and LR + medial imbrication. A high subjective satisfaction rate (VAS) was found in patients with ELHS (isolated LR), as well as in patients with EPD (LR + medial imbrication).

15.5 Summary Statement

Based on the conclusions of the available studies, it is acceptable to say that there is only one indication to perform an isolated lateral release: Excessive Lateral Hyperpression Syndrome. In case of Episodic Patellar Dislocation, there is no literature support for the role of an isolated lateral release. Hyperlaxity with hypermobility of the patella (medial and lateral patellar glide of three quadrants and more) is also an absolute contraindication for lateral release (isolated or associated). In Episodic Patellar Dislocation, lateral release may only be performed in combination with proximal realignment (associated lateral release). This should be done when there is physical exam evidence of a tight lateral retinaculum after patella relocation.

In Patellofemoral Osteoarthritis, lateral release only provides temporary benefit and delays the need for alternative surgical intervention such as patellofemoral resurfacing or total knee replacement. Lateral release (isolated or associated) in Patellofemoral Osteoarthritis is thus not indicated. In lateral PFOA, a partial lateral facetectomy results is a reliable clinical improvement and may be associated with a lateral release.

The lateral retinacular lengthening should also be kept in mind in patients with EHLS, because this has been also shown to provide significantly better clinical relief over LR in these patients, by preserving the lateral patellar muscle–capsuloligamentous continuity in an open but controlled fashion.

Before performing a lateral release, it is very important to know that the release has to be done judiciously and has to be gauged by the desired effect. Overreleasing can lead to a potentially devastating medial patellar instability.

References

1. Aderinto J, Cobb AG. Lateral release for patellofemoral arthritis. Arthroscopy. 2002;18(4):399–403.
2. Alemdaroglu KB, Cimen O, Aydogan NH, Atlihan D, Iltar S. Early results of arthroscopic lateral retinacular release in patellofemoral osteoarthritis. Knee. 2008; 15(6):451–5.
3. Bedi H, Marzo J. The biomechanics of medial patellofemoral ligament repair followed by lateral retinacular release. Am J Sports Med. 2010;38(7):1462–7.
4. Ceder LC, Larson RL. Z-plasty lateral retinacular release for the treatment of patellar compression syndrome. Clin Orthop Relat Res. 1979;144:110–3.
5. Christensen F, SØballe K, Snerum L. Treatment of chondromalacia patellae by lateral retinacular release of the patella. Clin Orthop Relat Res. 1988;234:145–7.
6. Dejour H, Walch G, Nove-Josserand L. Factors of patellar instability: an anatomic radiographic study. Knee Surg Sports Traumatol Arthrosc. 1994;2(1): 19–26.
7. Dzioba RB. Diagnostic arthroscopy and longitudinal open lateral release. A four year follow-up study to determine predictors of surgical outcome. Am J Sports Med. 1990;18(4):343–8.
8. Fithian DC, Neyret P. Patellar instability: the Lyon experience. Tech Knee Surg. 2007;6(2):112–23.
9. Fithian DC, Paxton EW, Post WR, Panni AS. Lateral retinacular release: a survey of the International Patellofemoral Study Group. Arthroscopy. 2004;20(5): 463–8.
10. Fulkerson JP. Diagnosis and treatment of patients with patellofemoral pain. Am J Sports Med. 2002; 30(3):447–56.
11. Fulkerson JP, Gossling HR. Anatomy of the knee joint lateral retinaculum. Clin Orthop Relat Res. 1980; 153:183–8.
12. Kolowich PA, Paulos LE, Rosenberg TD, Farnsworth S. Lateral release of the patella: indications and contraindications. Am J Sports Med. 1990;18(4):359–65.
13. Lattermann C, Drake G, Spellman J, Bach Jr BR. Lateral retinacular release for anterior knee pain: a systematic review of the literature. J Knee Surg. 2006;19(4):278–84.
14. Lattermann C, Toth J, Bach Jr BR. The role of lateral retinacular release in the treatment of patellar instability. Sports Med Arthrosc. 2007;15(2):57–60.
15. Merican AM, Amis AA. Anatomy of the lateral retinaculum of the knee. J Bone Joint Surg. 2008;4:527–34.
16. Merican AM, Kondo E, Amis AA. The effect on patellofemoral joint instability of selective cutting of lateral retinacular and capsular structures. J Biomech. 2009;42(3):291–6.
17. Merican AM, Sanghavi S, Iranpour F, Amis AA. The structural properties or the lateral retinaculum and capsular complex of the knee. J Biomech. 2009;42(14):2323–9.
18. O'Neill DB. Open Lateral retinacular lengthening compared with arthroscopic release. A prospective, randomized outcome study. J Bone Joint Surg Am. 1997;97(12):1759–69.
19. Osborne AH, Fulford PC. Lateral release for chondromalacia patellae. J Bone Joint Surg. 1982;64(2):202–5.
20. Pagenstert G, Wolf N, Bachmann M, Gravius S, Barg A, Hintermann B, Wirtz DC, Valderrrabano V, Leumann AG. Open lateral patellar retinacular lengthening versus open retinacular release in lateral hyperpression syndrome: a prospective double-blinded comparative study on complications and outcome. Arthroscopy. 2012;28(6):788–97.
21. Panni AS, Tartarone M, Patricola A, Paxton EW, Fithian DC. Long-term results of lateral retinacular release. Arthroscopy. 2005;21(5):526–31.
22. Ricchetti ET, Mehta S, Sennett BJ, Huffman GR. Comparison of lateral release versus lateral release with medial soft-tissue realignment for the treatment of recurrent patellar instability: a systematic review. Knee Surg Sports Traumatol Arthrosc. 2007;23(5):463–8.
23. Schneider T, Fink B, Niggemeyer O, Menke W, Rüther W, Schulitz KP. Critical evaluation of the indications for endoscopic lateral release of the patellofemoral joint. Eur J Orthop Surg Traumatol. 1997;7(4):271–5.
24. Scuderi G, Cuomo F, Scott WN. Lateral release and proximal realignment for patellar subluxation and dislocation. A long-term follow-up. J Bone Joint Surg. 1988;70(6):856–61.
25. Servien E, Verdonk PC, Neyret P. Tibial tuberosity transfer for episodic patellar dislocation. Sports Med Arthrosc. 2007;15(2):61–7.

Proximal Realignment: Medial Plication

16

Alberto Gobbi, Dnyanesh Lad, and Georgios Karnatzikos

Contents

A. Gobbi (✉) • D. Lad • G. Karnatzikos
Orthopaedic Arthroscopic Surgery International (O.A.S.I.) Bioresearch Foundation, Gobbi N.P.O., Milan, Italy
e-mail: gobbi@cartilagedoctor.it; giokarnes@gmail.com

A. Gobbi et al. (eds.), *The Patellofemoral Joint*,
DOI 10.1007/978-3-642-54965-6_16, © ISAKOS 2014

16.1 Introduction

Recurrent patellar dislocation and subluxation can be seen mostly in young and physically active individuals. The etiology of the disorder is varied, and numerous factors have been proposed [1, 2]. An acute lateral patellar dislocation is often postulated to be caused by the sudden thrust of valgus force on a plantigrade foot, along with the internal rotation of the femur or external rotation of the tibia [3]. More commonly, it occurs during normal activities as a result of dysplastic changes in the patellofemoral system. Genu recurvatum, patella alta, and increased Q-angle are some of the predisposing factors. Others include atrophy of the vastus medialis, lateral insertion of the iliotibial band, increased internal femoral rotation, increased femoral anteversion, and genu valgum. Constitutional connective tissue weakness also predisposes to patellar dislocation [4]. The spectrum of injury may range from elongation of the medial structures to complete rupture of medial para-patellar soft tissue. A combination of injury to the medial retinaculum and the medial patellofemoral ligament may be present in approximately 65 % of the cases [5]. Patients with lateral patellar dislocation have been treated conservatively with a 15–44 % recurrence rate [3, 6]. Historically, lateral retinacular release was performed with the belief that a tight lateral retinaculum is predisposed to lateral patellar subluxation or dislocation. Since the recognition of the importance of the medial patellofemoral ligament (MPFL), there has been increasing interest

Table 16.1 Modern management of patellar instability

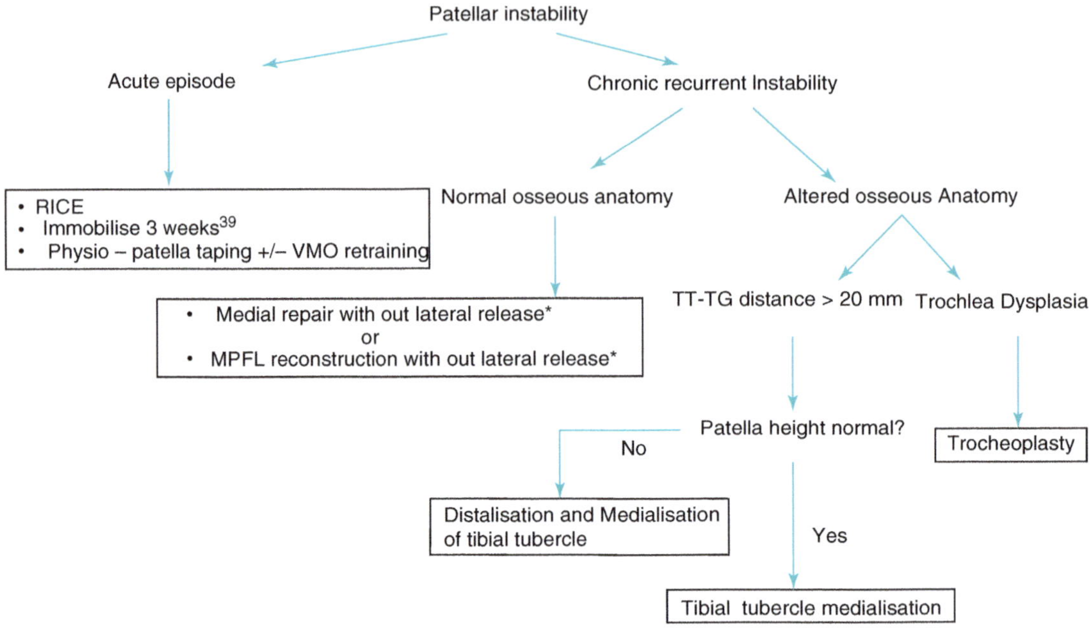

*Lateral release in isolation not supported by currecnt literature

Courtesy: Rhee et al. [1]

in different techniques for managing the medial stabilizer. Numerous surgical options have been evolved, ranging from repair, radio-frequency thermal reefing, imbrication (reefing), or plication of the medial retinacular structures (Table 16.1). Proximal realignment generally is addressed by a combination of lateral retinacular release and imbrications of medial retinaculum. With advances in arthroscopy and increased technical expertise, the concept of proximal realignment has yielded promising results. The aim of this chapter is to review the indications of this procedure, describe the surgical technique, and discuss the existing literature till date on its efficacy.

16.2 Indications

The surgical indications for medial plication in patellofemoral instability have been evolving. The accepted indications are:

- Failure of conservative treatment after acute dislocation or subluxation
- Recurrent dislocation with low-energy trauma
- Marked lateralization of the patella (persistent patellar subluxation)
- Slight lateralization accompanied by severe elongation of the medial retinaculum
- Severe patellofemoral pain with slight lateralization and hyper-lax capsule and ligaments
- Detachment of the vastus medialis or medial retinaculum from the medial aspect of the patella
- Patients with positive apprehension test on clinical examination and mild lateralization

16.3 Contraindications

- Presence of severe trochlear dysplasia (B, C, and D according to Dejour's classification) [7]
- Rupture of MPFL or medial retinaculum at femoral attachment
- Valgus knees with Q-angle greater than 20°
- Patella alta
- Congenital patellar dislocation
- History of previous surgery for patellar dislocation

The procedure can be performed in isolation or in combination with lateral release, distal realignment surgery depending on the biomechanical causes of instability.

16.4 Preoperative Workup

Symptomatology and clinical examination is an essential part of management and should be undertaken prior to any step regarding the treatment. Medial and lateral knee pain is common. The initial dislocation may present with hemarthrosis. Giving way and locking are characteristic of recurrent dislocations.

Ninety-eight percent of patellar dislocations undergo immediate and spontaneous reduction. If it remains dislocated, diagnosis is easy. Clinical examination of the reduced dislocation will show:

1. Tenderness or depression in the medial retinaculum (torn during the dislocation)
2. Tenderness over medial patellar facet (injured during reduction)
3. Tenderness over lateral femoral condyle (injured during the reduction)
4. Positive Fairbanks apprehension test (pathognomonic)

Always exclude a ruptured anterior cruciate ligament.

16.4.1 Radiographs and MRI

Anteroposterior and lateral views may show irregularities in the contour of the lateral femoral condyle. A displaced osteochondral fragment may be visible. Patellar sunrise view (Merchant view) must always be taken (Fig. 16.1). After initial dislocation, it may show lateralization of the patella and/or a sharp edged irregularity in the contour of the medial patellar facet consistent with osteochondral fracture. Recurrent dislocations are associated with rounded calcifications. When a free osteochondral fragment is suspected, an MRI is mandatory to detect its site of origin precisely, for better surgical planning.

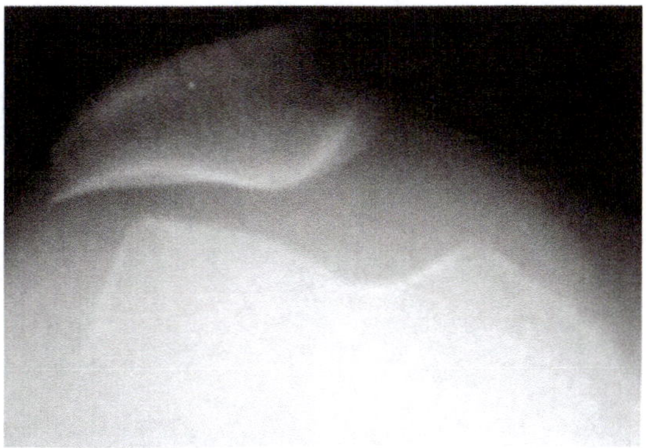

Fig. 16.1 Radiographic Merchant view of the patella (sunrise view)

16.5 Surgical Techniques for Medial Plication

The surgery is performed under spinal or general anesthesia. A tourniquet is applied to the extremity to be operated, and the patient is positioned in a standard manner for arthroscopic surgery on knee. After a sterile preparation and draping of the limb, standard anteromedial and anterolateral portals are made and a routine diagnostic arthroscopy is performed to identify possible concomitant pathologies.

A thorough assessment of the patellofemoral congruity, dysplasia, cartilage damage (especially at the medial patellar facet and lateral femoral condyle), and injury to medial retinaculum is performed. Particular attention should be paid to check the relationship of the patellar ridge with the femoral trochlear groove through range of motion of the knee. Assess the patellar tracking from both portals. Lateral tilt and overhang of the lateral patellar facet can be observed. The extent of laxity of the medial patellar capsuleligamentous complex and tightness of the lateral patellar retinaculum should be evaluated. Patient-appropriate treatments including partial meniscectomy, chondroplasty, and removal of chondral and osteochondral loose bodies not amenable to reattachment are performed. Depending on the alignment, a lateral release may be performed prior to the medial plication. The injured edges of the medial retinaculum are freshened by either

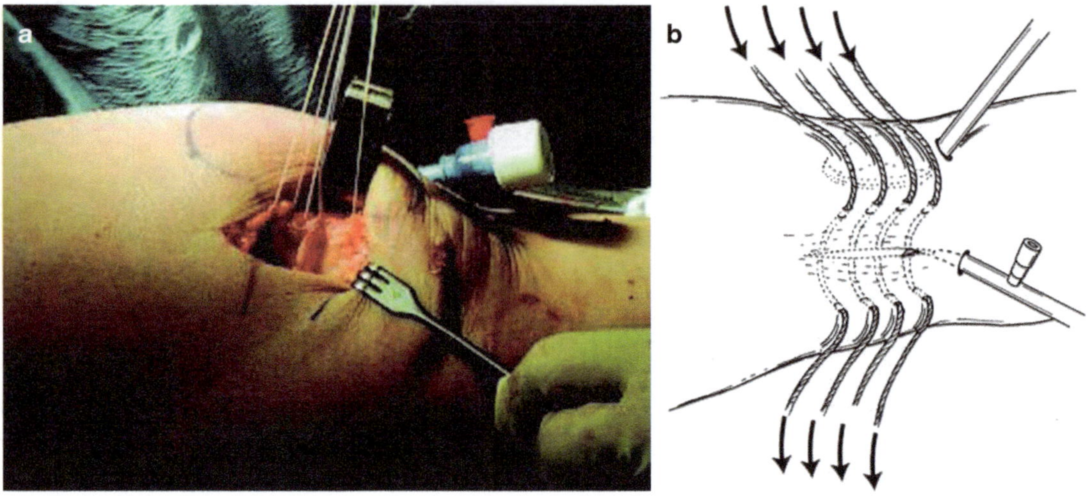

Fig. 16.2 (**a**) Mini open medial reefing. (**b**) Schematic view of access

rasping, gentle shaving, or using a thermal radiofrequency probe. This step helps in accelerated healing and strengthening of the repaired structures through better adhesion and approximation.

16.5.1 Mini-Open Medial Reefing

Nam and Karzel [8] use a 4-cm incision, beginning at the level of the superior pole of the patella, created 2 cm medial and parallel to the medial border of the patella extending distally. Dissection is carried down through the subcutaneous tissues. The vastus medialis and medial retinaculum are identified and carefully inspected for any areas of detachment. These structures are then grasped with a clamp and pulled laterally to assess the integrity at the adductor tubercle attachment, followed by lateral advancement to the patella. The vastus medialis and medial retinaculum are incised along the medial border of the patella down to but not through the level of the synovium. Using No. 2 Ethibond sutures (Ethicon Inc., Johnson & Johnson, Somerville, NJ), the medial retinaculum is advanced to the medial border of the patella using at least four mattress sutures (Fig. 16.2a, b). Before the sutures are tied, range of motion should be assessed to determine congruent tracking of the patella as well as to ensure at least 90° of knee flexion. The arthroscope is reintroduced to confirm centralization of the patella within the trochlear groove, and suture tension should be increased or decreased as necessary. The sutures are tied with the knee in full extension followed by meticulous closure.

16.5.2 Arthroscopically Assisted Medial Reefing

Miller et al. [9] described a technique with the knee maintained in 20° of flexion, and the arthroscope is placed in the anterolateral portal. From the superomedial edge of the patella, a No. 2 Vicryl suture is passed under arthroscopic visualization on a King needle, percutaneously through the joint. The needle must enter on the medial edge of the patella and exit the capsule approximately 25 mm posteromedially. This is repeated with a new suture at 1 cm intervals distally until the inferior pole is reached (Fig. 16.3). At least four sutures should be passed. A 5.5-mm Linvatec (Largo, Fl) cannula is placed in the anteromedial portal to protect the suture loops, and the capsule is then cut with a Linvatec meniscectomy electrode, from superior to inferior, bisecting between the suture loops. A 2-cm transverse incision is made in the Langer lines medial to the patella, midway between the superior and inferior pole. After blunt dissection between the subcutaneous layer and the deep fascia, the suture ends are

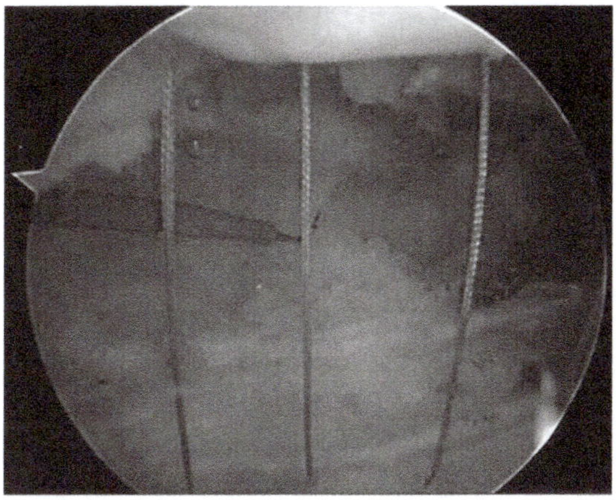

Fig. 16.3 Arthroscopically assisted medial reefing

retrieved into the incision and tagged. The medial capsular incision is palpated through the incision to ensure that it is complete. The medial and lateral capsular edges are grasped with separate 0-PDS sutures. A free King needle is placed on the two ends of the suture in the lateral limb and passed under the medial limb and out through the skin. While pulling the "pants over vest" (medial over lateral) with the PDS sutures, the surgeon ties the Vicryl sutures by hand through the incision. Capsular reefing and patellar tracking should be reevaluated arthroscopically and manually to confirm that initial contact is centralized and maintained throughout flexion.

16.5.3 Arthroscopic All-Inside Medial Plication

Currently, arthroscopic all-inside medial plication is the preferred treatment. It causes less morbidity and is associated with a better cosmetic result as well. Several methods have been described to perform this procedure entirely arthroscopically. We began using an all-inside medial plication in 1990 using an epidural needle (Tuohy needle, Rusch, Duluth, GA) because it does not cut the suture while passing. The sutures are placed starting at the superior border of the medial retinaculum and are gradually continued inferiorly spaced 1 cm apart; the Tuohy needle is passed adjacent to the patella, traversing the lateral border of the

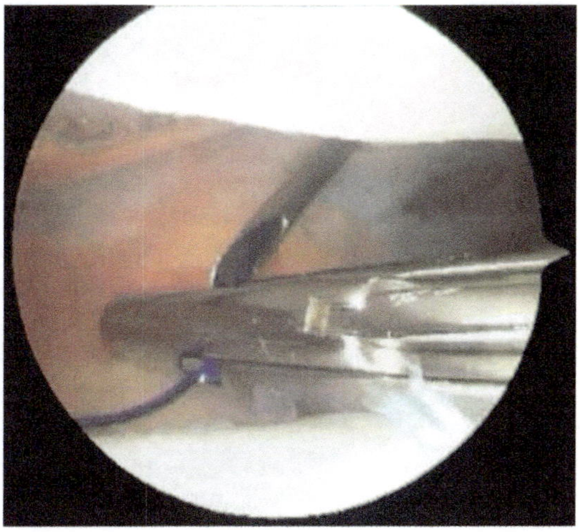

Fig. 16.4 Arthroscopic all-inside medial plication: A no 1 PDS suture is passed through the Tuohy needle and its end retrieved through the anterolateral portal

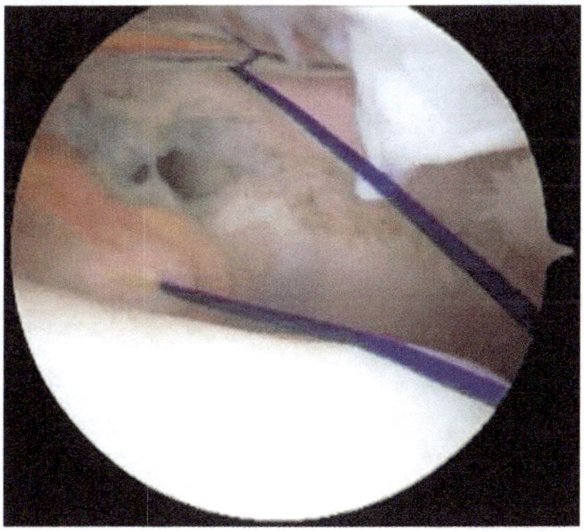

Fig. 16.5 Arthroscopic all-inside medial plication: The needle is then retracted out and is again passed along the medial border of the tear; once again, the suture is retrieved through the anterolateral portal

retinaculum tear and is visualized arthroscopically. A No. 1 PDS suture is then passed through the needle and its end retrieved through the anterolateral portal (Fig. 16.4). The needle is then retracted out of the retinaculum but not out of the skin and is again passed through the retinaculum along the medial border of the tear. Once again, the suture is retrieved through the anterolateral portal (Fig. 16.5). An arthroscopic knot is placed

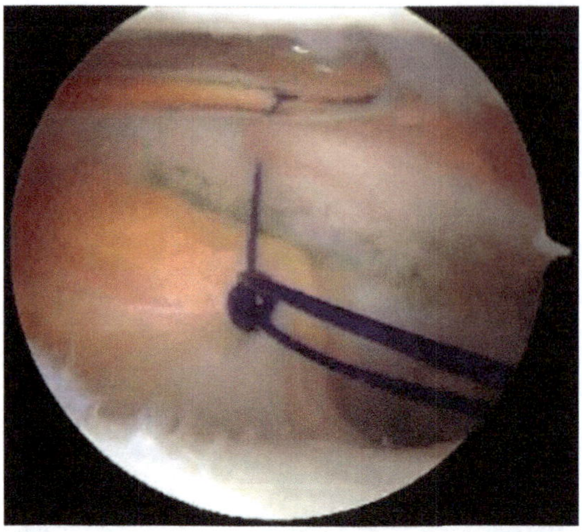

Fig. 16.6 Arthroscopic all-inside medial plication: An arthroscopic knot is placed to tighten the medial retinaculum

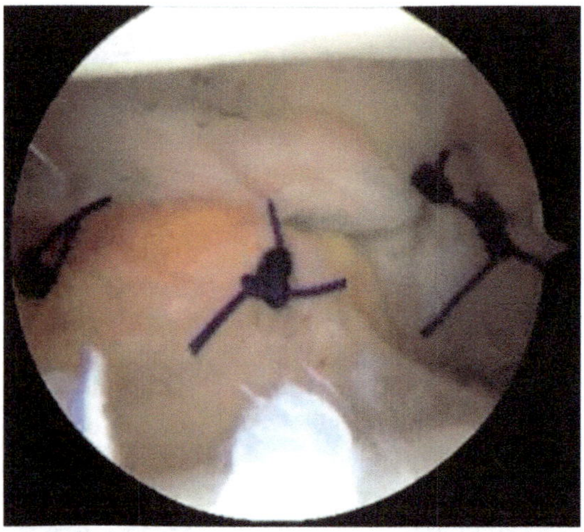

Fig. 16.7 Arthroscopic all-inside medial plication: Approximately 5 sutures are placed in this fashion

to tighten the medial retinaculum (Fig. 16.6). While tying the knot under arthroscopic vision, an assistant should manually displace the patella medially. The distance the suture is passed from the edge of the defect is chosen to affect the desired imbrication. This creates a pants-over-vest type of imbrication stitch. Approximately five sutures are placed in this fashion (Fig. 16.7). Tracking is then assessed from 0° to 90° flexion, both clinically and arthroscopically to confirm proper patellar tracking at the end of the procedure.

16.6 Postoperative Rehabilitation

Rehabilitation is a phased plan. Each patient moves from one phase to the next based on functional goals. Meeting the functional goal at each phase will allow the recovery to progress from normal walking to running and finally to sport activities.

16.6.1 Phase 1: Treat the Swelling and Inflammation

Severe postoperative pain can interfere with active muscle control. Pain can also impede progress with range of motion (ROM). Swelling, either as effusion or as soft tissue edema, also can interfere with joint motion. In addition, effusion inhibits quadriceps function and may be harmful to intra-articular structures, such as articular cartilage. Adequate pain relief should be ensured and swelling must be controlled. Use of ice packs and anti-inflammatory medications is advised.

16.6.2 Phase 2: Recovery of Range of Motion and Muscle Flexibility

Isometric quadriceps exercises maybe started under supervision, immediately postsurgery or the next day depending on the level of pain experienced. The knee joint is mobilized passively using a continuous passive motion (CPM) machine maintained between 0 and 30° in the first week. As the postoperative pain subsides, movement is increased in 5° increments per day. It is expected that the range of movement will reach up to 120° at the end of 6 weeks. Hydrokinesis (pool therapy) is useful in the first two phases where weight bearing needs to be avoided.

16.6.3 Phase 3: Recovery of Muscle Strength and Resistance

Quadriceps setting exercises should be started immediately after the surgery to keep the patellar tendon and infrapatellar fat pad stretched to their full length and to restore neuromuscular control. Resisted quadriceps and hamstring strengthening should be progressively used as the initial pain subsides.

With the knee fully extended, there is no contact between the patella and femur, limiting the chances for any chondral injury. Thus, straight-leg raises with the limb in a brace are usually safe and well tolerated early after surgery. The patient is advised a brace in complete extension during walking and at night for the first 3 weeks. Progressive exercises in closed kinetic chain (CKC) and open kinetic chain (OKC) can be introduced. At this point the patient may begin exercises in the gym in a controlled manner.

16.6.4 Phase 4: Recovery of Neuromuscular Control and Coordination

Facilitation of normal gait is an essential component of the overall treatment plan. This is particularly important for the returning athlete, in whom even a slight gait deviation can be compounded by repetitive loading. The clinician should pay particular attention to the quadriceps avoidance gait pattern (walking with the knee extended or hyperextended) because knee flexion during weight acceptance is critical for shock absorption. Taping or bracing of the patellofemoral joint may be done if pain is limiting the patient's ability to perform meaningful weight-bearing exercises. Partial squats can be introduced if not already begun earlier. Performing the exercises in front of a mirror provides a useful feedback. Ambulation postsurgery is nonweight bearing with the aid of two crutches followed by partial load bearing (10 % of the bodyweight) in the subsequent 3 weeks. Full weight bearing is allowed after 6 weeks.

16.6.5 Phase 5: Recovery of Specific Gestures

Single-leg activities must be initiated as the final step before returning to full unrestricted activity.

Return to sports activities are the long-term goal for amateur and professional athletes and can take up to 4–6 months once they can achieve satisfactory single limb dynamic control.

16.7 Complications

Potential complications with proximal realignment procedures include:

1. Recurrent instability
2. Medial patellar subluxation
3. Patellar malrotation which can lead to early-onset arthrosis
4. Arthrofibrosis

16.8 Discussion

Patellofemoral instability is a difficult condition to treat despite an improved understanding of the underlying biomechanics. The underlying complexity of the nature of the disorder may be one of the reasons. Clinical evidence demonstrates that medial structures are commonly damaged following patellar dislocation leading to the tendency for recurrent dislocation in affected knees if not managed appropriately [10, 11]. Considerable controversy continues on the management of a first-time patellar dislocation, with some researchers advocating conservative management and others stressing on the need for operative management. A review of the existing literature by White and Sherman [3] analyzed the existing treatment modalities and their efficacy and concluded that initial acute patellofemoral dislocations be treated with immobilization and rehabilitation and surgery best be reserved for patients with associated osteochondral fragments, persistent patellar subluxation, detachment of the VMO, and medial retinaculum from the medial aspect of the patella.

Although numerous proximal realignment surgeries have been described, there has been a lack of consensus on the appropriate technique. Since the recognition of MPFL as an important medial stabilizer of the patellofemoral joint (Warren and Marshall) [12], numerous surgical techniques have been described ranging from repair, imbrication (reefing), or plication of the medial retinaculum done by open or arthroscopic techniques. The earliest description of the use of arthroscopy for repair of medial retinaculum was given by Yamamoto [13] in 1986 in which he described arthroscopic lateral retinacular release and repair of the medial capsule in a group of 30 patients with a recurrence in one patient. Since then, advances in instrumentation and technical know-how and better understanding of the mechanisms have led to formulation of newer procedures. Recent progress provides the possibility of minimally invasive procedures for proximal realignment of the patella.

Nam and Karzel [8] reported their results (average 4.4 years) of a mini-open medial reefing with arthroscopic lateral release on 23 knees. Ninety-one percent of patients rated their results as good or excellent.

The efficacy of arthroscopic medial reefing has been extensively studied and fairly good results have been reported by researchers. Henry and Pflum [14] described their experience over 6 years with arthroscopic proximal realignment of the patella with lateral release in cases of acute instability, capsular defects, and recurrent subluxations. They reported significant improvement in patellofemoral pain and no recurrences and concluded that this technique was associated with decreased hospitalization, morbidity, and better cosmesis compared to the open method.

Halbrecht [15] described an arthroscopic all-inside medial reefing and lateral release in a group of 29 knees. In a retrospective evaluation of outcomes after 5 years, 93 % of the patients reported a significant improvement in pain and swelling and ability to climb stairs and return to sports. Furthermore, radiographic evaluation of the congruence angle, lateral patellofemoral angle, and lateral patellar displacement showed significant improvement postoperatively. He reported no complications or dislocations after the procedure and recommended that the procedure has the advantage of avoiding the violation of the vastus medialis obliquus.

Haspl et al. [16] also reported overall good results with no recurrent instability after a fully arthroscopic technique consisting of medial plication and lateral release. In another study involving arthroscopic stabilization of 30 acute patellar dislocations, Yamamoto reported only 1 redislocation at a 1- to 7-year follow-up.

The role of lateral release following medial reefing is an area of controversy with studies refuting its role in patellofemoral stability; moreover, a study by Desio et al. [17] has concluded that the lateral retinaculum contributes about 10 % to maintaining patellar stability. The proposed disadvantages of lateral release include hemarthrosis, recurrent effusions, adhesions, medial or lateral patellar instability, chondromalacia, decreased knee flexion, and decreased function. In supporting clinical study by Miller et al. [9], 24 patients (25 knees) were prospectively followed up for a period of 60 months and were evaluated clinically and radiologically after undergoing medial reefing without lateral release. At the end of the study period, 96 % of the patients were satisfied and showed good outcomes functionally and radiographically. Moreover, the researchers reported an improvement in congruence angle, lateral patellofemoral angle, and lateral patellar displacement.

Some recent randomized control trials have however questioned the efficacy of medial reefing on comparing it with other patella stabilizing soft tissue procedures. In a randomized controlled clinical trial, Zhao et al. [18] compared the clinical outcomes of arthroscopic medial plication with vastus medialis plasty (VMP) in a group of 60 patients and followed them for a mean follow-up period of 56.8 months (range 24–92 months). At the end of the study period, the VMP group had better clinical outcomes and less episodes of redislocation compared to the plication group. Another randomized control trial by the same author group [19] compared the clinical outcome of medial retinaculum plication with medial patellofemoral ligament reconstruction (MPFLR) in a group of 100 patients over 5 years follow-up showing that patients who underwent MPFLR had better functional outcome and static patellar position compared to those who underwent plication. However, these studies were not blinded and the assessment parameters used were purely subjective and no definitive objective parameters were used in assessing the clinical outcome in patients.

We began performing medial plication in 1992. Since then, we have performed more than 150 cases, starting with an open technique of medial reefing, gradually moving on toward a mini-open approach and finally to an all-arthroscopic technique of repair. Over the years we have had good results with this surgery, with no redislocations reported.

Although there are other anatomically based repairs and reconstructions of the medial patellofemoral ligament, these are more invasive and do not appear to yield better results. If a patient has insufficient retinacular tissue, consideration should be given to reconstruction of the medial patellofemoral ligament.

In our experience, we have treated patients with either medial plication alone or combined with a controlled lateral release when required and had good clinical outcomes. However, we believe that the ideal indication of medial reefing is a patient with painful patella syndrome (PPS) or potential patellar instability (PPI) with no evidence of MPFL rupture.

Conclusion

We conclude that arthroscopic medial reefing is a minimally invasive procedure indicated in cases of failed conservative management, requires a meticulous postoperative rehabilitation program, and has fairly good clinical outcomes.

References

1. Rhee SJ, Pavlou G, Oakley J, Barlow D, Haddad F. Modern management of patellar instability. Int Orthop. 2012;36(12):2447–56. doi:10.1007/s00264-012-1669-4. Epub 2012 Oct 7.
2. Hennrikus W, Pylawka T. Patellofemoral instability in skeletally immature athletes. J Bone Joint Surg Am. 2013;95(2):176–83. Review.
3. White BJ, Sherman OH. Patellofemoral instability. Bull NYU Hosp Jt Dis. 2009;67(1):22–9.
4. Strombel JM. Patellofemoral joint. Manual Arthrosc Surg. 2008;1:267–87.
5. Tuxoe JI, Teir M, Winge S, et al. The medial patellofemoral ligament: a dissection study. Knee. 2002;10: 138–40.
6. Cash J, Hughston JC. Treatment of acute patellar dislocation. Am J Sports Med. 1988;16:244–9.
7. Dejour D, Reynaud P, Lecoultre B. Douleurs et instabilite rotulienne: Essai de classification. Med Hyg. 1998;56:1466–71.
8. Nam EK, Karzel R. Mini-open medial reefing and arthroscopic lateral release for the treatment of recurrent patellar dislocation: a medium-term follow-up. Am J Sports Med. 2005;33:220–30.
9. Miller JR, Gregory JA, Marilyn MP, et al. Arthroscopically assisted medial reefing without routine lateral release for patellar instability. Am J Sports Med. 2007;35:622–9.
10. Basset H. Acute dislocation of the patella, osteochondral fractures, and injuries to the extensor mechanism of the knee. Instr Course Lect. 1976;25:40–9.
11. Sallay PI, Poggi J, Speer KP, et al. Acute dislocation of the patella. A correlative pathoanatomic study. Am J Sports Med. 1996;24:52–60.
12. Warren LF, Marshall JL. The supporting structures and layers on the medial side of the knee: an anatomical analysis. J Bone Joint Surg Am. 1979;61(1): 56–62.
13. Yamamoto RK. Arthroscopic repair of the medial retinaculum and capsule in acute patellar dislocations. Arthroscopy. 1986;2(2):125–31.
14. Henry JE, Pflum Jr FA. Arthroscopic proximal patella realignment and stabilization. Arthroscopy. 1995;11(4): 424–5.
15. Halbrecht JL. Arthroscopic patella realignment: an all inside technique. Arthroscopy. 2001;17(9):940–5.
16. Haspl M, Cicak N, Klobucar H, Pecina M. Fully arthroscopic stabilization of the patella. Arthroscopy. 2002;18:E2.
17. Desio SM, Burks RT, Bachus KN, et al. Soft tissue restraints to lateral patellar translation in the human knee. Am J Sports Med. 1996;26:59–65.
18. Zhao J, Huangfu X, He Y, Liu W. Recurrent patellar dislocation in adolescents: medial retinaculum plication versus vastus medialis plasty. Am J Sports Med. 2012;40(1):123–32.doi:10.1177/0363546511420551. Epub 2011 Sep 7.
19. Zhao J, Huangfu X, He Y. The role of medial retinaculum plication versus medial patellofemoral ligament reconstruction in combined procedures for recurrent patellar instability in adults. Am J Sports Med. 2012;40(6):1355–64. doi:10.1177/0363546512439193. Epub 2012 Mar 20.

Medial Patellofemoral Ligament Repair for Recurrent or Traumatic Patellar Dislocation

17

Masataka Deie and Mitsuo Ochi

Contents

17.1 Introduction

Medial patellofemoral ligament (MPFL) injury occurred when the patella dislocated laterally during sport activity and trauma. In the cadaver study, MPFL weighted 60 % to protect lateral force [1], and the MPFL is injured in more than 90 % of patellar dislocation [2]. Thus, MPFL function is one of most important factors in patellar stability.

When the patella is dislocated laterally, the MPFL can be ruptured at either the femoral or patellar sides or within the ligament itself. The majority of ruptures occur at the ligament's femoral origin [2] (Fig. 17.1); in such cases, we performed MPFL reconstruction as

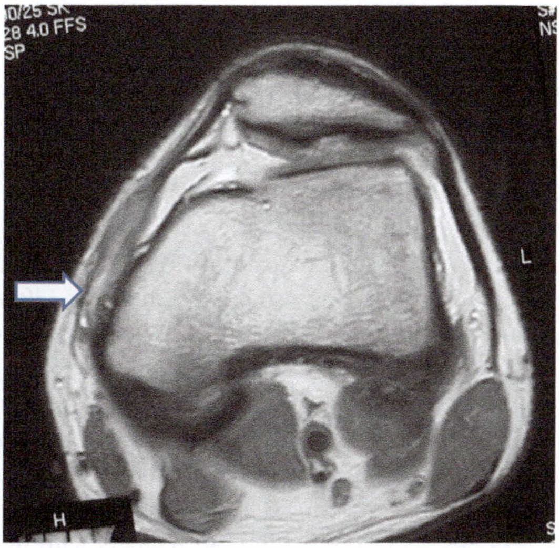

Fig. 17.1 MPFL ruptured at the femoral site. The *white arrow* indicates the position of the lesion

M. Deie (✉)
Laboratory of Musculoskeletal Functional Research and Regeneration, Institute of Biomedical & Health Sciences, Hiroshima University, Hiroshima, Japan
e-mail: masatakadeie@yahoo.co.jp

M. Ochi
Department of Orthopaedic Surgery, Institute of Biomedical & Health Sciences, Hiroshima University, Hiroshima, Japan

A. Gobbi et al. (eds.), *The Patellofemoral Joint*,
DOI 10.1007/978-3-642-54965-6_17, © ISAKOS 2014

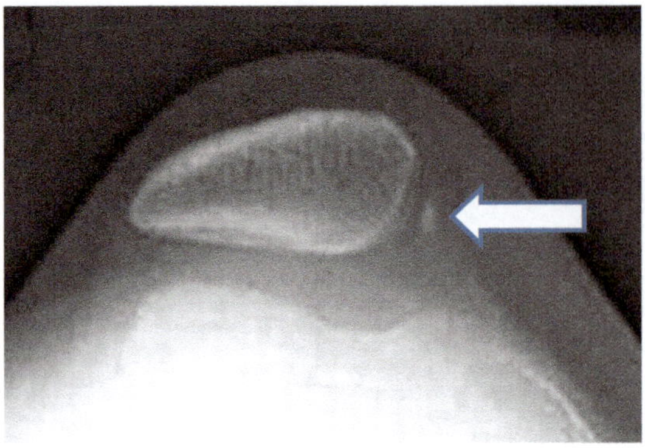

Fig. 17.2 Case of recurrent patellar dislocation. The *white arrow* shows the bony fragment of MPFL located at the medial patellar site

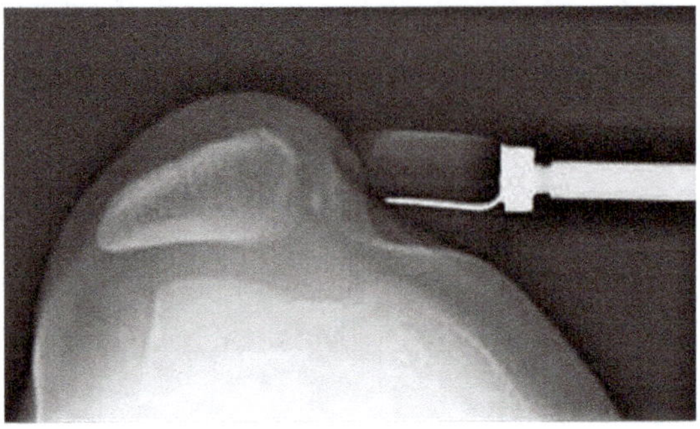

Fig. 17.3 Case of recurrent patellar dislocation. This view was obtained at 45° of knee flexion. A pushing apparatus was used to apply 20-N stress from the medial to lateral direction and from the lateral to medial direction

the first choice of treatment. We have reported the results of our MPFL reconstructions procedures previously [3–5]. In a few cases, we found that the rupture of the MPFL occurred at the patellar side, sometimes with small fragments at the rupture site (Fig. 17.2). Some researchers report that such lesions are indications for MPFL repair rather than reconstruction [6, 7]. We consider that the appropriate indications for MPFL repair are recurrent or traumatic patellar dislocation in which the patellar attachment of the MPFL has been damaged. Here, we report our surgical procedures and clinical results for MPFL repair following recurrent or traumatic patellar dislocation.

17.2 Surgical Procedures

17.2.1 Indications

We propose that the indications for MPFL repair are as follows: (a) instability remaining after traumatic patellar dislocation that has been in cases conservatively treated with muscle exercises, particularly for the vastus medialis, for more than 6 months; or (b) recurrent dislocations in which the MPFL is ruptured with an ossicle at the medial patellar site.

In physical examination, all patients with the above indications for MPFL repair showed a positive apprehension sign.

In x-ray examinations, the anterior-posterior, lateral, and skyline views at 30, 45, 60, and 90° of knee flexion were taken routinely, and axial views at 120° of knee flexion were sometimes taken. We also conducted the quantitative stress radiography at 45° of knee flexion with 20-N stress applied from the medial to lateral direction and from the lateral to medial direction, with the use of a pushing apparatus (Fig. 17.3). This method determined the indication of the lateral release to recognize the instability of medial and lateral directions. Magnetic resonance imaging (MRI) was always performed, because it revealed tear lesions of the MPFL at the site of femoral insertion, within the ligament itself or at the medial patellar site, as well as cartilage, meniscus injury, and malalignment of the patellofemoral joint (Fig. 17.4).

17.2.2 Operative Position and Anesthesia

MPFL repair was performed in the supine position with general or lumbar anesthesia or both.

17.2.3 Arthroscopy

The first step on the repair procedure was arthroscopy through the standard inferomedial and inferolateral portals (Fig. 17.5a). The patients were

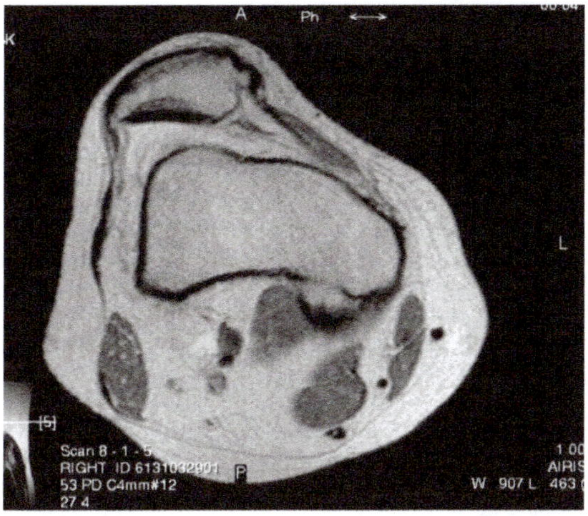

Fig. 17.4 MR imaging. This view shows malalignment of the patellofemoral joint

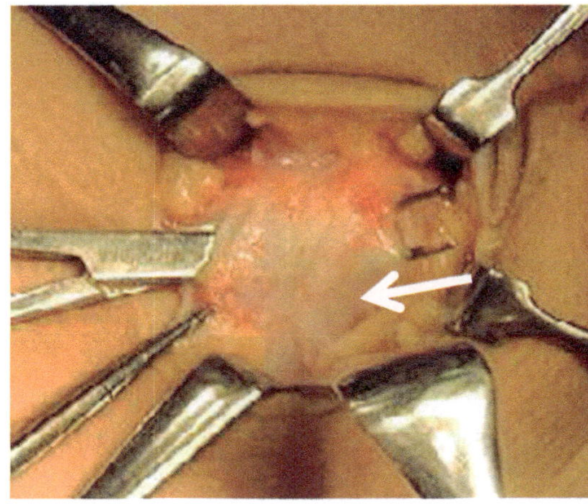

Fig. 17.6 The *white arrow* shows the identified MPFL substance

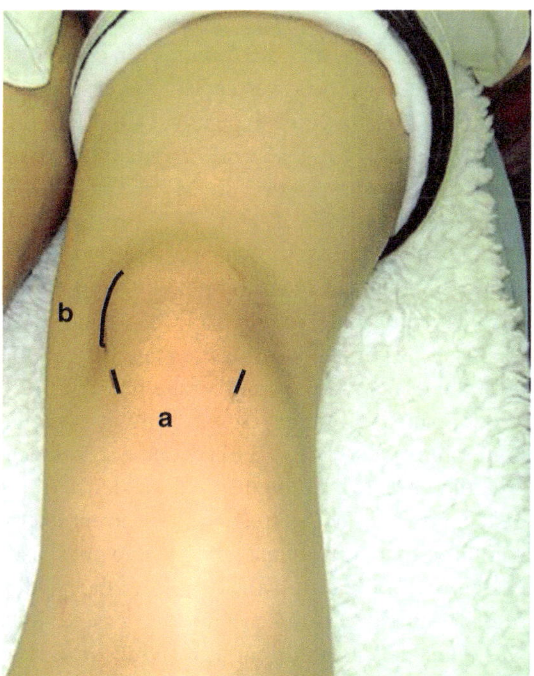

Fig. 17.5 Skin incisions required for the MPFL repair procedure. (*a*) Location of the standard inferomedial and inferolateral portals for arthroscopy and (*b*) incision site in MPFL repair procedure

checked for intra-articular lesions. When lateral release was required, it was performed arthroscopically by using an electrodevice or plastic scissors.

17.2.4 Repair of the MPFL

About 4-cm incision was applied along the medial border of the patella (Fig. 17.5b). The MPFL was carefully identified in the second layer of the knee because MPFLs are very thin and sometimes adhere to tissues within the third layer, such as the joint capsule (Fig. 17.6). When an ossicle was found at medial patellar ridge, the MPFL was easily identified because the MPFL was attached to this fragment. The MPFL substance was carefully removed from the adhered tissues. After the MPFL was identified, the operator checked the tension, volume, and strength of the ligament. Sometimes, even if the MPFL looked good, it did not show the tension required to function after the repair process, and we commenced the MPFL reconstruction procedure.

Next, the medial site at the MPFL insertion of the patella and one fourth of the patellar surface were exposed. With K-wire (diameter 1.2 mm), two small holes were made at the proximal one third and middle of the patellar medial lesions. The MPFL was sutured the patella medial edges by using a no. 1 sergiron through the small bone tunnels (Fig. 17.7). The surgeon checked the tension of the MPFL, the patellar alignment, and the position of patellofemoral joint. After repairing the MPFL, the patellar tracking course during knee flexion to extension was tested, and the patellofemoral alignment was reviewed with a Merchant's x-ray view (Fig. 17.8).

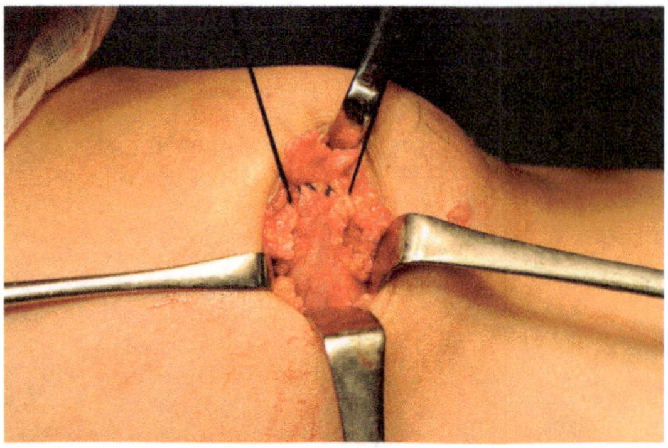

Fig. 17.7 The MPFL was sutured at the patellar site using the no.1 sergiron

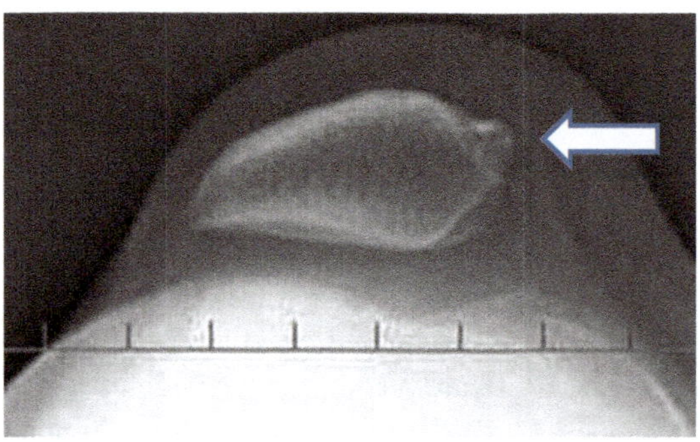

Fig. 17.8 Four years after MPFL repair, the ossicle (the *white arrow*) was adhered to the patella

17.2.5 Postoperative Rehabilitation

For 2 weeks after surgery, the knee was fixed with a knee brace at 20° of knee flexion. After 2 weeks, range of motion exercises were started, and the patients were allowed the partial weighted gait. Three months after surgery, the patients were allowed to jog if they could do the straight leg raises and had a normal patellar tracking course. Six months after surgery, the patients were allowed to return to the initial sports activity.

17.2.6 Clinical Results

From April 2002 to September 2012, we surgically treated 115 patellar dislocations. We performed MPFL repair procedure for nine cases of these cases, seven recurrent patellar dislocations and two traumatic patellar dislocations. Among these nine cases, one patient experienced a patellar dislocation 2 years after MPFL repair and subsequently underwent MPFL reconstruction using semitendinosus tendon. The other 8 patients experienced no re-dislocations or reappearance of apprehension sighs.

17.3 Discussion

It has been reported that MPFL injury most often occurs at the ligament's femoral insertion [2, 8, 9], when the patella is dislocated; however, some more recent studies based on MR imaging support the patella insertion as the more frequent point of injury [10, 11]. In a study of acute patellar dislocation, the principal injury site was the patellar insertion in 76 % of patients, the midsubstance in 30 % of patients, and the femoral attachment in 49 % of patients; however, in 39 % of cases there was more than one injury location [10].

There have been several reports describing surgical procedures used in MPFL repair and clinical results [6, 7, 12–15]. While these studies also include cases of medial reefing or plication, medial capsular imbrication, and VMO advancement with arthroscopic assistance, the MPFL repair of acute dislocation and dislocation with the ossicle at the patellar site had relatively good results. However, some authors describe high failure rates after MPFL repair: for instance, Arendt et al. [15] reported a high incidence of re-dislocation after MPFL repair for recurrent lateral patella dislocations. In our study, of the nine cases of traumatic or recurrent dislocation of the patella, re-dislocation occurred in only one case, which was diagnosed as recurrent dislocation with damage at the patellar insertion site, according to MR imaging. The MPFL is sometimes injured at more than one site, and the repaired MPFL can become dysfunctional even if it has been repaired at the patellar insertion sites. We suggest that surgeons should check the remaining MPFL tension during the operation and prepare to change to the MPFL reconstruction procedure if necessary.

The merits of our MPFL repair procedure were (1) no harvesting of hamstring tendon, (2)

no incision at the femoral attachment of the MPFL, and (3) less invasive surgery compared with MPFL reconstruction. We conclude that when the indications outlined here are met, MPFL repair is a very useful procedure for treatment of patellar dislocation.

References

1. Conlan T, Garth Jr WP, Lemons JE. Evaluation of the medial soft-tissue restrains of the extensor mechanism of the knee. J Bone Joint Surg Am. 1993;75:682–93.
2. Sallay PI, Poggi J, Speer KP, Garrett WE. Acute dislocation of patella. A correlative pathoanatomic study. Am J Sports Med. 1996;24:52–60.
3. Ochi M, Okada Y, Manabe H, Mochiduki Y, Sasashige Y, Jitsuiki Z, Yamanaka T, Ikuta Y. The medial tenodesis for patellar dislocation – from the view points of the isometricity – (In Japanese). The knee. 1989;15:45–50.
4. Deie M, Ochi M, Sumen Y, Adachi N, Kobayashi K, Yasumoto M. A long term follow-up after medial patellofemoral ligament reconstruction using the transferred semitendinosus tendon for patellar dislocation. Knee Surg Sports Traumatol Arthrosc. 2005;13:522–8.
5. Deie M, Ochi M, Adachi N, Shibuya H, Nakamae A. Medial patellofemoral ligament reconstruction fixed with a cylindrical bone plug and a grafted semitendinosus tendon at the original femoral site for recurrent patellar dislocation. Am J Sports Med. 2011;39:140–5.
6. Satterfield WH, Johnson DL. Arthroscopic patellar "bankart" repair after acute dislocation. Arthroscopy. 2005;21:627.
7. Torituska Y, Horibe S, Hiro-Oka A, Mitsuoka T, Nakamura N. Medial marginal fracture of the patella following patellar dislocation. Knee. 2007;14:429–33.
8. Nomura E. Classification of lesions of the medial patella-femoral ligament in patellar dislocation. Int Orthop. 1999;23:260–3.
9. Spritzer CE, Courneya DL, Burk DLJ, Garrett WE, Strong JA. Medial retinacular complex injury in acute patellar dislocation: MR findings and surgical implication. AJR Am J Roentgenol. 1997;168:117–22.
10. Elias DA, White LM, Fithian DC. Acute lateral patellar dislocation at MR imaging: injury patterns of medial patellar soft-tissue restraints and osteochondral injuries of the inferomedial patella. Radiology. 2002;225:736–43.
11. Guerrero P, Li X, Patel K, Brown M, Busconi B. Medial patellofemoral ligament injury patterns and associated pathology in lateral patella dislocation: a MRI study. Sports Med Arthrosc Rehabil Ther Technol. 2009;1:17.
12. Small NC, Glogau AI, Berezin MA. Arthroscopically assisted proximal extensor mechanism realignment of the knee. Arthroscopy. 1993;9:63–7.
13. Yamamoto RK. Arthroscopic repair of the medial retinaculum and capsule in acute patellar dislocation. Arthoscopy. 1986;2:125–31.
14. Henry JE, Pflum Jr FA. Arthroscopic proximal patella realignment and stabilization. Arthroscopy. 1995;11:424–5.
15. Arendt EA, Moeller A, Agel J. Clinical outcomes of medial patellofemoral ligament repair in recurrent (chronic) lateral patella dislocations. Knee Surg Sports Traumatol Arthrosc. 2011;19:1909–14.

MQTFL Reconstruction

18

John P. Fulkerson

Content

Deficiency of medial patellofemoral support structure may lead to recurrent lateral instability of the patella following a patellar dislocation, and therefore restoration of medial patella support is often necessary to reestablish permanent stable tracking of the patella. Most important in the process of achieving long-term patella stable function, however, is achieving balanced tracking initially before any retinacular reconstruction is undertaken. In patients with lateral tracking of the extensor mechanism, it is important to determine if medializing the extensor mechanism by tibial tubercle transfer is necessary for an optimal, stable result prior to medial reconstruction. In patients with a symptomatic articular lesion of the patella, anteriorization, anteromedialization, or articular resurfacing may be needed prior to medial patellofemoral reconstruction.

Once the decision has been made to do a medial patellofemoral retinacular reconstruction, one must decide if imbrication or restoration of medial structure by repair or plication will be sufficient (most typically in patients with minimal dysplasia and no evidence of malalignment). Medial patellofemoral ligament reconstruction using a tendon graft [1] however has been recommended for permanent restoration of medial patellofemoral support and may be the best option in many patients, particularly if the patient has trochlear dysplasia or malalignment factors putting the patient at added risk for a further recurrent dislocation.

Recently, Mochizuki et al. [2] has noted a prominent medial quadriceps tendon-femoral ligament. Careful study of the deep medial

J.P. Fulkerson
Orthopedic Associates of Hartford, P.C.,
Clinical Professor of Orthopedic Surgery,
University of Connecticut Medical School,
499 Farmington Avenue, Suite 300,
Farmington, CT 06032, USA
e-mail: patelladoc@aol.com

A. Gobbi et al. (eds.), *The Patellofemoral Joint*,
DOI 10.1007/978-3-642-54965-6_18, © ISAKOS 2014

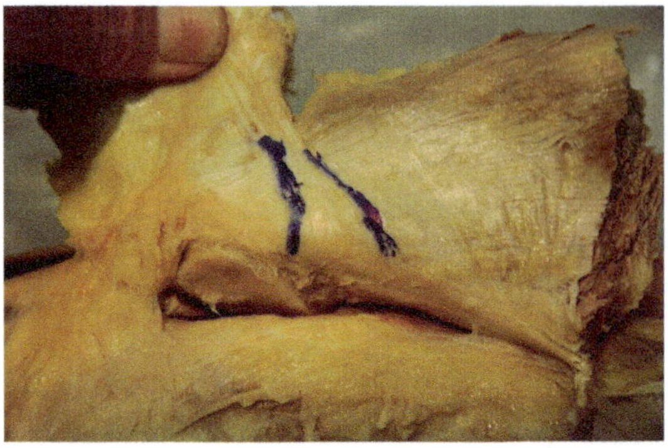

Fig. 18.1 Anatomic dissection showing deep side of the medial PF retinacular complex. A distinct medial quadriceps tendon-femoral ligament (MQTFL) component is readily noted and more distinct than MPFL

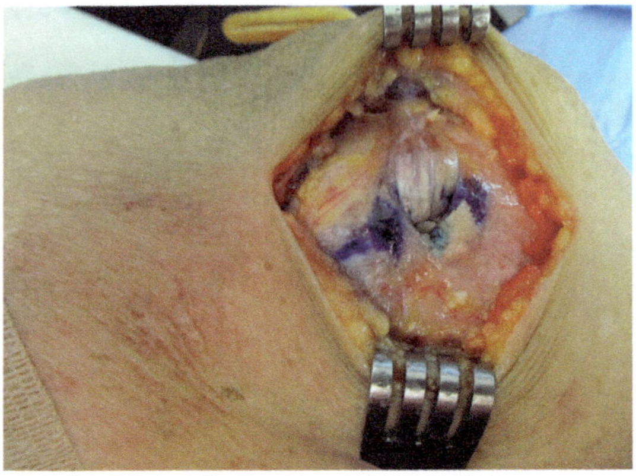

Fig. 18.2 The adductor magnus tendon is a foolproof guide to the adductor tubercle. The graft is then attached to a socket just at the distal end of the adductor tubercle. Make the incision as long as needed to be sure to accurately identify the adductor tendon in every case. This is more accurate than radiographic criteria

retinaculum in the subsynovial layer reveals that the medial support structure is predominantly a condensation of retinacular fibers that blend into the quadriceps (medialis and intermedialis) tendons rather than a distinct "medial patellofemoral ligament" (Fig. 18.1) Thus, the more appropriate reconstruction for many patients is a medial quadriceps tendon-femoral ligament. In any case, it is of paramount importance to reestablish a secure connection between the medial femur, anatomically, and the extensor mechanism.

Our choice for the stabilization currently is a medial quadriceps tendon-femoral ligament reconstruction, with or without specific attachment to the patella itself. The patient is placed supine on the OR table with a "bump" under the contralateral hip to give better exposure of the medial knee for surgery. Once arthroscopy has established the patella alignment and need for any additional surgery such as lateral release or tubercle transfer, those surgeries are completed prior to medial reconstruction in order to optimize balance of patella tracking prior to the medial restraint reconstruction.

To establish medial patellofemoral support, the tendon graft will typically be an autograft semitendinosus tendon or an allograft tendon. Our preference for allograft has been posterior tibialis tendon. The graft is prepared to fit in an 8-mm socket in the appropriate anatomic location on the medial femur.

The tendon graft fixation point on the medial femur is identified by first establishing the location and insertion of the adductor magnus tendon at the adductor tubercle. This is a key anatomic point and leads directly to the origin of the medial patellofemoral and medial quadriceps-femoral support structure consistently. This region is between the adductor tubercle and the medial epicondyle – an area called the "saddle region." It is identified most consistently by finding the adductor magnus tendon and following it to the adductor tubercle where it inserts. Once adductor tubercle is identified, a guide pin is placed just at the distal anterior aspect of the adductor tubercle (Fig. 18.2) and an 8 mm socket drilled in this location to 30-mm depth. The tendon graft is then secured in this location. We have not found radiographic criteria necessary when using precise anatomic criteria for placing the femoral fixation socket.

A variety of fixation devices may be used for fixation in this location.

A tendon graft is drawn anteriorly deep to the vastus medialis tendon through a 1-cm incision in the vastus medialis obliquus tendon just at the proximal pole of the patella. The graft may be secured there and may also be secured as desired into the patella itself although our preference is to make a slot 1-cm wide in the distal quadriceps

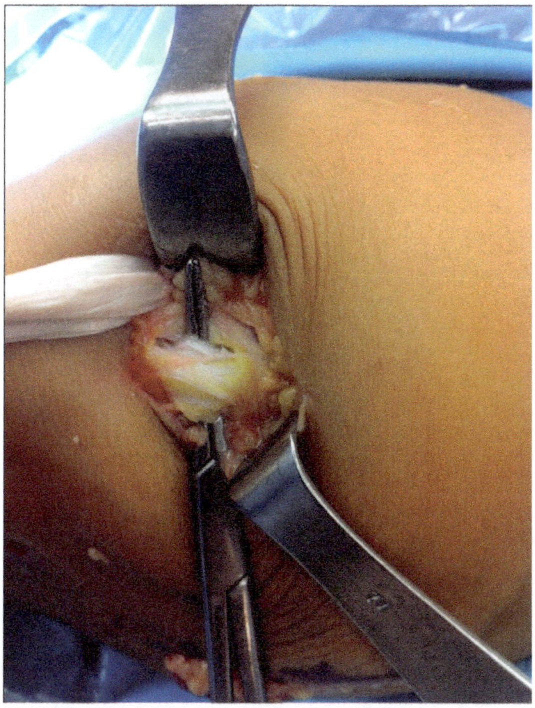

Fig. 18.3 Attachment of the MQTFL reconstruction graft into the quadriceps tendon immediately above the patella, thereby simulating and reconstructing the medial quadriceps tendon-femoral ligament surgically

tendon, 4–5-mm deep (Fig. 18.3), and draw the tendon graft through the slot in the distal quadriceps tendon where it should be secured in addition to suturing the tendon graft where it comes through the vastus medialis obliquus tendon. This reconstructs the medial quadriceps tendon-femoral ligament.

The knee is then cycled, and the patella is viewed arthroscopically as the graft is tensioned with the knee in approximately 45° flexion. The graft is first secured with a single heavy suture and then full motion applied to establish appropriate tensioning. The graft should never be too tight, and the patella should never be pulled medially but rather the patella should track accurately into the central trochlea without medializing and return to full extension in a slightly lateralized position as is normal. The graft may then be secured with additional sutures into the medialis tendon where it comes through and also into the quadriceps tendon slot. Again, range of motion assures proper patella tracking viewed arthroscopically. Excess tendon is removed or sutured back over the repair and the wound closed.

Rehabilitation involves immobilization for 6 weeks but starting a single flexion of the knee every day after the first week to assure early restoration of motion. Weight bearing may be as tolerated, or patients are given crutches until after the splint is removed and quadriceps function restored with physical therapy. Most patients can run after 4 months and play competitive sports after 6 months. The limiting factor for vigorous activity is usually articular damage.

References

1. Davis DK, Fithian DC. Techniques of medial retinacular repair and reconstruction. Clin Orthop Relat Res. 2002;402:38–52.
2. Mochizuki T, Nimura A, Tateishi T, Yamaguchi K, Muneta T, Akita K. Anatomic study of the attachment of the medial patellofemoral ligament and its characteristic relationships to the vastus intermedius. Knee Surg Sports Traumatol Arthrosc. 2013;21(2):305–10.

Medial Patellofemoral Ligament Reconstruction Based on Graft Tension Change and Anatomy

19

Hideyuki Koga, Tomoyuki Mochizuki, Tomohiko Tateishi, Keiichi Akita, Ichiro Sekiya, and Takeshi Muneta

Contents

H. Koga • T. Muneta (✉)
Department of Joint Surgery and Sports Medicine, Graduate School, Tokyo Medical and Dental University, Tokyo, Japan
e-mail: koga.orj@tmd.ac.jp; muneta.orj@tmd.ac.jp

T. Mochizuki
Department of Joint Reconstruction, Graduate School, Tokyo Medical and Dental University, Tokyo, Japan

T. Tateishi
Department of Orthopaedic Surgery, Douai Memorial Hospital, Tokyo, Japan

K. Akita
Unit of Clinical Anatomy, Graduate School, Tokyo Medical and Dental University, Tokyo, Japan

I. Sekiya
Department of Applied Regenerative Medicine, Graduate School, Tokyo Medical and Dental University, Tokyo, Japan

19.1 Introduction

The medial patellofemoral ligament (MPFL) has been recognized as the primary restraint against lateral dislocation of the patella. Biomechanical studies have shown that the MPFL contributes around 60 % of the medial stabilizing strength of the patella [2]. Therefore, restoration of medial stabilizing structure of the patella including MPFL is crucial after lateral patellar dislocation. Although conservative treatment or primary repair is normally performed as a first treatment for an initial lateral patellar dislocation, MPFL reconstruction has been widely accepted for the treatment of chronic patellar instability caused by dysfunction of medial stabilizing structure.

The Elmslie-Trillat procedure [1], one of the tibial tubercle transfer procedures, has been performed in our hospital as a standard procedure for patients with patellar instability. However, our first-choice procedure had been changed to MPFL reconstruction since 1998, because we have experienced some cases with residual apprehension after the tibial tubercle transfer alone, and satisfactory outcome was achieved after additional MPFL reconstruction in those cases [4]. Moreover, the importance of the MPFL has been well recognized as mentioned above.

Graft placement is one of the most crucial procedures for successful MPFL reconstruction. Knowledge of the precise attachment site of the MPFL is essential for the precise graft placement and thus restoration of patellar stability with MPFL reconstruction. We have recently investigated in

A. Gobbi et al. (eds.), *The Patellofemoral Joint*,
DOI 10.1007/978-3-642-54965-6_19, © ISAKOS 2014

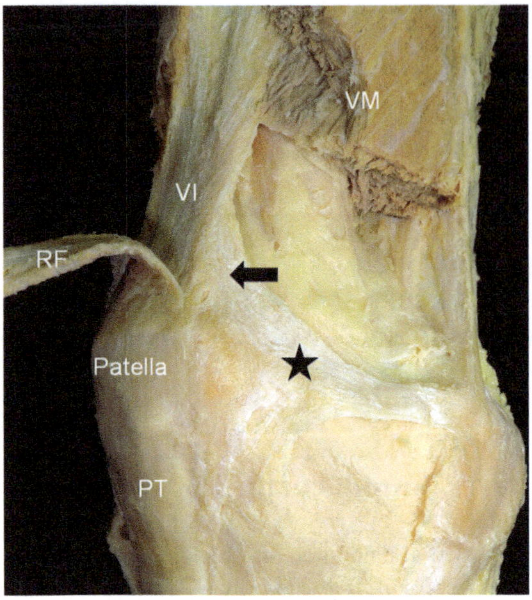

Fig. 19.1 Anatomy of MPFL. After removal of the distal part of the vastus medialis and peeling away the rectus femoris from the vastus intermedius, the MPFL (*black star*) was observed to be a fanlike shape. The proximal fibers of MPFL were mainly attached to the vastus intermedius tendon (*black arrow*), without tight adhesion to the vastus medialis. *PT* patellar tendon, *RF* rectus femoris, *VI* vastus intermedius, *VM* vastus medialis

detail the attachment of the patellar side of the MPFL, and it was revealed that the MPFL was directly attached to the vastus intermedius tendon and patella. In particular, the proximal fibers of the MPFL were mainly attached to the vastus intermedius tendon, without tight adhesion to the vastus medialis (Fig. 19.1) [3]. Based on this anatomic study, we positioned the graft to mimic its native attachment site on the patellar side during the MPFL reconstruction. On the other hand, for the femoral side graft placement, graft tension change was our first priority based on our previous study which investigated the relationship between graft length change and graft placement on the femoral side [6]. For the MPFL reconstruction, taking graft tension change into account is important as it would determine MPFL graft function such as patellar tracking and patellofemoral pressure by the graft, which would also be affected by preoperative patellar height, trochlear shape, patellar tilt, and subluxation. The length pattern also determines early knee range of motion recovery and postoperative patellar alignment. In this chapter, our MPFL reconstruction procedures based on graft tension change and anatomy are described.

19.2 Preoperative Assessment

Thorough clinical examinations are performed such as knee flexion angle where dislocation is most induced, the degree of apprehension, the degree of muscle contribution on instability, and active patellar tracking without anesthesia. Routine X-rays of anteroposterior, lateral, Rosenberg, and skyline (30, 60, and 90° of knee flexion) views are taken for both knees. Computed tomography for both lower extremities is taken to evaluate the alignment such as femoral anteversion, lateral patellar shift, lateral patellar tilt, dysplasia of the trochlea, patellofemoral joint congruity, and external rotation of tibia. Combined osteochondral fracture is also evaluated. MRI is taken to evaluate the status of residual MPFL, bone bruise, and other combined injuries. Under anesthesia, the degree of dislocation and passive patellar tracking are finally evaluated.

19.3 Surgical Procedures

A standard arthroscopic examination is performed, and combined injuries are managed according to the injury status. If any additional surgery such as lateral release or tibial tubercle transfer is required, those surgeries are performed before MPFL reconstruction in order to correct the patellar tracking. We think that MPFL reconstruction works for maintaining the restored patellar tracking and preventing the dislocation, but correcting abnormal tracking by MPFL reconstruction alone will put too much stress on the reconstructed MPFL, resulting in limited range of motion, osteoarthritis, and graft failure. Generally, patients with large tibial tubercle-trochlear groove distance (TT-TG) [5] and/or osteoarthritis of patellofemoral joint require tibial tubercle transfer, and patients with increased lateral dislocation along with knee flexion require extensive lateral release.

An oblique 2-cm incision is made on the anteromedial tibial surface at the level of the pes anserinus. Semitendinosus or gracilis tendon is harvested with an open-loop tendon stripper (Smith & Nephew Endoscopy, Andover, MA, USA). Graft selection is made depending on the tendon size and patient's height. The harvested tendon is folded, creating a double-stranded bundle with EndoButton CL (Smith & Nephew Endoscopy) with one bundle approximately 7 cm for the patellar bundle and the other at least 8 cm for the proximal bundle described later.

An oblique curve 3-cm incision is made at the proximal half of the patella along with medial edge is prepared. Another straight 2-cm incision is made at the femoral insertion of the MCL. A curved clamp is bluntly introduced between the second and the third layers from the medial patellar incision site in the direction of the femoral attachment of the MCL, creating the reconstruction route for the graft between the femur and patella.

A guidewire for the femoral drill hole is inserted at the center between the medial epicondyle and the adductor tubercle by palpating each prominence through the incision. Then, a folded polyester vessel tape is hooked onto the guidewire, introduced through the reconstruction route to the medial patellar site, and provisionally fixed to the proximal one-third of the patella with sutures at 70° of knee flexion. Tension of the vessel tape, patellar stability, and patellar tracking are confirmed from 0° to 120° of knee flexion. If those are not satisfactory, the guidewire is moved according to our previous study [6]. In this study, we have revealed that graft placement at the femoral site anterior-proximally is inclined to lengthen the graft length with knee flexion, whereas graft placement posterior-distally is inclined to lengthen the graft length with knee extension (Fig. 19.2). For instance, if the tension of the tape becomes too high with knee flexion or residual patellar instability is observed near knee extension, the guidewire is moved posterior-distally, and the same confirmation procedure is repeated.

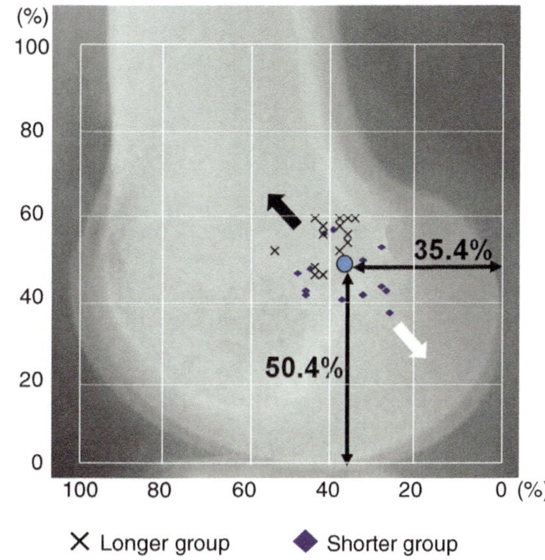

Fig. 19.2 Graft placement at the femoral site in 27 cases. The Isotac (Smith & Nephew Endoscopy) was placed at the center of the femoral drill hole, and the graft length change in the MPFL reconstruction route was measured. The X represents the Isotac position where the graft length became longer with knee flexion. The diamond represents the Isotac position where the graft length became shorter with knee flexion. The circle represents the mean Isotac position. The graft placement anterior-proximally was inclined to lengthen the graft length with knee flexion (*black arrow*), whereas graft placement posterior-distally was inclined to lengthen the graft length with knee extension (*white arrow*)

Once the optimal femoral tunnel position is determined, a 4.5-mm-diameter tunnel is created to the lateral cortex of the femur using an EndoDrill (Smith & Nephew Endoscopy). Thereafter, a femoral socket with a diameter matching the graft diameter is created according to the graft length inserted into the femoral tunnel. The graft is inserted into the femoral tunnel using a passing pin. The femoral side of the graft is fixed with the EndoButton CL.

For the patellar side fixation, we position the graft to mimic its native attachment site based on our anatomic study [3]. Namely, we reconstruct both the proximal bundle attached to the vastus intermedius tendon (proximal bundle) and the bundle attached to the proximal two-thirds of the patella (patellar bundle) (Fig. 19.3). After the femoral side fixation, the graft is introduced through the reconstruction route to the medial patellar site. A 4.5-mm-diameter bent tunnel is created at the

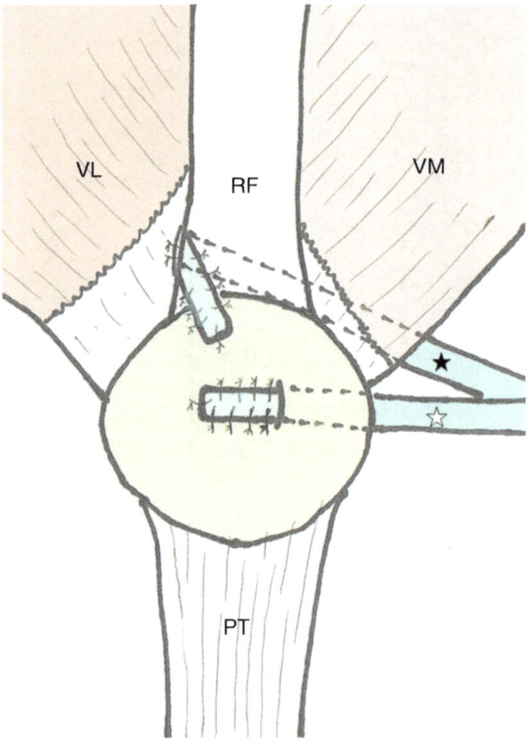

Fig. 19.3 MPFL reconstruction based on our anatomic study. The patellar bundle (*white star*) is introduced through the tunnel created at the proximal one-third of the patella to the patellar surface and sutured against the prepatellar fascia. The proximal bundle (*black star*) is introduced to the lateral border of the rectus femoris tendon under the vastus intermedius and vastus medialis tendon and sutured against the quadriceps fascia and the prepatellar fascia. *PT* patellar tendon, *RF* rectus femoris, *VM* vastus medialis, *VL* vastus lateralis

proximal one-third of the patella, and the patellar bundle is introduced through the tunnel to the patellar surface. We set the bent tunnel to avoid postoperative patellar fracture by the straight bone tunnel to the lateral side of the patella. The patellar bundle is sutured against the prepatellar fascia using No. 0 braided polyester sutures with adequate tension at 70° of knee flexion, where the patella is stabilized into the femoral groove. For the proximal bundle, a 1-cm longitudinal cut is made at the lateral border of the rectus femoris tendon, just proximal to the patellar insertion site. A curved clamp is bluntly introduced from the cut to the medial patellar site under the vastus intermedius and vastus medialis tendon to create the reconstruction route, and then the proximal bundle is introduced to the lateral border of the rectus femoris tendon. The proximal bundle is sutured against the quadriceps fascia and the prepatellar fascia using No. 0 braided polyester sutures with adequate tension at 70° of knee flexion. In this way, we expect not only static stabilization by the patellar bundle but also dynamic stabilization by the proximal bundle. Correction of patella alta could also be expected by tensioning the proximal bundle distally during its fixation.

19.4 Postoperative Management

Patients are encouraged to practice quadriceps setting and range of motion exercise from 2 days after the surgery. Static partial weight bearing is also permitted in knee extension with a knee brace. Wearing the knee brace is important until the recovery of knee extension muscle strength when the straight leg raising becomes possible. Walking with weight bearing on crutches is then started and gradually progressed. Jogging is allowed after 3 months, and patients progress to full activity after 6 months.

References

1. Cox JS. An evaluation of the Elmslie-Trillat procedure for management of patellar dislocations and subluxations: a preliminary report. Am J Sports Med. 1976; 4(2):72–7.
2. Desio SM, Burks RT, Bachus KN. Soft tissue restraints to lateral patellar translation in the human knee. Am J Sports Med. 1998;26(1):59–65.
3. Mochizuki T, Nimura A, Tateishi T, Yamaguchi K, Muneta T, Akita K. Anatomic study of the attachment of the medial patellofemoral ligament and its characteristic relationships to the vastus intermedius. Knee Surg Sports Traumatol Arthrosc. 2013;21(2):305–10.
4. Muneta T, Sekiya I, Tsuchiya M, Shinomiya K. A technique for reconstruction of the medial patellofemoral ligament. Clin Orthop Relat Res. 1999;359:151–5.
5. Schoettle PB, Zanetti M, Seifert B, Pfirrmann CW, Fucentese SF, Romero J. The tibial tuberosity-trochlear groove distance; a comparative study between CT and MRI scanning. Knee. 2006;13(1):26–31.
6. Tateishi T, Tsuchiya M, Motosugi N, et al. Graft length change and radiographic assessment of femoral drill hole position for medial patellofemoral ligament reconstruction. Knee Surg Sports Traumatol Arthrosc. 2011; 19(3):400–7.

Anteromedial Tibial Tubercle Transfer

20

John P. Fulkerson

Contents

J.P. Fulkerson, MD
Department of Othopedic Surgery,
Orthopedic Associates of Hartford, P.C.,
Clinical Professor of Orthopedic Surgery,
University of Connecticut Medical School,
499 Farmington Avenue, Suite 300,
Farmington, CT 06032, USA
e-mail: patelladoc@aol.com

Anteromedial tibial tubercle transfer (also called anteriorization of the tibial tubercle or AMZ) is most useful for unloading symptomatic patellofemoral articular lesions on the lateral and distal aspect of the patella while centering patella tracking [1]. The classic patient who benefits from anteromedial tibial tubercle transfer has Ficat's "excessive lateral pressure syndrome (ELPS) [2]." A patient who has had long-standing lateral tracking of the patella with overload of the lateral facet [2–7] generally breaks down the articular cartilage of the lateral patella and trochlear facets (Fig. 20.1), sometimes resulting in chronic pain. The typical pattern also breaks down distal patella articular damage as the patella courses from an abnormally lateral position in extension across the proximal lateral trochlea in early flexion, breaking down distal as well as lateral articular cartilage. Fortunately, the vast majority of patients with this aberration also are left with intact medial patella articular cartilage, particularly on the more proximal aspect of the medial patella.

Anteromedial tibial tubercle transfer may also be useful in patients with lateral patellar tracking who have damaged the distal patella medially as a result of relocation following patellar dislocation. An important prerequisite for success with anteromedial tibial tubercle transfer, however, is *intact* proximal medial patella articular cartilage. Anteromedial tibial tubercle transfer has also been useful in conjunction with patellofemoral resurfacing procedures.

A. Gobbi et al. (eds.), *The Patellofemoral Joint*,
DOI 10.1007/978-3-642-54965-6_20, © ISAKOS 2014

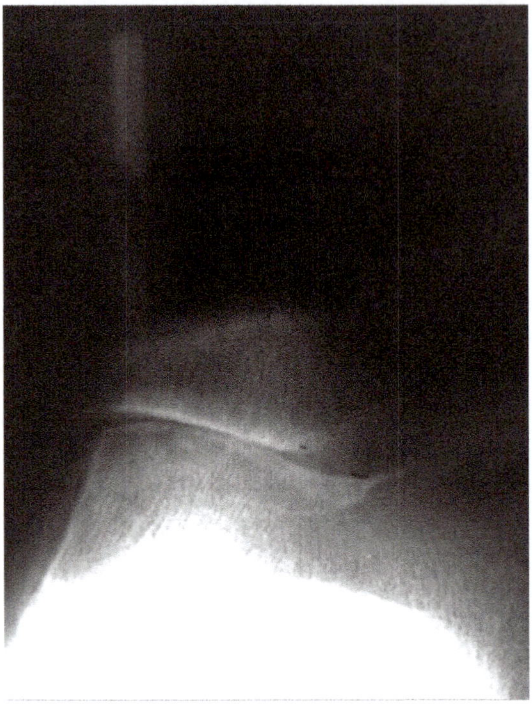

Fig. 20.1 Ideal patient for AMZ

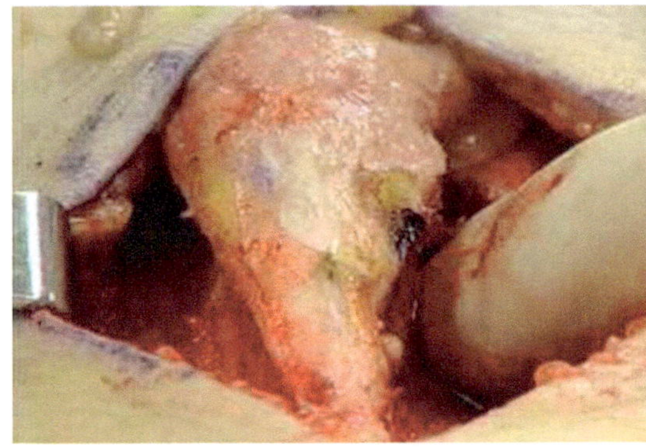

Fig. 20.2 Anteromedial tibial tubercle transfer before screw fixation. Note anterior taper at distal end and full exposure of lateral tibia

20.1 Technique

Anteromedial tibial tubercle transfer requires an incision of 6–10 cm along the tibial crest, extending from 2 cm proximal to the tibial tubercle to a point 4–7 cm distal to the tibial tubercle. After identifying the patella tendon insertion into the tibial tuberosity, the tibialis anterior muscle is reflected laterally. Using retractors to expose the entire lateral tibia, an oblique osteotomy is created from the medial patellar tendon insertion, extending to a point at the anterior tibial crest about 5–7 cm distally, such that the osteotomy is tapered anteriorly to exit at the anterior tibial crest distally. The osteotomy is designed to be oblique such that when the tibial tuberosity is moved, it will slide across the osteotomy plane both anteriorly and medially. The average osteotomy moves approximately 1 cm medially and 1 cm anteriorly. The tubercle may be moved slightly distally also to correct patella alta as needed (8) but this is usually not necessary or advised. To do this, some distal bone pedicle must be removed. *Most important in creating this oblique osteotomy is to watch the cutting blade laterally at the lateral tibia from distal to proximal.* The cut should be made through the distal aspect of the osteotomy first and then saw blade observed continuously as the blade moves proximally and posteriorly. The blade should never be out of the surgeon's direct vision, as the deep peroneal nerve and the anterior tibial artery are just behind the posterolateral tibia at this level.

To complete the osteotomy, an oblique cut is made on the lateral tibia from the most proximal lateral aspect of the osteotomy just above the patellar tendon. A third cut is then required to cut across the top of the osteotomy above the patellar tendon insertion to connect it to the starting point at the proximal medial tibial cortex just medial. After breaking loose the osteotomy shingle that has been created, it is shifted anteromedially (Fig. 20.2) and secured in its new location with two cortical screws into the posterior tibial cortex. After this is completed, lateral infrapatella releasing is done as needed to mobilize the patella, or the lateral release may be done preliminarily when necessary to relieve tilt, excessive tightness, or infrapatellar scarring. (Anteromedial tibial tubercle transfer is very helpful in many patients with infrapatellar contracture associated with excessive lateral pressure syndrome, as this osteotomy distracts any released infrapatellar contracture).

20.2 Rehabilitation

Following anteromedial tibial tubercle transfer, the patient should begin range of motion exercises as soon as the immediate surgical pain has abated. The patient should bend the knee once a day for the first 2–3 weeks, achieving 90° of knee flexion by 2 weeks following surgery. The patient should use crutches and protected weight bearing for 6 weeks following which the patient goes to physical therapy for further motion and strength exercises as well as weight bearing and progression off of crutches, usually by 8 weeks post-op.

20.3 Expectations

For properly selected patients, this procedure yields good and excellent results [1–9]. It is the best alternative to patellofemoral replacement particularly in younger patients with ELPS and for patients with lateral patella instability when the surgeon documents lateral patella tracking/overload and a need for patella medialization with unloading of the distal and lateral aspects of the patellofemoral joint.

References

1. Fulkerson JP, Becker GJ, Meaney JA, Miranda M, Folcik MA. Anteromedial tibial tubercle transfer without bone graft. Am J Sports Med. 1990;18(5):490–6.
2. Ficat P. The syndrome of lateral hyperpressure of the patella. Acta Orthop Belg. 1978;44(1):65–76.
3. Karamehmetoğlu M, Oztürkmen Y, Azboy I, Caniklioğlu M. Fulkerson osteotomy for the treatment of chronic patellofemoral malalignment. Acta Orthop Traumatol Turc. 2007;41(1):21–30.
4. Lin ZJ, Wu ZQ, Gao DW, Wu YF. Micro-modified Fulkerson osteotomy for the treatment of habituation patellar dislocation. Zhongguo Gu Shang. 2009;22(7):513–4.
5. Post WR, Fulkerson JP. Distal realignment of the patellofemoral joint. Indications, effects, results, and recommendations. Orthop Clin North Am. 1992;23(4):631–43. Review.
6. Farr J, Schepsis A, Cole B, Fulkerson J, Lewis P. Anteromedialization: review and technique. J Knee Surg. 2007;20(2):120–8. Review.
7. Cameron HU, Huffer B, Cameron GM. Anteromedial displacement of the tibial tubercle for patellofemoral arthralgia. Can J Surg. 1986;29(6):456–8.
8. Tjoumakaris FP, Forsythe B, Bradley JP. Patellofemoral instability in athletes: treatment via modified Fulkerson osteotomy and lateral release. Am J Sports Med. 2010;38(5):992–9.
9. Akgun U, Naran R, Karahan. Modified Fulkerzon osteotomy in recurrent patellofemoral dislocations. Acta Orthop Traumatol Turc. 2010;44(1):27–35.

Elmslie-Trillat Procedure: A Distal-Based Procedure for Patellar Stabilization

21

John Lane, Georgios Karnatzikos, and Alberto Gobbi

Contents

J. Lane (✉)
Coast Surgical Center, San Diego, CA, USA

Department of Orthopaedic Surgery,
University of California, San Diego, CA, USA
e-mail: jglane@san.rr.com

G. Karnatzikos • A. Gobbi
Orthopaedic Arthroscopic Surgery International
(O.A.S.I.) Bioresearch Foundation, Gobbi N.P.O.,
Milan, Italy
e-mail: giokarnes@gmail.com;
gobbi@cartilagedoctor.it

21.1 Introduction

Patellofemoral problems are a frequent common complaint in patients with knee pain. This is present in both adolescents and adults, and many terms have been used to describe this condition. Anterior knee pain and chondromalacia patella are two terms used to describe the knee pain which is noted to be present in the anterior aspect of the knee in the retropatellar area. In addition, many adolescents present with patellofemoral instability which can be with either frank dislocation or subluxation. Insall described this as patellofemoral malalignment in an attempt to explain patellofemoral tracking problems which lead to abnormal tracking, acute and chronic dislocation, and anterior knee pain.

Many physicians have avoided treating patients with patellofemoral disorders due to the difficulty in achieving good results. This is often due to the diffuse nature of their symptoms or the apparent psychological overlay which can contribute to pain in the retropatellar area.

Initial treatment needs to be focused on rehabilitation. Many of these patients have weakness of their extensor mechanism which often improves with strengthening and stretching of their quadriceps and hamstring muscle groups. Patellofemoral taping programs, as described by McConnell, have been an important adjunct to augment the therapeutic exercises which often improve patients with mild instability patterns or anterior knee pain [1].

A. Gobbi et al. (eds.), *The Patellofemoral Joint*,
DOI 10.1007/978-3-642-54965-6_21, © ISAKOS 2014

Many factors need to be considered in assessing patients with patellofemoral pain and instability as this is a multifactorial problem. Local factors in the knee which are primary determinants of patellar instability include trochlear dysplasia, patellar tilt, tibial tuberosity-trochlear groove (TT-TG) offset, and patellar height [2]. Overall structural alignments of the limb including hip rotation with femoral anteversion, distal rotation with tibial torsion, genu valgum, and genu recurvatum are also secondary contributory factors to patellar alignment and stability. Lateral retinacular tightness and poor muscular strength of the vastus medialis obliquus (VMO) are also contributory to instability and pain. Abnormalities in these structures and their relationships to the normal force vectors predispose the patient to develop patellar instability and/or patellofemoral pain syndrome.

After failure of conservative care, surgical intervention is focused on treatment of recurrent patellar instability due to the significant disability which occurs from recurrent dislocation of the patella. As a result, numerous operative procedures and modifications which number over 100 have been described to try to prevent a recurrence of the instability episode. These procedures have focused on different factors which were thought to be the primary cause of instability, and authors differed in their opinion as to the importance of proximal soft tissue reconstructions or distal realignment procedures which often included osteotomies. The Elmslie-Trillat procedure was developed to treat patellofemoral instability as a combination of soft tissue and bone realignment. This procedure evolved from a procedure initially noted by Roux in 1888 in a surgical procedure to stabilize patellar instability which combined a distal osteotomy procedure with a proximal soft tissue repair. The distal tibial tubercle osteotomy involved medialization of the tubercle with medial rotation of the patellar tendon insertion site into the tibial tubercle. This decreased the Q angle present and was successful in improving patellar stability. Golthwaite modified this procedure in 1895.

Further modifications were proposed in an attempt to improve the results of treatment of patellar instability. Hauser changed the site of fixation of the tibial tubercle osteotomy, and in addition to medialization, the patellar tendon insertion was moved distally. Posteriorization of the tubercle was associated with increased patellofemoral pain and increased levels of osteoarthritis of the patellofemoral joint [3]. Unfortunately, due to the altered biomechanics of the extensor mechanism and altered forces on the patellofemoral joint, numerous studies have demonstrated the development of patellofemoral arthritis following the Hauser modification [4, 5]. Subsequently, Trillat et al., in 1964, introduced a modification in which the tibial tubercle was only medialized without distalization [6].

During this procedure, a lateral retinacular release was performed. Modifications of the Elmslie-Trillat procedure have been successful in restoring stability to the knee following patellar dislocation.

21.2 Surgical Indications

Patients with recurrent patellar instability and patellar malalignment who have failed conservative treatment are appropriate for evaluation for surgical treatment. Assessment of the need for proximal realignment with medial patellofemoral ligament reconstruction or medial plication with lateral release versus distal realignment with possible distalization of the tibial tubercle for patella alta is needed. Patients with recurrent patellar instability or subluxation with an abnormally increased Q angle and increased TT-TG translation without patella alta are good candidates for surgical reconstruction. Patients with significant patellofemoral disease are not good operative candidates for this procedure due to poor results which appear related to the patellofemoral arthrosis [7]. For this reason, performance of this procedure is limited often to patients less than 35 years of age or to those undergoing a combined patellofemoral compartment biologic resurfacing procedure in conjunction with the realignment procedure in our practice. Patients with open physis of the tibial tubercle are also excluded due to the risk of growth plate arrest.

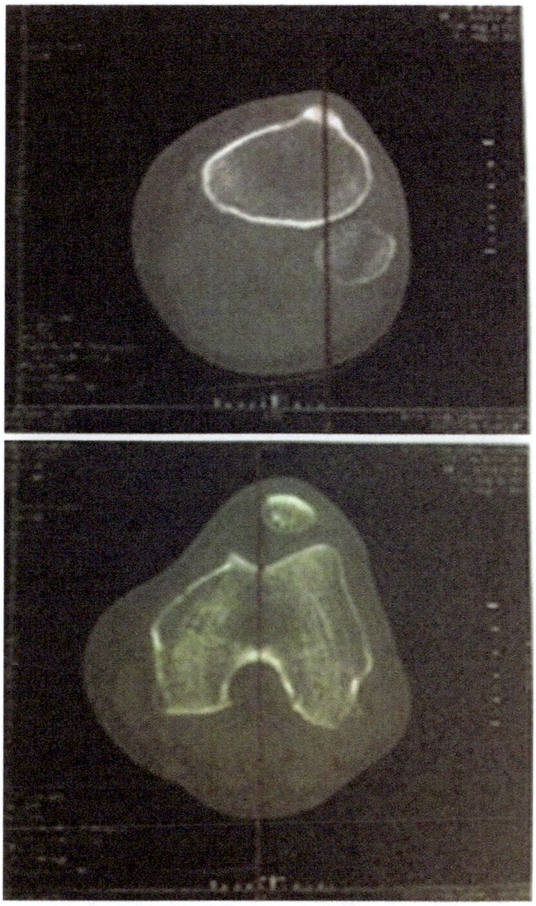

Fig. 21.1 Calculation of the TT-TG distance is made by measuring the difference between two projections made on simultaneous axial CT scans with the extremity in extension and the patella directed anteriorly. The first projection is drawn perpendicular to the posterior aspect of the knee through the center of the tibial tubercle. This line projects onto a grid superimposed onto the scan below. The second projection is drawn perpendicular to the posterior condyles of the femur (in the same plane as the line through the tibial tubercle) in the central aspect of the trochlea, extending to the grid below. The grid is calibrated to measure distance in millimeters between the two scans through which the two projections pass

A CT scan is obtained preoperatively, and the TT-TG distance is calculated by using the ruler on the scan in the posterior aspect of the knee and measuring the position of the center of the trochlear groove and determining the distance to the central portion of the tibial tubercle (Fig. 21.1).

TT-TG distances greater than 20 mm are present in patients who are candidates for the Elmslie-Trillat procedure. Determination of the need for advancement of the medial retinaculum is based on the amount of translation of the patella at time of surgery, and the need for a lateral release is dependent upon the amount of patellar subluxation seen on x-ray and the tightness noted after performance of the tubercle osteotomy with rotation. Reduction of the TT-TG distance to 10–15 mm is optimal [8].

21.3 Surgical Technique

In patients with patellofemoral instability, preoperative assessment includes physical examination and diagnostic studies. These studies include a CAT scan to assess the tibial tuberosity-trochlear groove distance. This is important to calculate the distance that the tibial tubercle needs to be medialized to improve the alignment. The patient is placed on the operating table in the supine position and a diagnostic arthroscopy is performed to assess the patellofemoral joint surfaces as well as the patellofemoral alignment and stability. To perform the distal osteotomy, an incision is made over the anterior fascia and a dissection of the muscles off of the tubercle is performed in preparation for the osteotomy. Multiple drill holes are made in the coronal plane starting at the superior aspect of the tibial tubercle in the coronal plane with a drill or K wires which can be left in place. Proximally, the depth of the insertion of the wires is approximately 5–7 mm. This is sequentially decreased to 1–2 mm at the distal aspect of the osteotomy site 4–5 cm distal to the patellar tendon insertion to the tubercle. An osteotome or saw is then used to complete the proximal aspect of the osteotomy. Distally where the osteotomy fragment is only 1–2 mm thick, an osteotome is directed distally and is used to crack the anterior cortex with the soft tissue hinge present (Fig. 21.2). This then allows distal attachment to remain intact, but the tubercle can be rotated medially on this soft tissue attachment. This facilitates postoperative therapy as proximal migration of the tubercle is prevented by the soft tissue and periosteum attachment of the distal osteotomy site.

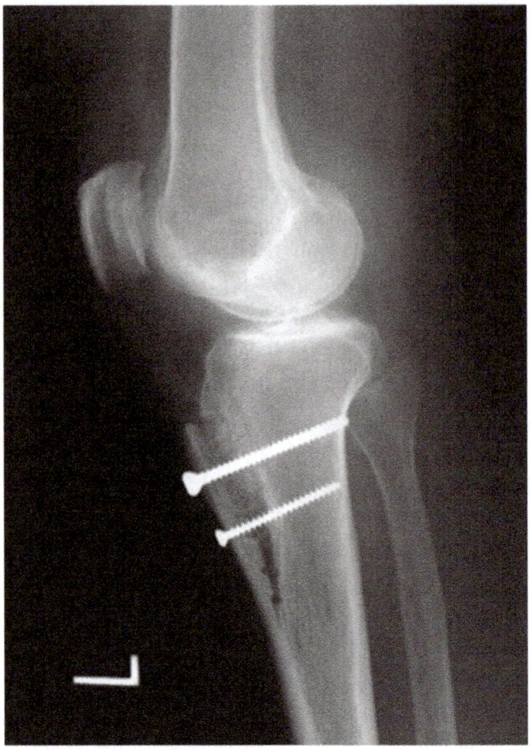

Fig. 21.2 Lateral radiograph of the knee demonstrating the medialization of the tibial tubercle which is rotated anteromedially with an intact distal periosteal hinge. Multiple drill holes are made to initiate the plane of the osteotomy, and then osteotomes are used to complete the cut and crack the distal periosteal hinge

The amount of medialization of the proximal aspect of the tibial tubercle at the patellar tendon insertion site is determined with consideration of the preoperative studies. Arthroscopic evaluation of the patellar tracking is determined prior to definitive fixation to assess appropriateness of the translation. The tubercle is then fixed with two screws. This can be performed with either 3.5 or 4.5 mm screws based on the integrity of the soft tissue attachment distally. Medial plication of the medial retinaculum can be performed either open with an arthrotomy or arthroscopically if needed to help improve stability dependent upon patellar stability based on the integrity of the medial structures assessed after transfer of the tibial tubercle. Closure of the skin is sometimes performed with use of a drain if needed following the procedure to prevent hematoma formation.

Postoperative care includes early passive range of motion exercises in a range of motion brace through a safe range of motion determined during the surgical procedure. Early weight bearing is encouraged in an extended position with the range of motion brace locked in full extension. Quadriceps strengthening with the knee in a limited range of motion from 0 to 60 is encouraged. Range of motion with full weight bearing is dependent upon the amount of elevation of the tubercle and the need for medial reefing. In patients with only medialization and an intact hinge at the distal edge of the osteotomy, full weight bearing with limited flexion angles can begin as early as 6 weeks postoperatively depending upon radiographic evaluation and symptoms with weight bearing.

21.4 Discussion

Patellofemoral instability and patellofemoral pain syndrome continue to be difficult problems to treat. Improved physical therapy and exercise programs have shown to be successful in resolving many of the symptoms. The many types of patellofemoral instability procedures can be classified into proximal, distal, or combined procedures. Patients that continue to have significant instability or pain are often responsive to surgical treatment.

Proximal repair or reconstruction procedures with repair or reconstruction of the medial patellofemoral ligament have been shown to be successful [9, 10]. Advancement of the medial retinaculum has long been an option for restoring stability to an unstable patella [11, 12]. Lateral retinacular release has been shown not to be effective in restoring stability, and studies have shown that release of the lateral retinaculum in knees with injury to the medial structures causes the development of additional instability [13–15]. The importance of the medial patellofemoral ligament has been shown to be an important factor in patellar instability and this has improved our ability to obtain good results in restoring patellar stability.

Distal realignment procedures can be done in isolation. Fulkerson has popularized anteromedialization of the tibial tubercle with good results in patients with both instability and anterior knee pain [16].

Many studies have demonstrated the benefits of the Elmslie-Trillat procedure in achieving stability with a combination of distal and proximal procedures. Barber and McGarry evaluated 35 knees in patients with an average of 27.7 years and noted that 91.4 % of the patients had stable knees at follow-up [17]. The results of the procedure have been shown to be long-standing as noted in a 26-year follow-up by Carney et al. in which there was no increase in instability over time [18].

The procedure however has not been successful in treating patellofemoral pain or patellofemoral arthritis. Nakagawa showed that 74 % of patients that underwent the Elmslie-Trillat procedure developed degenerative changes at 10–20-year follow-up [19]. 13 of 31 knees (42 %) developed definitive osteoarthritis. When stratified for success in treatment of patellar instability versus treatment of patellofemoral pain, the results are noted to vary widely. Kumar noted that all patients that underwent distal realignment for patellar instability received a good or excellent result. The results were much worse when performed for patellofemoral pain and patellofemoral pain with instability as the percentages of good results in these scenarios were 40 and 44 %, respectively (functional evaluation of the modified Elmslie-Trillat procedure for patellofemoral dysfunction) [7]. In response to results as noted in these clinical scenarios, alternate treatment options have been suggested as noted with different osteotomies including anteromedialization as described in a surgical technique by Fulkerson [16].

There still remain patients with trochlear dysplasia which remain problematic. Dejour has shown the importance of trochlear dysplasia, and performance of trochleaplasty is effective in helping to restore stability in experienced hands. For patients with patella alta, consideration of distalization of the tubercle insertion of the patellar tendon is important to help the patella engage in the trochlea earlier in the flexion cycle of the knee.

In patients with increased tibial tuberosity-trochlear groove distances and intact articular surfaces, medialization via the Elmslie-Trillat technique can be very effective in providing patellofemoral stability. In knees with patellar chondral defects, the combination of anteromedialization as noted by Fulkerson helps improve results in patients with patellar subluxation/instability.

References

1. McConnell J. The physical therapist's approach to patellofemoral disorders. Clin Sports Med. 2002;21:363–87.
2. Dejour H, Walch G, Nove-Josserand L, Guier C. Factors of patellar instability: an anatomic radiographic study. Knee Surg Sports Traumatol Arthrosc. 1994;2:19–26.
3. Hauser EDW. Total tendon transplant for slipping patella: new operation for recurrent dislocation of the patella. Surg Gynecol Obstet. 1938;66:199–214.
4. Hampson WGJ, Hill P. Late results of transfer of the tibial tubercle for recurrent dislocation of the patella. J Bone Joint Surg Br. 1975;57-B:209–13.
5. Crosby EB, Insall J. Recurrent dislocation of the patella: relation of treatment to osteoarthritis. J Bone Joint Surg Am. 1976;58-A:9–13.
6. Trillat A, Dejour H, Couette A. Diagnostic et traitement des subluxations r'ecidevantes de la rotule. Rev Chir Orthop Reparatrice Appar Mot. 1964;50:813–24.
7. Kumar A, Jones S, Bickerstaff DR, Smith TW. Functional evaluation of the modified Elmslie-Trillat procedure for patello-femoral dysfunction. Knee. 2001;8(4):287–92.
8. Fithian DC, Nyret P, Servien E. Patellar instability: the Lyon experience. Tech Knee Surg. 2007;6(2):112–23.
9. Howell NR, et al. Medial patellofemoral ligament reconstruction. J Bone Joint Surg. 2012;94-B:1202–8.
10. Buckens CF, Saris DB. Reconstruction of the medical patellofemoral ligament for treatment of patellofemoral instability: a systematic review. Am J Sports Med. 2010;38:181–8.
11. Abraham E, Washington E, Huang TL. Insall proximal realignment for disorders of the patella. Clin Orthop Relat Res. 1989;(248):61–5.
12. Halbrecht J. Arthroscopic patella realignment: an all inside technique. Arthroscopy. 2001;17:940–5.
13. Lattermann C, Toth J, Bach BR. The role of lateral retinacular release in the treatment of patellar instability. Sports Med Arthrosc. 2007;15:57–60.
14. Aglietti P, Pisaneschi A, De Biase P. Recurrent dislocation of the patella: three kinds of surgical treatment. Ital J Orthop Traumatol. 1992;18:25–36.

15. Dainer RD, Barrack RL, Buckley SL, et al. Arthroscopic treatment of acute patellar dislocations. Arthroscopy. 1988;4:267–71.
16. Fulkerson JP, Becker GJ, Meaney JA, Miranda M, Folcik MA. Anteromedial tibial tubercle transfer without bone graft. Am J Sports Med. 1990;18: 490–7.
17. Barber FA, McGarry JE. Elmslie-Trillat procedure for the treatment of recurrent patellar instability. Arthroscopy. 2008;24(1):77–81.
18. Carney JR, Mologne TS, Muldoon M, Cox JS. Long-term evaluation of the Roux-Elmslie-Trillat procedure for patellar instability: a 26-year follow-up. Am J Sports Med. 2005;33(8):1220–3. Epub 2005 Jul 6.
19. Nakagawa K, Wada Y, Minamide M, Tsuchiya A, Moriya H. Deterioration of long-term clinical results after the Elmslie-Trillat procedure for dislocation of the patella. J Bone Joint Surg Br. 2002; 84-B:861–4.

Open Proximal Trochleaplasty (Grooveplasty)

22

Lars Peterson and Haris S. Vasiliadis

Contents

L. Peterson (✉)
Department of Orthopaedics, University of Gothenburg, Gothenburg, Sweden
e-mail: peterson.lars@telia.com

H.S. Vasiliadis
Department of Orthopaedics, University of Gothenburg, Gothenburg, Sweden

Department of Orthopaedics, School of Medicine, University of Ioannina, Ioannina, Greece

A. Gobbi et al. (eds.), *The Patellofemoral Joint*,
DOI 10.1007/978-3-642-54965-6_22, © ISAKOS 2014

22.1 Passive and Active Stability of the Patellofemoral Joint

Skeletal, ligamentous (passive), and muscular (active) stabilizers of the patella and the knee joint all work together for the stability of the patellofemoral joint (see Fig. 22.1).

The skeletal geometry of the lower limb creates a passive stability during knee motion. The knee alignment measured, for example, by the Q angle or the femoral anteversion plays important roles for the stability of the patella in the trochlear groove but above all for the congruity between the patellotrochlear articular surfaces.

The geometry of the trochlear groove itself is of the highest importance during the first 30° of flexion (30–0° of extension) affecting the skeletal stability. Most of the dislocations occur at the beginning of the flexion after which the patella is more stably trapped between the femoral condyles at the lower part of the trochlear groove. The skeletal stability can be affected by patellofemoral dysplasia, where the geometric configuration of the trochlea is primarily not developed. The patella shape in comparison with the trochlea usually plays a secondary role even if the patella is found dysplastic [1].

The ligamentous stabilizers of the patella (retinacula, medial patellofemoral ligament) also play an important role. The shape of the trochlear surface affecting the skeletal stability and the ligaments of patellofemoral joint preserve the passive stability which cannot be affected by training or any voluntary activity. Ligamentous

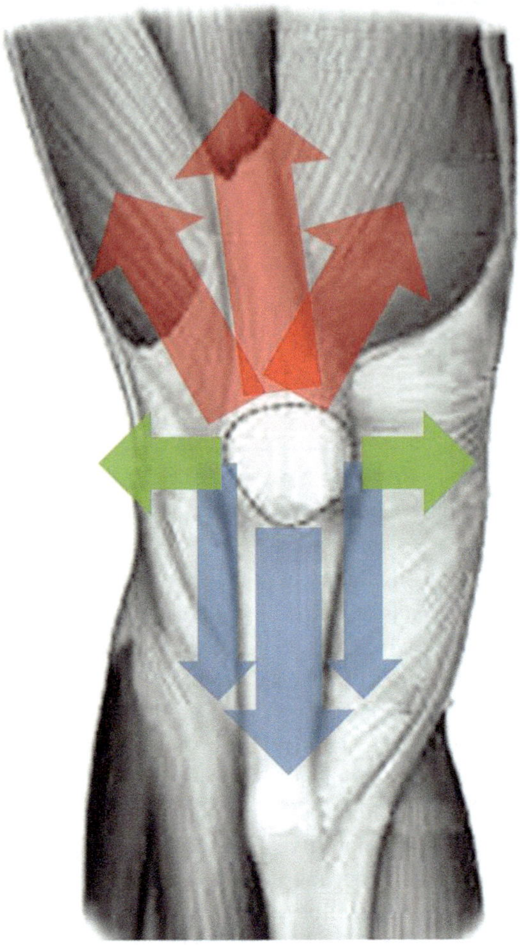

Fig. 22.1 Active and passive stabilizers of the patella. The *red arrows* show active muscle stabilizers, VMO, rectus femoris, and VL. The *green arrows* show passive stabilizers, medial patellofemoral ligament (medial transverse retinaculum), and lateral patellofemoral ligament (lateral transverse retinaculum). The *blue arrows* show passive and active stabilizers, patellar ligament, and longitudinal medial, and lateral retinacula

stability works by stabilizing the joint throughout the whole range of motion.

The medial patellofemoral ligament and the lateral patellofemoral ligament (transverse retinaculum) are also passive stabilizers medially and laterally of the patella. They work synergetically with the VMO and VL muscles and add to the active stability [2].

The quadriceps, acting over the joint during contraction, contributes to the patellofemoral stability as active stabilizers. The active stability of the patellofemoral joint can be improved by training.

The longitudinal medial and lateral retinacula act both through passive support to the patellar ligament. However, they are also active stabilizers acting as tendon-like aponeuroses of the vastus medialis obliquus (VMO) and vastus lateralis muscles (VL).

The patellar ligament (tendon) has a double role. It mainly acts as a passive stabilizer by limiting the proximal movement of the patella but also secondarily as a tendon of the quadriceps muscle functioning as an active stabilizer.

22.2 Trochlear Dysplasia

Trochlear dysplasia can be classified into 3 grades depending on the shape of the articular surface. In grade 1, trochlea is flat or shallow, while in grade 2 it is convex shaped and extends more proximal. In grade 3, the lateral trochlea is flat with the lateral patellar facet articulating on its lateral surface and not to the dysplastic medial trochlea (Figs. 22.2 and 22.3).

In patients with dysplasia of the trochlea, there is usually a lateral tracking and tilting of the patella, especially during muscle contraction. Depending on the grade of the dysplasia, the patella may be slightly tracked or tilted laterally, recurrently subluxating or dislocating during knee motion [3].

In trochlear dysplasia, the contact areas of the patella and trochlea are altered along with the contribution of the applied forces on the articular surfaces. The forces are applied in different areas than in normal joints (Fig. 22.4). Subsequent tilting or subluxation or dislocation of the patella also subjects the articulating areas to increased shearing forces on the medial-lateral axis. The abnormal lateral tracking causes a stress concentration in a small contact area of the patellofemoral joint resulting in kissing cartilage lesions on lateral trochlear surface and lateral patellar facet. These lesions may be the result of the repeated abnormal forces or of an acute episode of traumatic subluxation or dislocation causing damage to the articular cartilage surfaces in the contact areas.

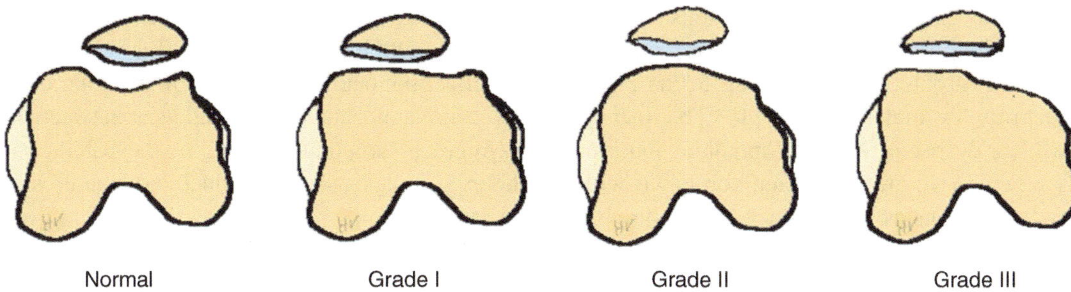

Fig. 22.2 Schematic drawings of trochlear dysplasia. Grade I, flat or shallow trochlea; grade II, convex-shaped trochlea that extends more proximally; and grade III, lateral trochlea is flat with the patella articulating on the lateral trochlear surface and medial trochlea dysplastic

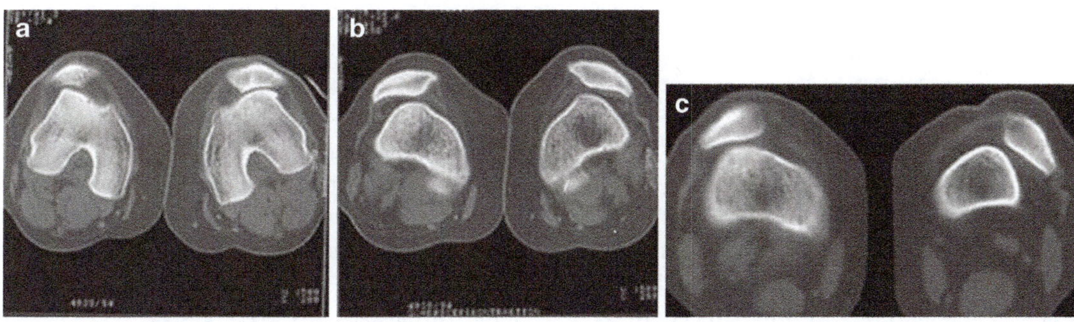

Fig. 22.3 Computerized tomography with the knees in extension. (**a**) Patient with grade III trochlear dysplasia showing flat lateral trochlear articular surface articulating with lateral patellar facet, with slight subluxation on the right knee. (**b**) Patient with extended knees and quadriceps contraction showing subluxation of both patellas. (**c**) Patient with extended knees during quadriceps contraction showing right knee with subluxation and left knee with dislocation. Both (**b**) and (**c**) show flat trochlear grooves

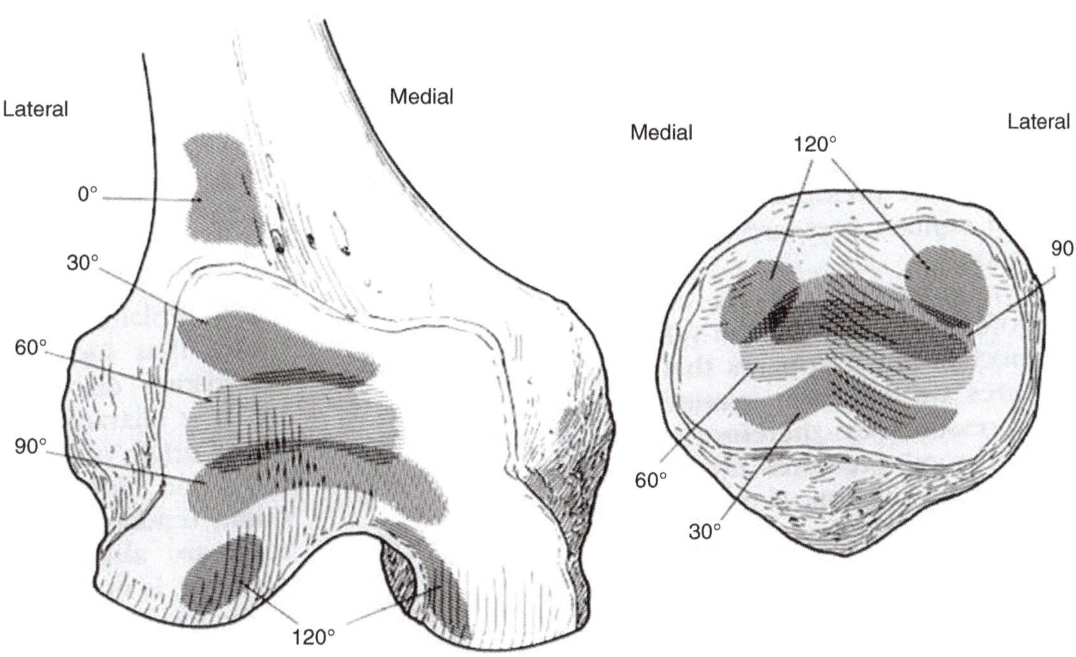

Fig. 22.4 The different contact areas of normal trochlea and patella in different angles of knee flexion

In these cases, the cartilage lesions should be treated along with the trochlear dysplasia or any other background factors leading to the patellar instability or malalignment [4]. The fact that trochlear dysplasia is a developmental abnormality affecting young ages often combined with a patella alta makes the contact of patella to the trochlea important for the development of the trochlear groove (sulcus). This increases the importance of prompt identification and treatment of this problem in order to develop and preserve the congruity and integrity of the joint.

22.3 Open Proximal Trochleaplasty (Grooveplasty)

The open proximal trochleaplasty was first described in 1988 [1]. It is indicated in cases with trochlear dysplasia causing symptomatic patellofemoral instability or cartilage patellofemoral lesions. The purpose of this procedure is to reconstruct to a close to normal trochlear groove and to subsequently stabilize the patella during the first 30° of knee flexion (extension 30–0°). The technique aims to avoid and/or minimize interfering with the patella-trochlea congruity.

The philosophy of the procedure is based on that patellar dislocations initiate during the first 30° of knee flexion. In cases with trochlear dysplasia, where the skeletal stability is already insufficient and the patella is not stable lying in the center of the trochlea, the quadriceps contraction is usually the cause of lateral tracking, dislocation, or subluxation of the patella. If the patella passes the first 30° of knee flexion without being subluxated, then in further flexion it will remain stable by the passive and active stabilizers. At flexion angles exceeding 30°, the patella is trapped between the femoral condyles at the notch (passive stabilization), and the stretched ligaments (especially the medial and lateral retinacula) will effectively stabilize the patellofemoral joint. According to this theory, only the proximal part of the trochlea, which is in contact with the patella during the first 30°, needs to be deepened in order to achieve the desired patellofemoral joint stability. Therefore, there is no functional need for altering the congruity of the patella-trochlea distally as it does not affect the patellar tracking. On the contrary, it will cause abnormal loading and wear between the incongruent articular surfaces of the patella and the existing sulcus with gradual cartilage erosion as a result.

Other trochlear grooveplasties have also been used, such as lateral trochlear open-wedge osteotomy and subchondral trochlear burring [5–8]. Although these techniques may anatomically restore the proximal skeletal stability, they interfere with the patella-trochlea congruity. Although the dislocation of the patella may be resolved, the alteration of the congruity of the patellofemoral joint may increase the forces applied on the cartilage and may also lead to eccentric loading of the joint resulting in a progressive osteoarthritis.

The open proximal trochleaplasty should almost always be performed as a part of more extended strategy. The aim should be the total restoration of the patellofemoral alignment along with the repair of the cartilage lesions when present.

The whole surgical protocol starting from the preoperative planning up to the grooveplasty and rehabilitation is described below.

22.3.1 Preoperative Planning

The diagnostic approach of any patient with patellofemoral instability should include the identification of the factors that may contribute to the malalignment and instability of the patella. Based on the findings of the diagnostic approach, a preoperative plan should be scheduled, including the corrections needed to be performed. The therapeutic approach should be individualized, for any single patient, depending on the factors that need to be corrected.

A meticulous clinical examination is of high importance and should be performed prior to any laboratory exams. In grade I instability, the patella will usually not dislocate but track laterally before moving proximally on slow quadriceps contraction with the knee in extension; slow VMO activation delays proximal traction. Apprehension test is negative. In grades II and

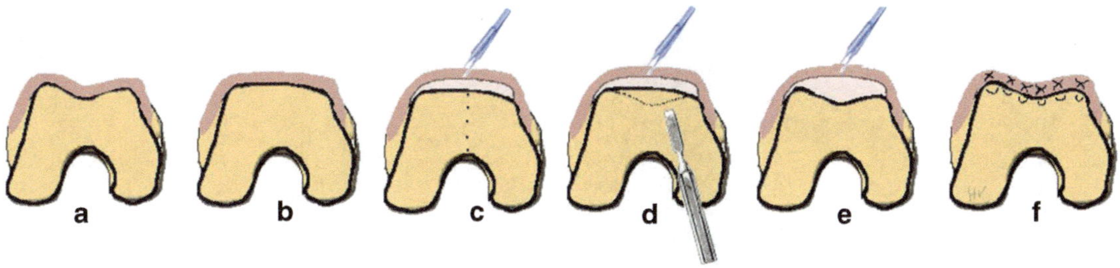

Fig. 22.5 Proximal trochlear grooveplasty. (**a**) Normal configuration of the trochlea groove, (**b**) flat trochlear dysplasia, (**c**) synovial lining released from the trochlea articular border, (**d**) with a curved osteotome perform distal resection of about 10–12 mm of the cartilage and bone, aiming to the top of the intercondylar notch. Then, complete the trochleaplasty by about 15 mm medial and lateral enlargement of the groove, (**e**) restoration of the trochlea groove after completed resection, (**f**) re-suturing of the synovial lining to the cartilage border of the trochlea using mattress sutures

III, the apprehension test is positive. Patellar lateral subluxation and patellar tilt are found. Quadriceps contraction on an extended knee may occasionally dislocate the patella in grade II. A persisting instability with recurrent subluxation and sporadic dislocations is present in 0–30° of flexion, in grade III instability.

Long plain X-rays of the limb (hip-knee-ankle) on AP plane should be performed in order to measure the mechanical axis, assess the Q angle, and identify a potentially valgus knee. Assessment of the tibial tubercle-trochlear groove distance (TT-TG) could be helpful. CT scan of the patellofemoral joint in extension is used, identically with and without quadriceps contraction in order to reveal lateral tracking or instability and trochlear dysplasia.

Skyline views of the patella are usually performed as the first imaging approach in order to evaluate the trochlear sulcus angle and assess the patellofemoral congruity. However, it only gives a static, nonfunctional image of the joint and lacks the ability of the active evaluation of the joint in full extension or 20° of flexion and fails to demonstrate the tendency of subluxation under the traction of quadriceps. Besides, this is a less accurate method of evaluating the exact anatomy comparing with CT scan.

MRI can be added to the CT scan, also demonstrating the integrity of the ligamentous as well as muscular stabilizers of the patella. The integrity of medial patellofemoral ligament (MPFL) can also be examined, potentially indicating patellar dislocations in the past. MRI (conventional of more advanced techniques like dGEMRIC) is also used for the evaluation of the cartilage [9]. Dynamic MRI is expected to be used in the future giving the ability of a more functional assessment of the patellofemoral joint.

MRI arthrogram and arthroscopy are valuable tools for assessing the patellofemoral joint in details especially regarding articular cartilage lesions and dysplasia.

22.3.2 Surgical Technique

A central skin incision is usually performed, followed either by a medial or lateral arthrotomy (depending on other potentially performed operations or the location of cartilage lesions). The trochlea is exposed and an exploration of the patellofemoral joint is performed. The surgeon should first identify and confirm the type of dysplasia and the location and extent of any articular cartilage injuries (Figs. 22.5b and 22.6a).

Then, the synovial lining (posterior wall of the suprapatellar bursa) should be released from the articular border of the proximal trochlea (on the anterior femoral cortex) (Fig. 22.5c).

Then, the surgeon should aim for the top of the intercondylar notch and mark an imaginary line from the top of the notch to the top of the articular surface (Fig. 22.5c). With the use of a curved osteotome, he or she should remove articular cartilage and bone from the center of the trochlea, about 10 mm distal from the proximal edge of the cartilage and 15 mm medial and lat-

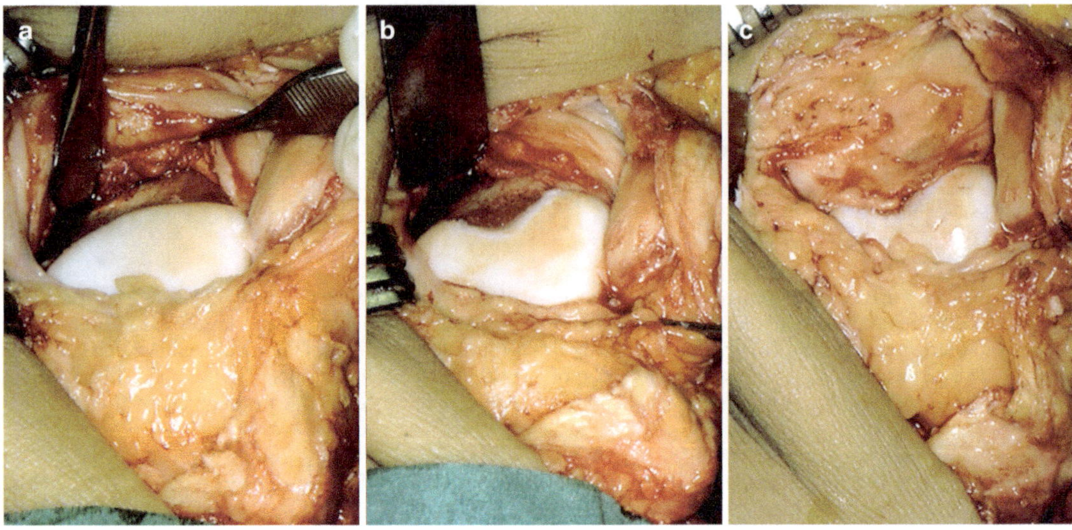

Fig. 22.6 Surgical pictures of different stages of proximal open trochleaplasty. (**a**) Dysplasia of the trochlea with a convex groove, (**b**) after creating the desired concavity with the curved osteotome, (**c**) after advancing distally and suturing the synovial lining to the proximal cartilage border

eral of the previously marked line, into a maximum depth of about 5 mm. The aim is to create a concavity on the most proximal part of the trochlea (Figs. 22.5d, e and 22.6b). If the bone is also flat or convex proximal to the trochlea, it should be removed with the osteotome or a burr and a continuous concavity should be created. Then the patella sliding through the groove should be checked and the Q angle should be adjusted if necessary to centralize the patella.

Then the surgeon should re-suture the synovial lining back to the cartilage border using mattress reabsorbable 3-0 sutures, to cover the raw bone surface of the distal femur preventing postoperative bleeding (Figs. 22.5f and 22.6c). Fibrin glue should usually be used to assist the fixation to the synovial lining to the roughened bone (Fig. 22.6) (as described below).

The surgeon should always evaluate the trochlear dysplasia in cases with patellofemoral instability. He should consider trochlear dysplasia an important part of the instability and correct it if symptomatic. However, isolated dysplasia of the trochlea is rare, and other background factors are usually identified after a meticulous diagnostic approach. These factors should be assessed and are equally important to be corrected along with the trochleaplasty.

It is important that the proximal location of the concavity created is aiming and directed to the top of the intercondylar notch. Then, when deciding the degree of medialization if needed,it should be adjusted so that the extensor mechanism with the patella is entering the trochleaplasty without angulation, not to compress laterally or dislocate medially but to run centrally. Sometimes in the patella alta, a distalization is needed and performed by an oblique medial to lateral distal osteotomy where 10–30–50° angle give a 1–3–5 mm distalization when sliding the tibial tuberosity medially.

Care should be taken so that a step from the bone to the cartilage border is avoided; the cartilage borders should be cut oblique with the use of a knife.

For the reattachment of the synovial lining to the cartilage border, care should be taken so that a step from the bone to the cartilage border is avoided; the cartilage border should be cut oblique with the use of a knife. Then, the surgeon should start through the synovial lining, continue deep into the cartilage penetrating 10 mm to the surface, start the return 3 mm aside, then go through the cartilage back to the border, and catch the synovial lining 2–3 mm apart from the initial entrance. Then, he or she should tighten the suture so that the synovial lining adapts to the cartilage border. As many sutures as needed should be placed in order to adapt the synovial lining to the whole border with about 6–8 mm intervals.

Sometimes, it is hard to advance the synovium to cover the whole bony defect. In this case, a transverse incision to the synovium proximally is needed to release it and to allow more distal advancement and reattachment to the cartilage.

Then, injecting the fibrin glue, between the rough bone and the synovium, it should be compressed to the bone with a dry sponge for 2 min to adhere the synovium to the bone surface and avoid bleeding.

The proximal trochleaplasty can also be performed arthroscopically. Correction of the concavity and deepening of the trochlear groove can be performed with a curved osteotome and an arthroscopic burr. Initially an elevator (raspartorium) can be used for the release of the synovial lining, while it can be advanced and fixed distally to the cartilage border with the use of fibrin glue (Tisseel). Arthroscopic proximal trochleaplasty may be advised to experienced arthroscopists, when minor concomitant arthroscopic surgery is required (e.g., medial soft tissue plication and the restoration of the patellofemoral alignment) and when no open cartilage surgery is required. In these cases, arthroscopically performed grooveplasty including minor procedures will minimize the morbidity and allow for a more accelerated postoperative rehabilitation.

22.4 Additional Surgeries

The trochlear dysplasia is usually one part of the problems that need to be assessed and corrected for the treatment of patellar instability and cartilage lesions of the patellofemoral joint. There are usually coexisting factors leading to patellar instability, some of them being the result of the recurrent subluxations or dislocations of the patella. The subsequent elongation of the medial passive stabilizing elements (medial retinaculum, medial patellofemoral ligament) along with shrinkage of the lateral retinaculum is usually the result of recurrent patellar subluxations. However, this ligamentous insufficiency may also be a part in these patellar subluxations or dislocations along with the active muscular stabilizer (the VMO) which may be stretched out, weakened, and insufficient.

Besides that, there is usually a dysfunction of the whole musculoskeletal arrangement of the knee joint; that dictates the need for anteromedial-distal transfer of the tibial tuberosity for the correction of the forces and unloading of the patellofemoral joint. The restoration of the trochlear dysplasia by the grooveplasty should be accompanied by the correction of all the concomitant background factors. Only with the above-mentioned factors corrected, the treatment of the potential cartilage lesions, which should preferably be with autologous chondrocyte implantation, will lead to satisfactory clinical outcomes.

So, the whole surgery is described in the following text, keeping the order that all the stages should be performed. In the absence of trochlear dysplasia, depending on the underlying background factors, some of those stages can be enough to restore the patellar alignment. Whenever trochlear dysplasia is present, it should be corrected.

The operation is performed with the patient supine using a tourniquet. A midline skin incision starting 1–2 cm proximal to the base of the patella down and distal to the tibial tuberosity should be performed; then, a medial parapatellar arthrotomy starting 1 cm proximal to the patella, between the rectus femoris and the vastus medialis obliquus, running 5–7 mm medial to the patellar insertion of the VMO and patellofemoral ligament should be performed. The surgeon should continue down to the tibial condyle and incise the joint capsule and inspect the joint, especially the patella and trochlear groove. He or she should evaluate the articular cartilage lesions along with the patellofemoral joint incongruity or any other malfunctions and address the surgical plan for his/her treatment. A meticulous assessment of the Q angle should be performed. The patellar tracking in the groove should be checked along with the inspection of the groove, and the identification of a dysplastic trochlea is apparent.

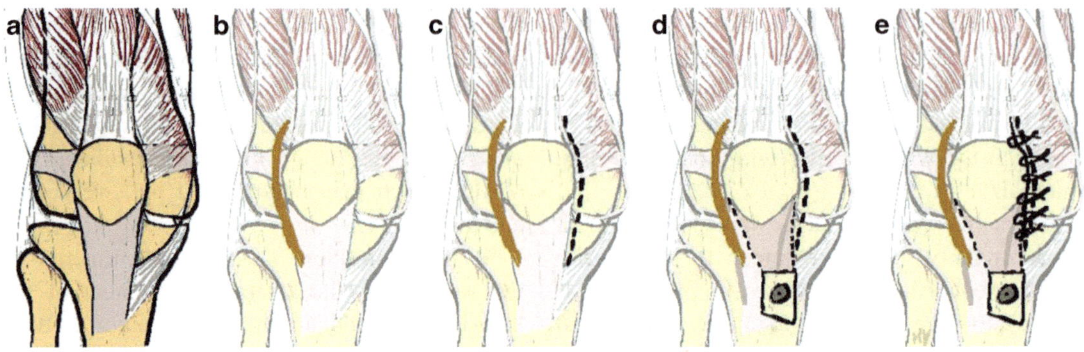

Fig. 22.7 The row of surgeries for the reconstruction of the extensor mechanism (realignment procedure). (**a**) Important structures to address. (**b**) Release of the lateral transverse retinaculum and distal VL insertion. (**c**) Anteromedial incision starting 1 cm proximal to the base of the patella between the VMO and rectus femoris down through the VMO and MPFL and capsule 5–7 mm medial to the insertion in the patella down to the tibial tuberosity. (**d**) Osteotomy and medial-distal transfer of the tibial tuberosity and screw fixation. Trochleaplasty or ACI should be performed at that stage, before the screw fixation. (**e**) Plication of the VMO and MPFL en bloc. For details of the plication technique, see Fig. 22.8

- *Lateral Release*
  The lateral release (release of the lateral transverse retinaculum and distal VL insertion) and the medial arthrotomy incising the medial retinaculum and VMO should be performed first (Fig. 22.7b, c). However, the suturing (plication) of the medial soft tissues en bloc (the MPFL, the VMO, and the retinaculum) should be preserved as the last stage of the surgery (Fig. 22.7e).
- *Tibial Tuberosity Transfer: Unloading Procedures*
  In case of an increased Q angle, a tibial tuberosity transfer may be indicated in order to correct it. A straight medial transfer should be undertaken in order to correct the Q angle and an anteromedial transfer to correct the Q angle along with unloading the patellofemoral joint. An additional distal transfer should be performed in the case of patella alta.

  Simultaneous anteriorization is achieved by using an oblique osteotomy starting from posterior laterally aiming to anterior medially. Increased angle of the osteotomy increases the anteriorization.

  For both medial and anteromedial transfers, a distalization can be added by an oblique distal osteotomy angulating by 10° for every mm of distal transfer needed.

  Isolated tibial tuberosity anteriorization (ventralization) is used to unload the patellofemoral joint in kissing patella-trochlea lesions or large uncontained patellar and trochlear lesions. A straight proximal osteotomy of the tibial tuberosity, keeping the attachment to the bone distally, is elevated about 10 mm, and a 10 mm wedge of bone is taken from the lateral tibial plateau and pressed into the osteotomy and fixed with a screw.

  The surgeon should first open the infrapatellar bursa, medial and lateral, and dissect free the tibial condyle if you need an anteriorization. He should use a saw or an osteotome and go from posterior-lateral to anterior-medial through the tibial tuberosity and then perform an oblique osteotomy 3–5 cm distal to the patellar tendon insertion after predrilling for later screw fixation. He should check the degree of medialization for correcting the Q angle and maltracking; usually 10–14 mm of medial transfer is needed. When a distal transfer of 3–5 mm is needed to correct a patella alta, an oblique distal osteotomy is used (Fig. 22.7).

  The fixation of the transferred tubercle in its new place should be performed at the end of surgery, after the trochleaplasty or ACI (if performed), just before the medial plication (Fig. 22.7d, e).
- *Trochlear Grooveplasty*
  The trochlear grooveplasty is performed (as described before) as the next step after the

tibial tuberosity osteotomy, if dysplasia is present.

- *Autologous Chondrocyte Implantation (ACI)*
  After the tibial tuberosity osteotomy and the grooveplasty, the ACI for the treatment of cartilage lesions follows. In that case of course, a cartilage biopsy and chondrocyte culturing should have been preceded in a previous arthroscopy. When a trochlear grooveplasty is scheduled, the cartilage biopsy may be retrieved from the proximal central part of the dysplastic trochlea; this area will anyway be removed during the grooveplasty.
- *Medial Plication*
  The shortening of the medial patellofemoral ligament (medial retinaculum) and the VMO should be performed at the last stage of the surgery. The medial patellofemoral ligament and VMO should be shortened en bloc (Figs. 22.7a and 22.8). The bone between the soft tissue insertion to the medial side of the patella and the articular cartilage should be roughened with the use of a curette. Sutures should then be passed through the soft tissue flap close to the patellar bone from outside to inside. They should then pass through the medial end of the soft tissue block, including the medial patellofemoral ligament and the VMO tendon, from anterior to posterior. Finally, sutures should return from inside to outside through the soft tissue close to the patella. After the passage of 3–4 of these sutures, they should be tightened, and the surgeon should check the stability of the patella. Then, he should overlap the lateral flap by interrupted sutures, continue with the suturing of the medial arthrotomy, and finally close the skin and bandage the knee.

The row of the above-described interventions is of high importance. The lateral release should be performed first and then the incision to the medial soft tissue (retinaculum-MPFL and VMO) (not the suturing) and then the trochleaplasty and the ACI (if they are also performed), preserving for the end the plication of the medial retinaculum-MPFL and VMO (Fig. 22.7).

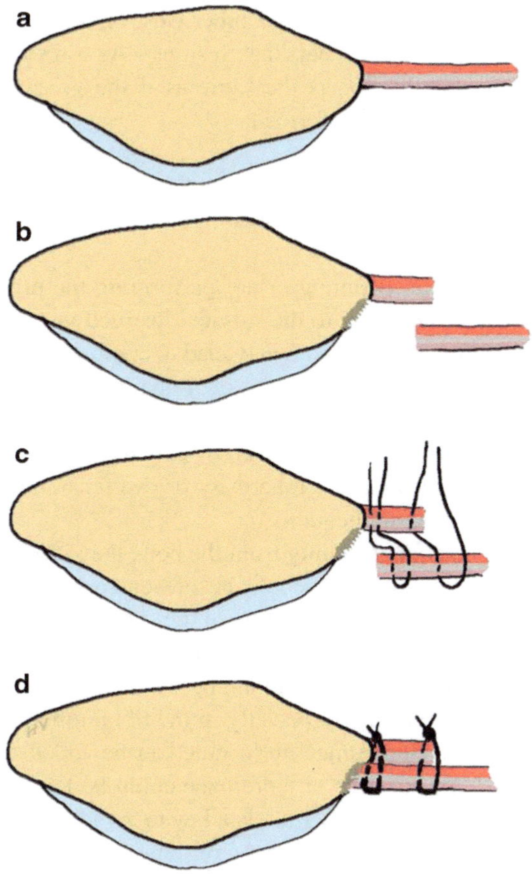

Fig. 22.8 Medial soft tissue plication including VMO (*red color*), MPFL (*gray color*), and synovial capsule (*violet color*). (**a**) Medial structures (VMO, MPFL, synovial capsule). (**b**) Transaction of soft tissue en bloc and roughening the area between the soft tissue insertion and the articular cartilage border for better ingrowth. (**c**) Technique of reinsertion and fixation of the soft tissue plication to the roughened surface of the patella using mattress sutures. (**d**) Tightening and knotting of the sutures

22.5 Rehabilitation

No bone to bone healing is required after the open proximal grooveplasty, in contrary to other trochleaplasty procedures. This offers an increased ability for an accelerated postoperative rehabilitation, as the integrity of the operation is not jeopardized. Therefore, the rehabilitation is mainly dependent on other concomitant surgeries performed along with the trochleaplasty, such as the transfer of tibial tuberosity or the ACI (if performed for cartilage lesions). As a result, the rehabilitation is the same as if the grooveplasty was not performed.

22.6 Complications

Not careful planning or not performing the procedure according to the surgical instructions and rehabilitation protocol may lead to complications like medial subluxation or remaining lateral instability. The medial subluxation may occur if the Q angle is overcorrected by excessive medialization of the tibial tuberosity or overtensioning of the medial plication.

Persisting bleeding from the bone may lead to arthrofibrosis. This could be prevented by adequate coverage and fixation of the synovium with sutures and fibrin glue. Arthrofibrosis may also be the result of concomitant procedures and subsequent bleeding especially in the tibial tuberosity area. Sometimes fibrin glue can be added to the bony surfaces or a drainage could be left for 24 h. Early mobilization is a key to avoid arthrofibrosis, and the progress of range of motion should be followed closely in the early postoperative period.

22.7 Summary

The open proximal trochlear grooveplasty is an effective treatment option for trochlear dysplasia. It aims to the reconstruction of the affected area to a close to normal trochlear concavity, stabilizing the patella in the critical first 30° of knee flexion. In the same time, the technique achieves not to extensively interfere with the patella-trochlea congruity, thus avoiding complications connected to an excessive and eccentric loading.

As for all trochleaplasties, the open grooveplasty should almost never be performed alone. The other background factors contributing to the patellofemoral malalignment should also be addressed. If concomitant background factors need to be addressed and could be managed arthroscopically (like MPFL reconstruction), an arthroscopic proximal trochleaplasty is a good alternative.

References

1. Peterson L, Karlsson J, Brittberg M. Patellar instability with recurrent dislocation due to patellofemoral dysplasia. Results after surgical. Bull Hosp Jt Dis Orthop Inst. 1988;48(2):130–9.
2. Andrikoula S, Tokis A, Vasiliadis HS, Georgoulis A. The extensor mechanism of the knee joint: an anatomical study. Knee Surg Sports Traumatol Arthrosc. 2006;14(3):214–20.
3. Arendt EA, Dejour D. Patella instability: building bridges across the ocean a historic review. Knee Surg Sports Traumatol Arthrosc. 2013;21(2):279–93.
4. Vasiliadis HS, Lindahl A, Georgoulis AD, Peterson L. Malalignment and cartilage lesions in the patellofemoral joint treated with autologous chondrocyte implantation. Knee Surg Sports Traumatol Arthrosc. 2011;19(3):452–7.
5. Bereiter H, Gautier E. The trochleaplasty as a surgical therapy of recurrent dislocation of the patella in dysplastic trochlea of the femur. Arthoscopie. 1994;7: 281–6.
6. Brattstroem H. Shape of the intercondylar groove normally and in recurrent dislocation of the patella. A clinical and X-ray-anatomical investigation. Acta Orthop Scand Suppl. 1964;68:1–148.
7. Dejour H, Neyret P, Walch A. Factors in patellar instability. In: Aichroth PM, Cannon WD Jr, Patel DV, editors. Martin Dunitz (ed); Knee Surg Curr Pr. London, United Kingdom 1992. p. 408–12.
8. Koëter S, Pakvis D, van Loon CJM, van Kampen A. Trochlear osteotomy for patellar instability: satisfactory minimum 2-year results in patients with dysplasia of the trochlea. Knee Surg Sports Traumatol Arthrosc. 2007;15(3):228–32.
9. Vasiliadis HS, Danielson B, Ljungberg M, McKeon B, Lindahl A, Peterson L. Autologous chondrocyte implantation in cartilage lesions of the knee: long-term evaluation with magnetic resonance imaging and delayed gadolinium-enhanced magnetic resonance imaging technique. Am J Sports Med. 2010;38(5): 943–9.

Sulcus-Deepening Trochleoplasty for the Treatment of Recurrent Patellar Dislocation with High-Grade Trochlear Dysplasia

23

David Dejour and Panagiotis G. Ntagiopoulos

Contents

23.1 Introduction

Trochlear dysplasia is a developmental condition where the femoral trochlea loses its normal concave shape to an abnormal flat or even convex geometry [1, 2]. It has been identified as the most consistent anatomic factor present in patients with recurrent patellar dislocation [3, 4]. For its treatment, various surgical procedures have been published and proposed to reshape the abnormal trochlea [5]. Trochleoplasty procedures involve a certain amount of technical difficulty [6], and they were initially reserved as salvage options [2]. Deepening trochleoplasty procedures are a relatively rare option in the surgical treatment of recurrent patellar dislocation [5]. But recent literature contains studies with encouraging results on the treatment of patients with patellar dislocation due to high-grade trochlear dysplasia [7–18].

Four basic trochleoplasty procedures have been proposed; Albee's pioneer procedure involved the elevation of the lateral facet in order to restore normal anatomy [19]. The second procedure is the 'sulcus-deepening trochleoplasty', which was first proposed by Masse [20] and later modified by Henri [21] and David Dejour [22]. The third procedure was introduced by Bereiter and Gautier in 1994 [23], was later followed by von Knoch [13] and is known as the 'Bereiter procedure'. An osteochondral flake with only 2 mm of subchondral bone is elevated from the trochlea without osteotomy of the condyle, and the distal femoral subchondral bone is deepened

D. Dejour (✉) • P.G. Ntagiopoulos
Clinique de la Sauvegarde, Lyon-Ortho-Clinic, Lyon, France
e-mail: corolyon@wanadoo.fr

A. Gobbi et al. (eds.), *The Patellofemoral Joint*,
DOI 10.1007/978-3-642-54965-6_23, © ISAKOS 2014

with osteotomes and a high-speed burr. The same procedure has been performed arthroscopically by Blønd and Schöttle [15]. The fourth procedure is the 'recession wedge trochleoplasty' which was introduced by Goutallier [24] in 2002, and its results were later published by Beaufils [6]. In this technique, the trochlea's shape is not changed, but the removal of a proximally based wedge results in a more posteriorly placed trochlea.

On one hand, trochlear dysplasia has been identified as the most recognized factor in patients with patellofemoral dislocation [3]. On the other hand, there are only a small number of surgeons who perform trochleoplasty for the treatment of patellar dislocation [6], in comparison to tibial tuberosity osteotomy for realignment procedures or patella alta, or more recently MPFL reconstruction. Subsequently, there is a group of patients with patellofemoral instability, in whom the key aetiological factor of instability, i.e., trochlear dysplasia, is underestimated or not addressed. Furthermore, the evaluation of the results from the application of trochleoplasty presents with certain difficulties, because there is no agreement on the appropriate clinical (patellofemoral pain [12, 24], patellar dislocation [7, 8, 13] or both) and radiographic (type of dysplasia, height of prominence, sulcus depth [8, 9, 13]) indications for trochleoplasty. For sulcus-deepening trochleoplasty [22], authors agree that is indicated to severe cases of Dejour types B and D of dysplasia (high-grade dysplasia), where the presence of trochlear prominence is amenable to correction with this technique [8, 9, 22].

There is controversy on the indications of trochleoplasty in the primary treatment of recurrent patellar dislocation, but some authors state that it remains a primary surgical option for a subgroup of patients with high-grade trochlear dysplasia and a revision option for specific patients with unsuccessful previous surgeries on patellofemoral instability where the trochlear dysplasia was the main factor for the dislocation and was underestimated [2, 16].

23.1.1 Surgical Indications

Inclusion criteria for sulcus-deepening trochleoplasty are:

- *Patients with more than three documented episodes of patellar dislocation with a high-grade trochlear dysplasia type B or D according to Dejour classification* [1].

Patients with open growth plates, patellofemoral arthritis or patellofemoral pain syndrome with no true episodes of patellar dislocation are not suitable candidates for trochleoplasty. Sulcus-deepening trochleoplasty is always combined with additional bone and/or soft-tissue surgery as described below.

23.1.2 Preoperative Evaluation

Preoperative patient evaluation includes clinical examination (apprehension test, lateral patellar glide test and patellar tracking) and subjective evaluation (patellofemoral pain and/or feeling of instability). Radiographic examination includes the study of the type of trochlear dysplasia in true lateral x-rays, measurement of sulcus angle in patellar axial views in 30° of knee flexion and measurement of patellar height according to Caton–Deschamps index [25] in lateral views in 20° of knee flexion. Tibial tuberosity–trochlear groove (TT-TG) distance and lateral patellar tilt (without quadriceps contraction) are measured in axial computed tomography (CT) slices.

Patients with recurrent patellar dislocation and trochlear dysplasia are treated according to the presence of concomitant predisposing anatomic abnormalities, and trochleoplasty is combined with other procedures when needed [1, 3, 5, 26–28]:

- *If the TT-TG distance is excessively increased (e.g. over 25 mm), a tibial tuberosity medialization osteotomy is performed in order to obtain postoperatively a TT-TG between 10 and 15 mm* [1, 3]. *In cases where the TT-TG distance is between 20 and 25 mm, a 'proximal realignment procedure' was performed by lateralizing the new trochlear groove during trochleoplasty in a more lateral procedure, thus reducing the slightly increased TT-TG.*
- *If patella alta is recorded (Caton–Deschamps index >1.3), a distalization osteotomy is done to obtain a normal patellar index of 1.0* [1, 3]. *In cases where a slightly elevated Caton–Deschamps index is recorded (e.g. 1.2–1.3), trochleoplasty alone was performed, even though this resulted in a slight patella alta.*

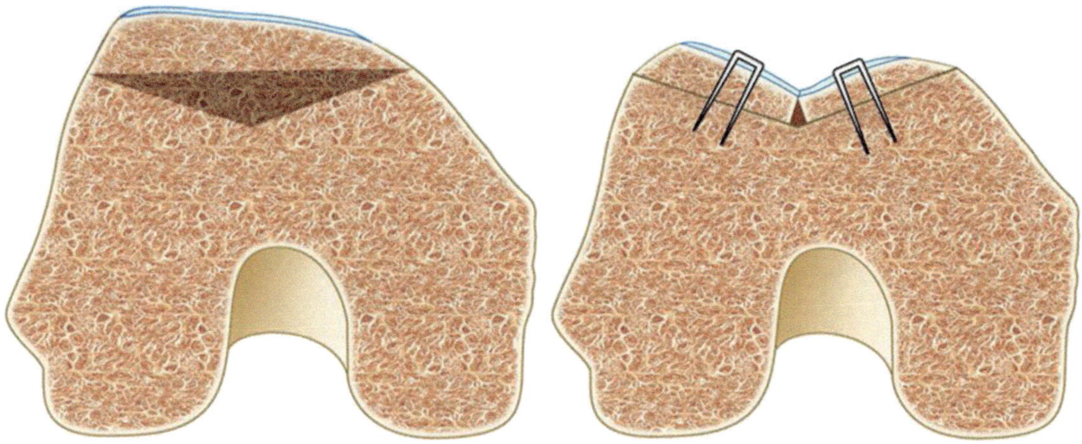

Fig. 23.1 The principle of sulcus-deepening trochleoplasty is to remove subchondral cancellous bone under the trochlea with a high-speed burr. The amount of bone removal is determined to have a new trochlear groove flushed with the anterior femoral cortex and to make the prominence disappear. The goal is to decrease the trochlear bump and deepen the sulcus angle

- *If the lateral patellar tilt is over 20°, the correction is achieved with a reconstruction of the medial patellofemoral ligament (MPFL) using a double-looped gracilis graft* [1, 3, 29, 30] *and lateral retinaculum release/lengthening.*

23.1.3 Surgical Technique

Trochleoplasty is performed according to Dejour's 'sulcus-deepening trochleoplasty' in all patients [22]. The principle is to remove subchondral cancellous bone under the trochlea with a high-speed burr. The amount of bone removal is determined to have a new trochlear groove flushed with the anterior femoral cortex and to make the prominence disappear (Fig. 23.1). The goal is to decrease the trochlear bump and deepen the sulcus angle.

Trochleoplasty can be performed under general or regional anaesthesia. Patient is positioned supine, and the lower limb is prepared and draped in a standard fashion. A tourniquet is applied and the knee is flexed at 90°. A straight midline skin incision is carried out from the superior patellar margin to the tibiofemoral articulation. The arthrotomy is performed through an adapted mid-vastus approach: the medial retinaculum is sharply dissected 2 cm of the medial border of the patella, and the vastus medialis oblique (VMO) is bluntly dissected within its fibres starting proximally 3–4 cm into the muscle belly and ending at the superomedial pole of the patella. The patella is then briefly everted only to inspect and document chondral injuries, and then it is retracted laterally.

The trochlea is now exposed, and the peritrochlear synovium and periosteum are incised along their osteochondral junction and reflected from the field using a periosteal elevator. It is very important to visualize the level of the anterior femoral cortex in order to evaluate the exact height of the prominence and determine the amount of required deepening, by removing with an osteotome the prominent bone superiorly to the trochlear and up to the anterior femoral cortex (e.g. supratrochlear spur) (Fig. 23.2).

Once the trochlea is fully exposed, the native trochlear groove and medial and lateral facets are marked with three lines starting point from the intercondylar notch. The new trochlear groove is marked in a more lateral position according to the preoperative TT-TG value (e.g. 'proximal realignment' in order to get a postoperative value of 10–15 mm) (Fig. 23.3).

With the use of a burr with a 5 mm offset, undermining the cancellous bone posteriorly of the trochlear cartilage is performed (Fig. 23.4).

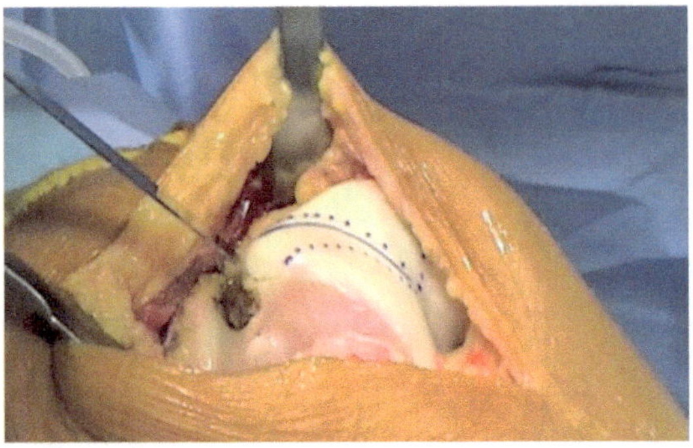

Fig. 23.2 The trochlea is now exposed and the peritrochlear synovium and periosteum are incised along their osteochondral junction and reflected from the field using a periosteal elevator. It is very important to visualize the level of the anterior femoral cortex in order to evaluate the exact height of the prominence and determine the amount of required deepening, by removing with an osteotome the prominent bone superiorly to the trochlear and up to the anterior femoral cortex (e.g. supratrochlear spur)

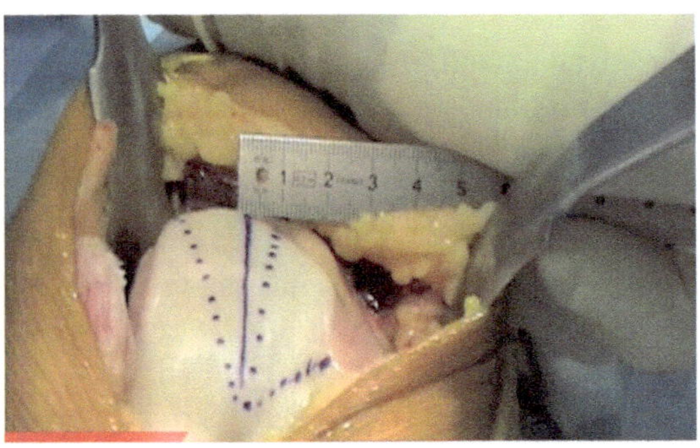

Fig. 23.3 The native trochlear groove, medial and lateral facets are marked with three lines starting from the intercondylar notch. The new trochlear groove is marked in a more lateral position according to the pre-operative TT-TG value (e.g. "proximal re-alignment" in order to get a post-operative value of 10–15 mm)

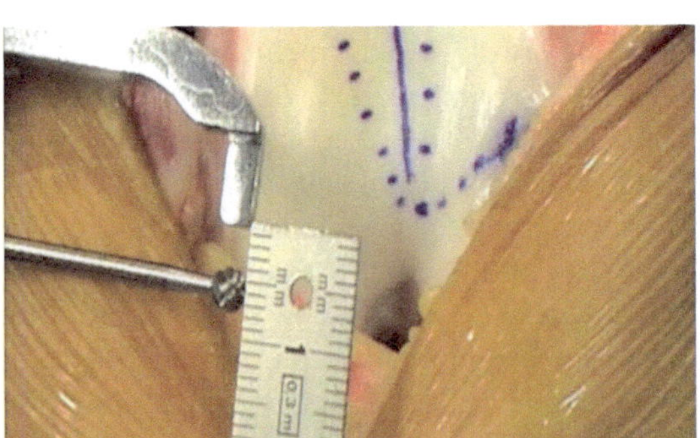

Fig. 23.4 With the use of a burr with a 5 mm offset, undermining the cancellous bone posteriorly of the trochlear cartilage is performed

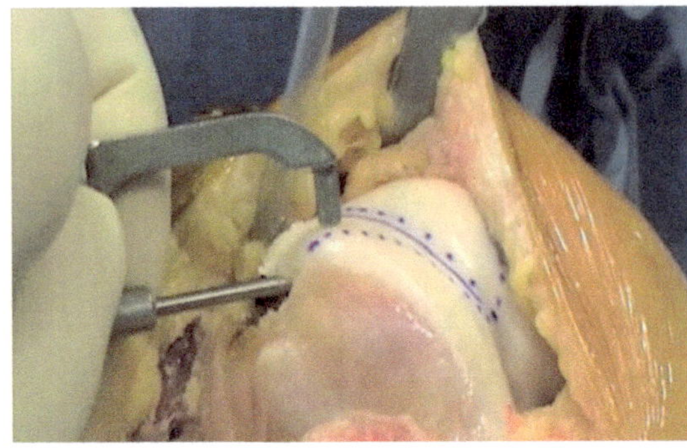

Fig. 23.5 The burr must reach down to the intercondylar notch, but with care not to damage the cartilage

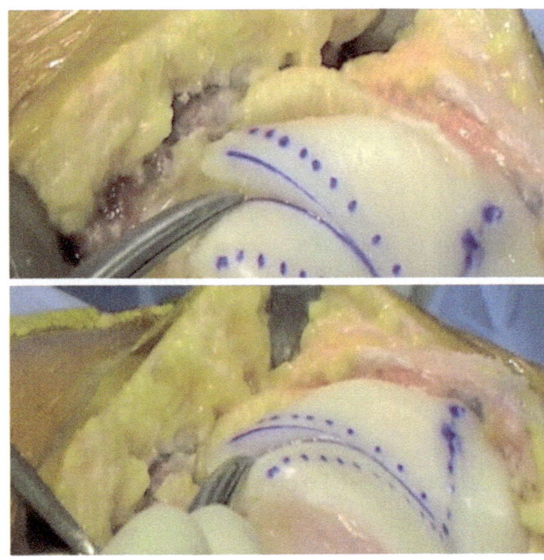

Fig. 23.6 When the trochlea is pressed down to the level where it is flush with the anterior femoral cortex (e.g. removal of the prominence) and moves backwards like a "trampoline", after we remove the pressure, sufficient bone has been removed

The burr must reach down to the intercondylar notch, but with care not to damage the cartilage (Fig. 23.5).

When sufficient cancellous bone has been removed, the trochlea must be elastic when applying light pressure with the fingers. When the trochlea is pressed down to the level where it is flushed with the anterior femoral cortex (e.g. removal of the prominence) and moves backwards like a 'trampoline', when we remove the pressure, sufficient bone has been removed (Fig. 23.6).

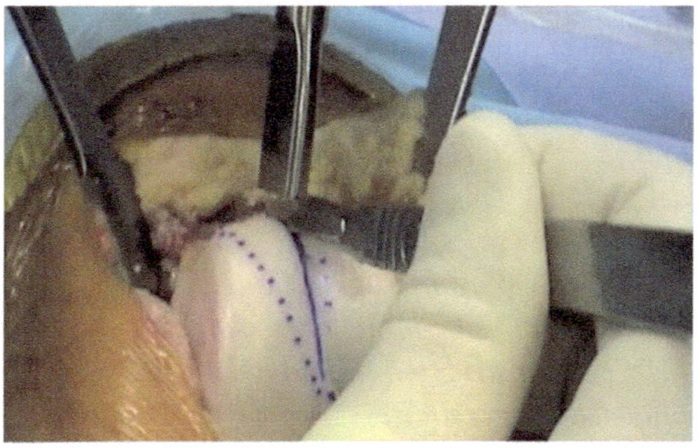

Fig. 23.7 After sufficient cancellous bone has been removed, the trochlea must be elastic when applying light pressure with the fingers. The trochlea is then osteotomized with a scalpel carefully over the position of the desired new groove

The trochlea is then osteotomized with a scalpel carefully over the position of the desired new groove (Fig. 23.7).

At this point, if the trochlear facets are too rigid and it is difficult to suppress them, a further osteotomy over the medial and/or the lateral facet may be required. The trochlear facets are fixed with two absorbable anchor sutures from the intercondylar notch while applying pressure posteriorly and evaluating that the prominence is removed and they are flushed with the anterior femoral cortex (Fig. 23.8).

Lateral retinaculum release is performed after the trochleoplasty, and it is mostly dependent on the ability of the examiner's to evert the patella inwards (e.g. no lateral release is required) or not (e.g. lateral structures are tightened and lateral release is required). The joint capsule is then closed. MPFL reconstruction is performed last, with the surgeon's preferred technique. No drain is used and the knee is put in an extension brace with no restriction in weight bearing. Prophylactic anticoagulants are advised for 2–3 weeks.

23.1.4 Rehabilitation Protocol

A common postoperative rehabilitation protocol includes full weight bearing with the use of crutches in an extension knee brace for 15 days and continuous passive motion between 0° and 100°. In the first 6 weeks, the patients are encouraged to perform exercises for early range of motion and isometric quadriceps and hamstring strengthening. From the 6th to 12th postoperative week, the protocol also includes closed-chain and weight-bearing proprioception exercises. After the 12th postoperative week, patients start running and gradually training in their sport of preference and are allowed to full sports activities after 6 months. During this period (3–6 months postoperatively), running can be initiated on a straight line. Closed kinetic chain muscular reinforcement between 0 and 60o with minor loads but long series is also allowed. Stretching of the anterior and posterior muscular chains is continued as during previous phases. The patient is encouraged to proceed with the rehabilitation on his or her own. After 6 months, sports on a recreational or competitive level can be resumed.

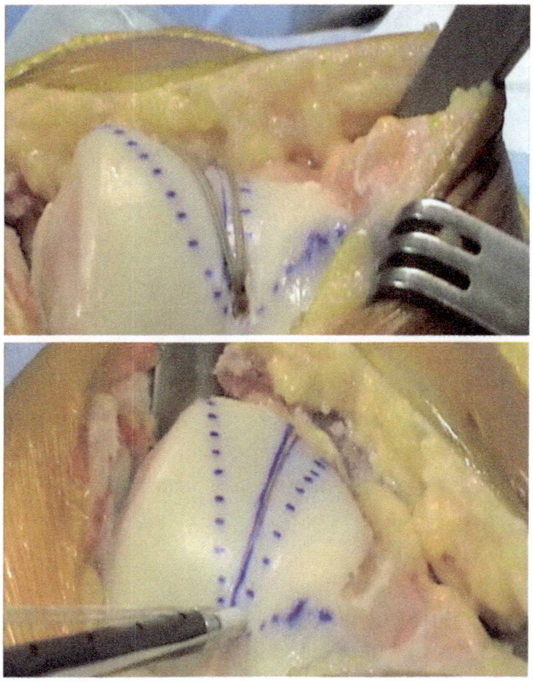

Fig. 23.8 The trochlear facets are fixed with two absorbable anchor sutures from the intercondylar notch while applying pressure posteriorly and evaluating that the prominence is removed and they are flush with the anterior femoral cortex

23.2 Discussion

The rationale of sulcus-deepening trochleoplasty has three key elements: First, in cases of a flat or convex trochlea, it tries to bring back to a more anatomic and concave geometry by deepening the proximal sulcus, so that it engages the patella in early degrees of knee flexion [22]. *Second, in cases of a convex trochlea, trochleoplasty removes the sulcus prominence that the patella needs to override during flexion and that leads to patella dislocation over the lateral facet. Last, trochleoplasty creates a new trochlear groove in a more lateral position than the dysplastic one, thus decreasing the excessive TT-TG distance and serving as a 'proximal realignment procedure'* [22].

The biomechanical effects of the sulcus-deepening trochleoplasty have been studied by Amis et al. who reported that the mediolateral flattening of the anterior surface of the trochlear facets results predominantly from an excess of bone centrally in the groove. This can also form a supratrochlear prominence anterior to the femoral shaft; during knee flexion, the patella has to override this prominence. The authors showed that 'simulated' trochlear dysplasia led to significant reduction in lateral stability, and by re-creating a deep trochlear groove with a trochleoplasty procedure, lateral stability increased significantly similarly to the intact knee [31].

The goal of sulcus-deepening trochleoplasty is to reshape the abnormal trochlear shape, but the instability of the patella is also caused by the presence of coexistent anatomic factors that must be addressed (e.g. tuberosity osteotomy for patella alta or increased TT-TG distance), and its treatment almost always requires a combined soft-tissue procedure like MPFL reconstruction [1, 22, 32]. This is because patellar stability during early flexion is accomplished by MPFL which is tight in full knee extension and acts as a dynamic stabilizer during early flexion (15–20°), brings the patella into the trochlear grove which is necessary for the normal further tracking of the patella and in greater degrees of flexion (>30°) is loose and the normal concave trochlear geometry acts as a static stabilizer [3, 5, 33, 34]. *MPFL rupture is the consequence of pathologic lateral patellar translation, while the actual causes are the underlying bony abnormalities like trochlear dysplasia and patella alta* [3, 32], *and most importantly, a normal trochlea with a deep groove and an elevated lateral facet is necessary for MPFL to provide stability* [32, 35]. *This is the reason why, although MPFL has a fundamental role in patellar stability, isolated MPFL reconstruction should not be considered a panacea for patellar dislocation, without excluding the presence of other factors contributing to instability* [32] *that need associated correction* [1, 5–9, 13, 14, 22, 24].

This is helpful in eliminating the need for an additional procedure (medial transfer of the tuberosity) because the groove position could be lateralized while doing the trochleoplasty, similar to a 'proximal realignment procedure'. The effect of trochleoplasty alone in reducing the TT-TG distance without additional tibial tuberosity surgery was on average 5 mm.

In the case of a convex trochlea, there is a centrally located sulcus prominence that the patella needs to override during flexion and that leads to patella dislocation off the lateral facet [31, 36]. This prominence is present in type B and D trochlear dysplasia, where the central groove is elevated. *In this scenario of underlying high-grade trochlear dysplasia, if patellar dislocation is approached with no regard to dysplasia and the elevated groove and is addressed by medial soft-tissue augmentation alone, the patella will obtain firm medially stabilizing structures and a medial tilt, but the elevated and abnormal trochlear groove will still be present* (Fig. 23.9). *This could lead to undesirable impingement between the patella and the dysplastic trochlea prominence and subsequent increased contact pressures and graft loosening, effects similar to overmedializing the patella after MPFL reconstruction* [26, 37] *or to the pattern of MPFL rupture recorded in patients with trochlear dysplasia* [38].

In a different scenario, the isolated correction of patella alta and the underestimation of a concomitant high-grade trochlear dysplasia would have similar inefficient results to treat patellar dislocation; a distalization osteotomy of the tuberosity would normalize patellar height but would only lower the patella in front of an

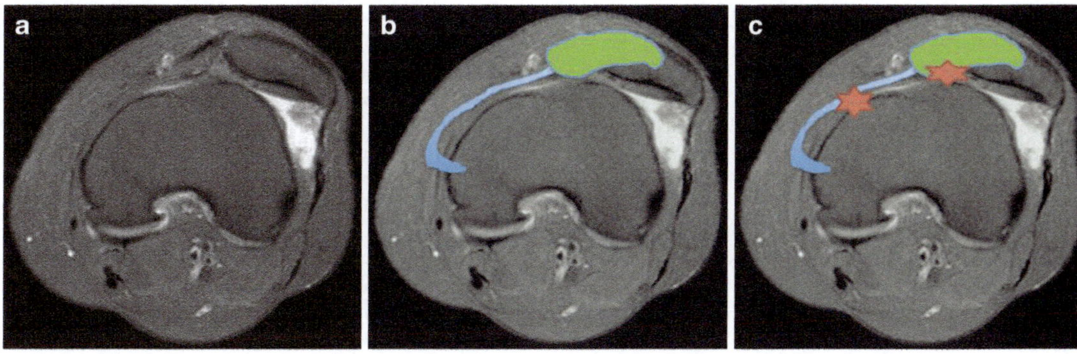

Fig. 23.9 If patellar dislocation is approached with no regard to dysplasia and the elevated groove (**a**), and is addressed by medial soft-tissues augmentation alone (**b**), the patella will be positioned medially in front of an untreated elevated groove that would lead to impingement between the patella and the trochlea (**c**)

abnormally shaped trochlear groove, if trochlear dysplasia were not corrected. The absence of a deep trochlear groove would not provide to the patella the required lateral stability from further lateral dislocation.

In the clinical setting, other authors report success rates from 90 to 100 % after trochleoplasty procedures [7, 9, 12–14, 24]. After trochleoplasty, recurrence of dislocation is highly uncommon, and patient satisfaction is high [5]. The results of trochleoplasty as a revision option in selected patients who had previous failed surgery for patellofemoral instability have been recently recorded [16].

In conclusion, trochleoplasty is an important surgical option both as primary and as a revision option in the case of previously operated patients with persistent patellar dislocation and undiagnosed or underestimated trochlear dysplasia. The combination of the procedure with soft-tissue surgery, such as MPFL reconstruction, is necessary to achieve patellar stability. There is a need for agreement on the choice of the right candidate for trochleoplasty, the surgical technique, the rehabilitation protocol and the postoperative parameters that must be corrected. Sulcus-deepening trochleoplasty requires careful attention to detail. *It is a technically demanding procedure that addresses a rare condition with satisfactory results and acceptable level of complications and is more suitable for severe cases of types B and D of dysplasia, where the presence of trochlear prominence is amenable to correction with this technique.*

References

1. Dejour D, Le Coultre B. Osteotomies in patello-femoral instabilities. Sports Med Arthrosc. 2007;15(1): 39–46. doi:10.1097/JSA.0b013e31803035ae, 00132585-200703000-00006 [pii].
2. Bollier M, Fulkerson JP. The role of trochlear dysplasia in patellofemoral instability. J Am Acad Orthop Surg. 2011;19(1):8–16. 19/1/8 [pii].
3. Dejour H, Walch G, Nove-Josserand L, Guier C. Factors of patellar instability: an anatomic radiographic study. Knee Surg Sports Traumatol Arthrosc. 1994;2(1):19–26.
4. Balcarek P, Jung K, Ammon J, Walde TA, Frosch S, Schuttrumpf JP, Sturmer KM, Frosch KH. Anatomy of lateral patellar instability: trochlear dysplasia and tibial tubercle-trochlear groove distance is more pronounced in women who dislocate the patella. Am J Sports Med. 2010;38(11):2320–7. doi:10.1177/0363546510373887, 0363546510373887 [pii].
5. Mulford JS, Wakeley CJ, Eldridge JD. Assessment and management of chronic patellofemoral instability. J Bone Joint Surg Br. 2007;89(6):709–16. doi:10.1302/0301-620X.89B6.19064. 89-B/6/709 [pii].
6. Beaufils P, Thaunat M, Pujol N, Scheffler S, Rossi R, Carmont M. Trochleoplasty in major trochlear dysplasia: current concepts. Sports Med Arthrosc Rehabil Ther Technol. 2012;4:7. doi:10.1186/1758-2555-4-7, 1758-2555-4-7 [pii].
7. Utting MR, Mulford JS, Eldridge JD. A prospective evaluation of trochleoplasty for the treatment of patellofemoral dislocation and instability. J Bone Joint Surg Br. 2008;90(2):180–5. doi:10.1302/0301-620X.90B2.20017, 90-B/2/180 [pii].
8. Fucentese SF, Zingg PO, Schmitt J, Pfirrmann CW, Meyer DC, Koch PP. Classification of trochlear dysplasia as predictor of clinical outcome after trochleoplasty. Knee Surg Sports Traumatol Arthrosc. 2011; 19(10):1655–61. doi:10.1007/s00167-011-1410-7.
9. Donell ST, Joseph G, Hing CB, Marshall TJ. Modified Dejour trochleoplasty for severe dysplasia: operative

technique and early clinical results. Knee. 2006;13(4):266–73. doi:10.1016/j.knee.2006.01.004, S0968-0160(06)00006-8 [pii].
10. Fucentese SF, Schöttle PB, Pfirrmann CW, Romero J. CT changes after trochleoplasty for symptomatic trochlear dysplasia. Knee Surg Sports Traumatol Arthrosc. 2007;15(2):168–74. doi:10.1007/s00167-006-0140-8.
11. Schöttle PB, Schell H, Duda G, Weiler A. Cartilage viability after trochleoplasty. Knee Surg Sports Traumatol Arthrosc. 2007;15(2):161–7. doi:10.1007/s00167-006-0148-0.
12. Verdonk R, Jansegers E, Stuyts B. Trochleoplasty in dysplastic knee trochlea. Knee Surg Sports Traumatol Arthrosc. 2005;13(7):529–33. doi:10.1007/s00167-004-0570-0.
13. von Knoch F, Bohm T, Burgi ML, von Knoch M, Bereiter H. Trochleaplasty for recurrent patellar dislocation in association with trochlear dysplasia. A 4- to 14-year follow-up study. J Bone Joint Surg Br. 2006; 88(10):1331–5. doi:10.1302/0301-620X.88B10.17834, 88-B/10/1331 [pii].
14. Thaunat M, Bessiere C, Pujol N, Boisrenoult P, Beaufils P. Recession wedge trochleoplasty as an additional procedure in the surgical treatment of patellar instability with major trochlear dysplasia: early results. Orthop Traumatol Surg Res. 2011;97(8):833–45. doi:10.1016/j.otsr.2011.07.013, S1877-0568(11)00235-0 [pii].
15. Blønd L, Schöttle PB. The arthroscopic deepening trochleoplasty. Knee Surg Sports Traumatol Arthrosc. 2010;18(4):480–5. doi:10.1007/s00167-009-0935-5.
16. Dejour D, Byn P, Ntagiopoulos PG. The Lyon's sulcus-deepening trochleoplasty in previous unsuccessful patellofemoral surgery. Int Orthop. 2013;37(3):433–9. doi:10.1007/s00264-012-1746-8.
17. Nelitz M, Dreyhaupt J, Lippacher S. Combined trochleoplasty and medial patellofemoral ligament reconstruction for recurrent patellar dislocations in severe trochlear dysplasia: a minimum 2-year follow-up study. AmJSportsMed.2013.doi:10.1177/0363546513478579, 0363546513478579 [pii].
18. Blond L, Haugegaard M. Combined arthroscopic deepening trochleoplasty and reconstruction of the medial patellofemoral ligament for patients with recurrent patella dislocation and trochlear dysplasia. Knee Surg Sports Traumatol Arthrosc. 2013. doi:10.1007/s00167-013-2422-2.
19. Albee FH. Bone graft wedge in the treatment of habitual dislocation of the patella. Med Rec. 1915;88:257.
20. Masse Y. Trochleoplasty. Restoration of the intercondylar groove in subluxations and dislocations of the patella. Rev Chir Orthop Reparatrice Appar Mot. 1978;64(1):3–17.
21. Dejour H, Walch G, Neyret P, Adeleine P. Dysplasia of the femoral trochlea. Rev Chir Orthop Reparatrice Appar Mot. 1990;76(1):45–54.
22. Dejour D, Saggin P. The sulcus deepening trochleoplasty-the Lyon's procedure. Int Orthop. 2010;34(2):311–6. doi:10.1007/s00264-009-0933-8.
23. Bereiter H, Gautier E. Die trochleaplastik als chirurgische Therapie der rezidivierenden Patellaluxation bei Trochleadysplasie des Femurs. Arthroskopie. 1994;7:281–86.
24. Goutallier D, Raou D, Van Driessche S. Retro-trochlear wedge reduction trochleoplasty for the treatment of painful patella syndrome with protruding trochleae. Technical note and early results. Rev Chir Orthop Reparatrice Appar Mot. 2002;88(7): 678–85.
25. Caton J, Deschamps G, Chambat P, Lerat JL, Dejour H. Patella infera. Apropos of 128 cases. Rev Chir Orthop Reparatrice Appar Mot. 1982;68(5): 317–25.
26. Colvin AC, West RV. Patellar instability. J Bone Joint Surg Am. 2008;90(12):2751–62. doi:10.2106/JBJS.H.00211, 90/12/2751 [pii].
27. Dejour D, Saggin P. Disorders of the patellofemoral joint. In: Scott NW, editor. Insall & scott surgery of the knee, Insall & scott surgery of the knee, vol. 1. 5th ed. Philadelphia: Elsevier; 2012. p. 592–623.
28. Schoettle PB, Zanetti M, Seifert B, Pfirrmann CW, Fucentese SF, Romero J. The tibial tuberosity-trochlear groove distance; a comparative study between CT and MRI scanning. Knee. 2006;13(1):26–31. doi:10.1016/j.knee.2005.06.003, S0968-0160(05)00076-1 [pii].
29. Arendt EA. MPFL reconstruction for PF instability. The soft (tissue) approach. Orthop Traumatol Surg Res. 2009;95(8 Suppl 1):S97–100. doi:10.1016/j.otsr.2009.09.002, S1877-0568(09)00135-2 [pii].
30. Saggin PR, Saggin JI, Dejour D. Imaging in patellofemoral instability: an abnormality-based approach. Sports Med Arthrosc. 2012;20(3):145–51. doi:10.1097/JSA.0b013e3182553cfe, 00132585-201209000-00004 [pii].
31. Amis AA, Oguz C, Bull AM, Senavongse W, Dejour D. The effect of trochleoplasty on patellar stability and kinematics: a biomechanical study in vitro. J Bone Joint Surg Br. 2008;90(7):864–9. doi:10.1302/0301-620X.90B7.20447, 90-B/7/864 [pii].
32. Arendt EA, Fithian DC, Cohen E. Current concepts of lateral patella dislocation. Clin Sports Med. 2002;21(3):499–519.
33. Arendt E. Anatomy and malalignment of the patellofemoral joint: its relation to patellofemoral arthrosis. Clin Orthop Relat Res. 2005;436:71–5.
34. Senavongse W, Amis AA. The effects of articular, retinacular, or muscular deficiencies on patellofemoral joint stability: a biomechanical study in vitro. J Bone Joint Surg Br. 2005;87(4):577–82. doi:10.1302/0301-620X.87B4.14768, 87-B/4/577 [pii].
35. Bicos J, Fulkerson JP, Amis A. Current concepts review: the medial patellofemoral ligament. Am J Sports Med. 2007;35(3):484–92.
36. Biedert RM, Bachmann M. Anterior-posterior trochlear measurements of normal and dysplastic trochlea by axial magnetic resonance imaging. Knee Surg

Sports Traumatol Arthrosc. 2009;17(10):1225–30. doi:10.1007/s00167-009-0824-y.

37. Panagopoulos A, van Niekerk L, Triantafillopoulos IK. MPFL reconstruction for recurrent patella dislocation: a new surgical technique and review of the literature. Int J Sports Med. 2008;29(5):359–65. doi: 10.1055/s-2007-965360.

38. Weber-Spickschen TS, Spang J, Kohn L, Imhoff AB, Schottle PB. The relationship between trochlear dysplasia and medial patellofemoral ligament rupture location after patellar dislocation: an MRI evaluation. Knee. 2011;18(3):185–8. doi: 10.1016/j.knee.2010.04.002, S0968-0160(10)00075-X [pii].

Role of Rotational Osteotomy in the Treatment of Patellofemoral Dysfunction

24

Ricardo Bastos Filho, Munehiro Naruo, Elvire Servien, Sebastien Lustig, and Philippe Neyret

Contents

R. Bastos Filho (✉)
Department of Surgery, Antônio Pedro Hospital, Fluminense Federal University, Niterói, Rio de Janeiro, Brazil
e-mail: rpsbastos@hotmail.com

M. Naruo
Department of Orthopaedics Surgery, Tomei Atsugi Hospital, 232 Hunako, Atsugishi, Kanagawa, Japan

E. Servien • S. Lustig • P. Neyret
Department of Orthopaedics Surgery, Centre Albert Trillat, Croix Rousse Hospital, 69004, Lyon, France

24.1 Biomechanic Considerations

The most important function of the patella is to improve quadriceps efficiency by increasing the lever arm of the extensor mechanism. It centralizes the divergent forces from the four heads of the quadriceps and transmits tension around the femur to the patellar tendon and the tibial tuberosity. These forces must be as frictionless as possible and equally distributed in the patellofemoral joint. The skeleton dictates where the load will cross the patellofemoral joint, and any variation from optimal skeleton alignment may result in abnormal forces.

Patellofemoral dysfunctions include a large number of pathologies such as instability, anterior knee pain, chondromalacia, and osteoarthrosis. Some of these problems are caused by variations from the normal anatomy involving ligaments, patellar and trochlear articular surfaces, subchondral bone, tendons articular geometry, and finally lower limb alignment. These variations may alter the ability of the joint to accept load and, consequently, cause pain, instability, locking, and swelling.

If the knee is viewed in the coronal plane, the muscle loads tend to displace the patella laterally when the knee is near extension, because of the Q (quadriceps) angle, which is essentially the axis of pull of the quadriceps muscle group relative to the axis of the patellar tendon. Hvid and Andersen [1] found a significant correlation between the Q angle and hip rotation, thus suggesting that the Q angle is in fact

A. Gobbi et al. (eds.), *The Patellofemoral Joint*,
DOI 10.1007/978-3-642-54965-6_24, © ISAKOS 2014

increased because of excessive femoral neck anteversion.

Concerning this observation, torsional abnormalities of the femur and tibia are possible factors leading to patellar pain or instability. The pattern would involve increased femoral neck anteversion, so that the trochlear groove faces inward, the patellae are squinting, patellar tracking is compromised, and the patella tends to track more laterally than normal. Compensatory external tibial torsion is required to produce a foot aligned in the sagittal plane with consequent knee external rotation and lateralization of the tibial tuberosity, and this way, the Q angle is maximized. This combination of increased femoral anteversion and increased external tibial torsion has been termed *miserable malalignment syndrome* [2], a spectrum that includes squinting patellae, genu varum, genu recurvatum, patella alta, and pronated feet.

Brattstom [3] defined the "Q angle" as reflecting the direction of the quadriceps vector and explained how this quadriceps vector is increased by an inward rotation of the knee joint and would be altered by rotational femoral osteotomy.

Summarizing, there may be rotational dysfunction at various levels in the lower limb, resulting in increased forces to which the body cannot adapt and which can manifest clinically at the knee joint as patellar maltracking, anterior knee pain, and patellofemoral osteoarthritis, and all of them may share a common underlying malrotation in one or more segments of the hip, femur, tibia, or knee.

24.2 Anatomic Considerations

Femoral torsion is a twist of the proximal femur relative to the distal femur.

Kingstey and Olmstead [4], on the basis of 630 dry specimens, reported that the mean femoral anteversion angle was 8.0° (range,−20° to 38°), with females having a minor increase in mean anteversion angle compared with males. In infants and children, the mean femoral anteversion was higher but, with growth, this angle decreases. This observation can explain the fact that most femoral torsional issues in childhood resolved or are accommodated with age, leaving only a few with functional or cosmetic problems [5].

Dejour et al. [6] correlated femoral anteversion and tibial lateral torsion with the incidence of patellar instability. They found that femoral neck anteversion was increased in patients with patellofemoral instability (15.6°) as compared with controls (10.8°). Tibial torsion was a less important factor (33° in the control group and 35° in the knees with patellar instability).

Less is known about tibial version. It develops in childhood and it has been suggested that tibial version is largely a reflection of femoral anteversion. There is a weak correlation of tibial torsion with anterior knee pain and patellofemoral osteoarthritis. Tibial torsion is defined as the anatomic twist of the proximal versus distal articular axis of the tibial bone around the longitudinal axis [7]. The external tibial torsion is always greater than the angle of femoral anteversion. Lerat et al. [8] described the relationship between the angles for external tibial torsion and femoral anteversion, noting that the larger the difference, the more the correlation with patellofemoral pathology. When the knee is in extension, the tibia is externally rotated in relation to the femur. Hyperextension of 15°, as in a ligamentously lax individual, is associated with a 13° increase in femorotibial rotation and increases varus. Turner and Smillie [9] found that an average of 19° of lateral tibial torsion is considered normal. They noted an average of 24.5° in knees with patellofemoral instability and 24° in those with chondromalacia.

24.3 Physical Exam

Clinical measurements based on physical examination are precise and reliable [10] but do not quantify the true rotational profile of the hip–knee–ankle axis.

Examination should be sequentially performed with the patient while standing, walking, sitting supine, and prone.

With the patient standing with the feet together, rotatory malalignment can be observed. The patient shows an infacing or "squinting" of

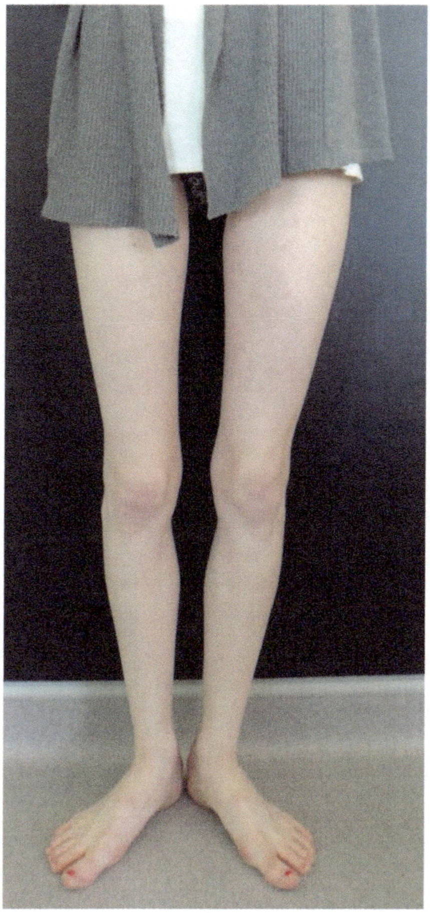

Fig. 24.1 Patient in standing position with the patellae pointed forward. Observe the external rotation of the feet

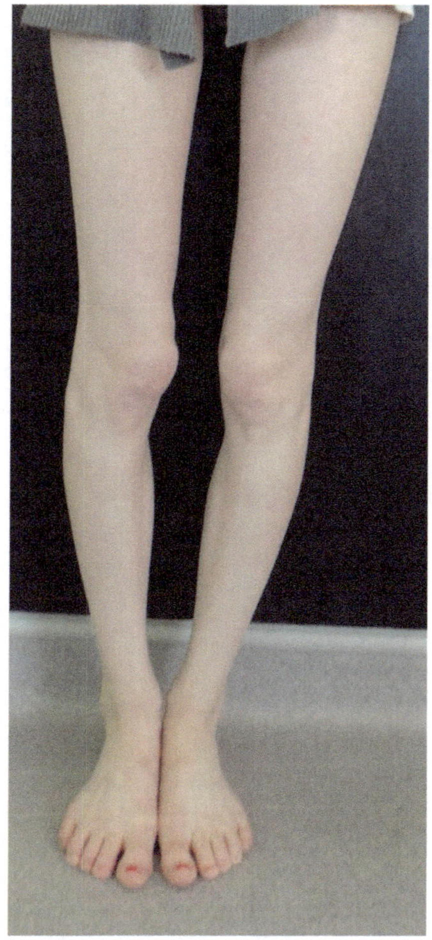

Fig. 24.2 Patient in standing position with the feet together, rotatory malalignment can be observed. The patient shows an infacing or "squinting" of the patellae, an apparent varus with a hyperpronation of the foot. This deformity is noted on the right leg. Asymmetric rotational deformity

the patellae, an apparent varus with a hyperpronation of the foot (Figs. 24.1 and 24.2). It is very important to observe if the torsional malalignment has a characteristic asymmetric (unilateral). Unilateral deformity leads us to consider a surgical treatment.

The Q angle is assessed by drawing a line from the anterior superior iliac spine to the center of the patella and represents the line of pull of the quadriceps muscle. A second line is drawing from the center of the patella to the tibial tubercle and indicates the line of the patellar tendon (Fig. 24.3).

According to Aglietti, the Q angle should be considered normal when around 14° in men and 17° in women [11]. In fact, we should consider an abnormal value when Q angle is greater than 20°.

In sitting position, the amount of tibial external rotation is estimated by observing and comparing the external rotation of the feet (Fig. 24.4).

In prone position, the amount of femoral neck anteversion is indirectly estimated by measuring the proportion of internal to external rotation of the hips in extension. If internal rotation of the hip in extension exceeds external rotation by more than 30°, femoral anteversion is increased [13] (Fig. 24.5).

Staheli [12] established age- and sex-related values for rotational profile. The following values were measured in 1,000 normal limbs (Tables 24.1 and 24.2).

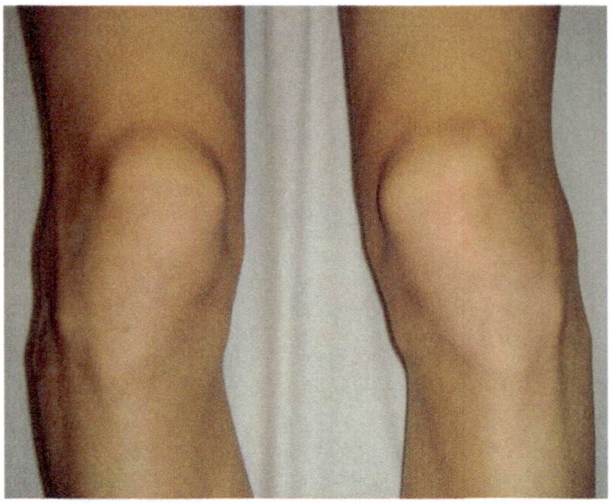

Fig. 24.3 Physical exam with the patient in supine position. Observe the Q angle increased

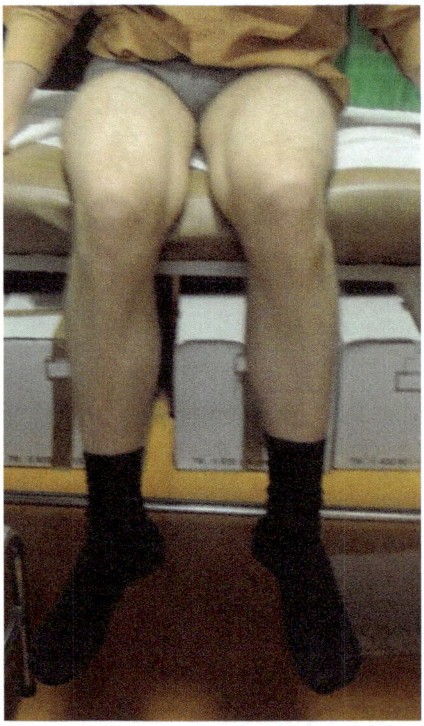

Fig. 24.4 Patient in sitting position. The amount of tibial external rotation is estimated by observing and comparing the external rotation of the feet (Fig. 24.3)

We believe that the lack of clinical method accuracy in determining the true rotational profile and the Q angle of the lower limb is due to error in positioning, variability of surface landmarks, anatomic variations among individuals, and the subjective nature of the technique. For this reason, we consider

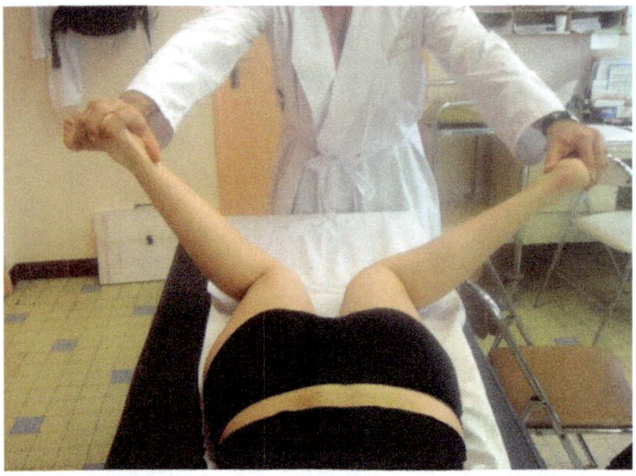

Fig. 24.5 Patient in prone position. The amount of femoral neck anteversion is indirectly estimated by measuring the proportion of internal to external rotation of the hips in extension

Table 24.1 Values for rotational profile

	Mean (degrees)	Range (degrees)
Foot progression angle	10 external	–3–20
Medial rotation of hip (males)	50	25–65
Medial rotation of the hip (females)	40	15–60
Lateral rotation of hip	45	25–65
Thigh–foot angle	10 outward	–5–30
Transmalleolar axis angle	20	0–45

Table 24.2 Severity of femoral torsion

	Medial hip rotation (degrees)	Lateral hip rotation (degrees)
Mild	70–80	10–20
Moderate	80–90	0–10
Severe	>90	<0

that these values are not accurate and the computed tomography is fundamental to accurately analyze the tibial tubercle–trochlear groove distance (TT–TG) and rotational lower limb alignment.

24.4 Imaging the Lower Limb Torsional Abnormalities

24.4.1 Radiography

Radiographic limb alignment is obtained with an entire limb radiograph with the patient in standing position and the feet aligned in the sagittal plane.

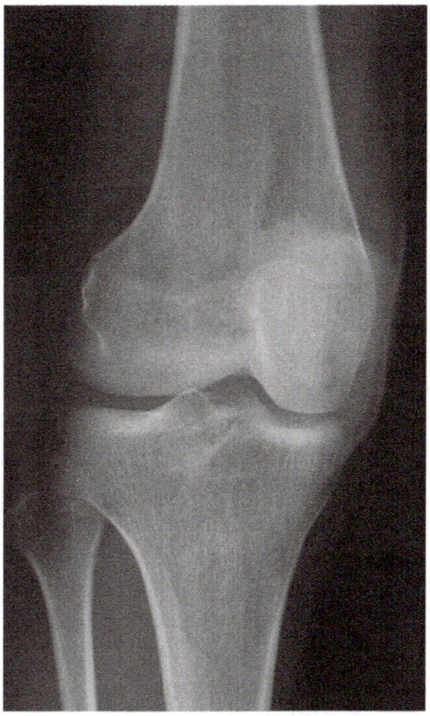

Fig. 24.6 Knee radiography in the frontal plane. The patella should be centered in the middle of the distal femur. If patella is centered in the trochlear, malrotation should be considered

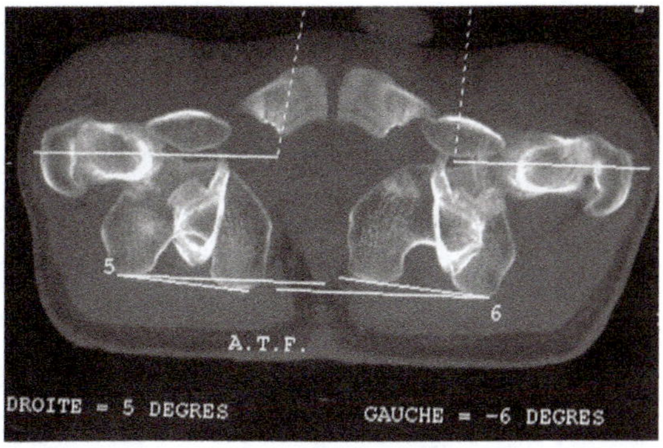

Fig. 24.7 CT scan. Femoral anteversion is measured as the angle between the femoral neck axis and the line tangent to the posterior aspect of the femoral condyles

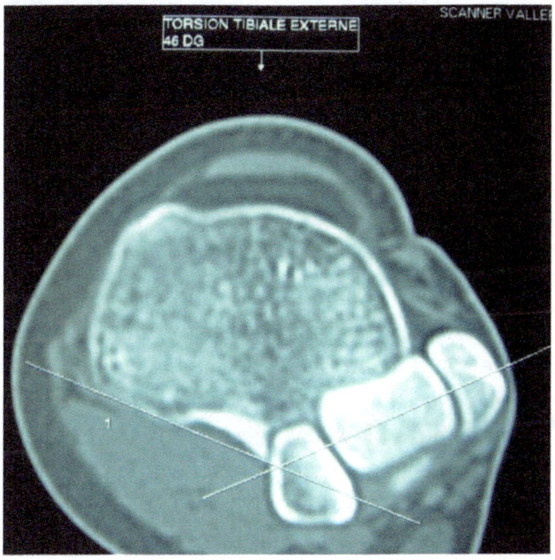

Fig. 24.8 CT scan. External tibial torsion is measured as the angle between the posterior aspect of the tibial metaphysis and the ankle joint line

The patella position is evaluated. In the frontal plane, the patella should be centered in the middle of the distal femur. If patella is centered in the trochlear, malrotation should be considered (Fig. 24.6). In this view, we observe the lesser trochanter. In excessive femoral anteversion, the lesser trochanter is excessively prominent. Although these findings can suggest a rotational deformity, we cannot measure the rotational alignment.

24.4.2 CT Scan

Torsional deformities in the limb are measured with CT scan.

Femoral anteversion is measured as the angle between the femoral neck axis and the line tangent to the posterior aspect of the femoral condyles [14]. The measurement must always be bilateral and comparative (Fig. 24.7).

External tibial torsion is measured as the angle between the posterior aspect of the tibial metaphysis and the ankle joint line (Fig. 24.8). This measurement is not reproducible and it is inaccurate. This is explained by the fact that there is an important influence of the cut level in the

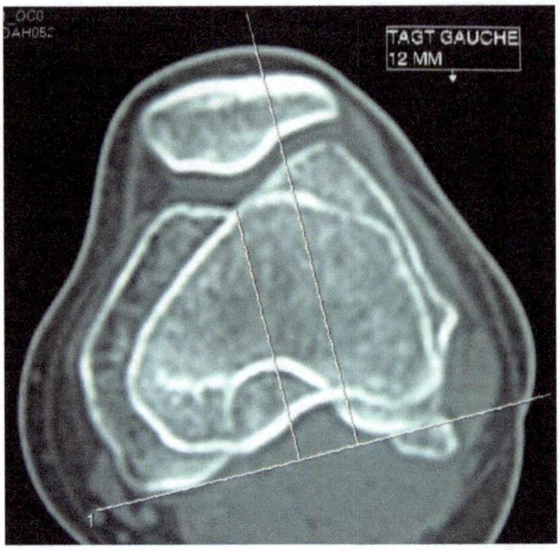

Fig. 24.9 CT scan. TT–GT can be measured by superposition of axial CT images

CT scan in the epiphysis. With slight variations in the tibial cut level, we can observe differences in the angle result.

The TT–GT quantifies in millimeters the distance between the tibial tuberosity and the middle of the condylar groove as defined by Goutallier et Bernageau. It is used to evaluate the anatomic relationship between the femoral trochlear groove and the anterior tubercle. It can be measured by superposition of axial CT images, and the measurement of tibial tubercle lateralization is more accurate than the clinical assessment of the Q angle. Dejour et al. [6] defined 20 mm of offset as the pathologic threshold (Fig. 24.9).

24.5 Surgical Indications

Concerning torsional osteotomies, it is very difficult to achieve a precise correction with a low incidence of morbidity. Moreover, the absolute mechanical importance of different maltorsions also is unclear. Femoral anteversion should only be considered when femoral anteversion is greater than 20°. Nonetheless, femoral and tibial rotational osteotomy represents an important option in the management of patellar dysfunctions. It seems to be logical to place the trochlear groove under the patella rather than force the patella under the trochlear groove. Compared with frontal-plane and sagittal-plane deformities of the lower limb, which are apparent on clinical examination and conventional radiographs, transverse-plane rotational deformities are often missed or ignored because of difficulties in their assessment [12].

In the presence of rotational malalignment, the increase in femoral anteversion produces high lateral-direct patellofemoral joint forces and pain not relieved by performing a proximal or distal realignment procedure. Moreover, there is an increase in the Q angle, leading to an increase in the lateral-directed force on the patella. A medial displacement osteotomy of tibial tubercle increases the external tibial torsion in patients with an increase of the Q angle and, in the presence of underlying malalignment, exacerbates the symptoms [15].

This way, we consider a rotational osteotomy procedure in patients with surgical indication for patellofemoral dysfunction in the presence of lower limb rotational malalignment, especially if the deformity is asymmetric and unilateral.

Patients with medial hip rotation >85° and lateral rotation <10° are considered to be candidates for surgical intervention. External tibial torsion >30° based on thigh–foot angle measurements may need surgical correction [16].

When there is an excessive TT–GT, over 20 mm, a medialization of the tibial tubercle is indicated.

24.6 Surgical Considerations

Despite the fact that there is no evidence that the location of the rotational osteotomy is preferable, we are used to performing the osteotomy at the

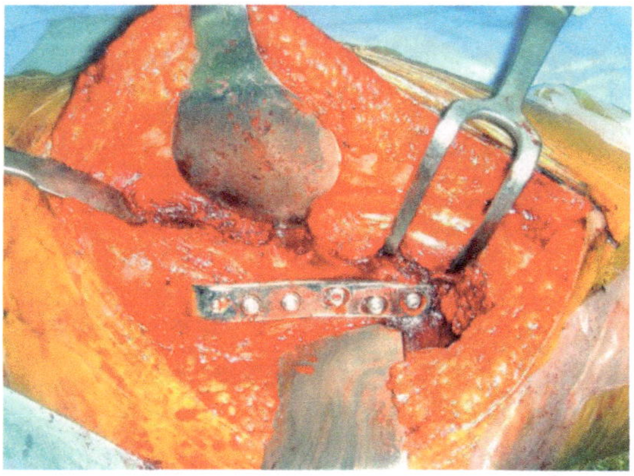

Fig. 24.10 Surgical image. Rotational osteotomy of the proximal femur

intertrochanteric level in the femur and in the proximal tibia.

In the intertrochanteric level, the proximal femur is more cylindrical so control of varus–valgus and flexion–extension is easier than when dealing with the distal flare of the femur. Another advantage is that, in this level, the quadriceps muscle rotates with the diaphysis. If there is a varus or valgus deformity, the correction must be made near the knee joint usually in the supracondylar region. We prefer using an angled blade plate as the distal fragment is more easily aligned to the plate (Figs. 24.10 and 24.11). We have to take care and consider the fact that a large rotational correction in the proximal femur can change significantly the supratrochlear anatomy.

In the tibia, we perform the tibial tubercle osteotomy before the rotational osteotomy. We raise the tibial tubercle with the patellar tendon attached. We perform the transversal rotational osteotomy at the metaphysis at the level of the tibial tubercle. After the rotational correction, the transverse osteotomy fixation is fixed with two agrafes; the tibial tubercle is repositioned according to the correct TT–GT and fixed with one or two screws (Figs. 24.10, 24.11, and 24.12).

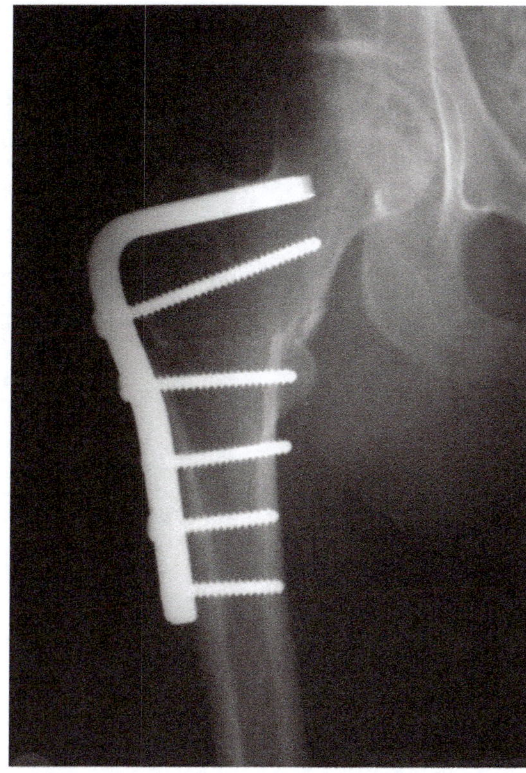

Fig. 24.11 Radiographic image. Rotational osteotomy of the proximal femur

When there is the necessity to perform both femoral and tibial rotational osteotomy, we should avoid performing the two procedures at the same time. First femoral osteotomy is performed and a delay of 1 or 2 weeks between the surgeries is indicated. After femoral osteotomy, a new CT scan is realized to observe the tibial correction need in relation to the correction performed previously at the femur.

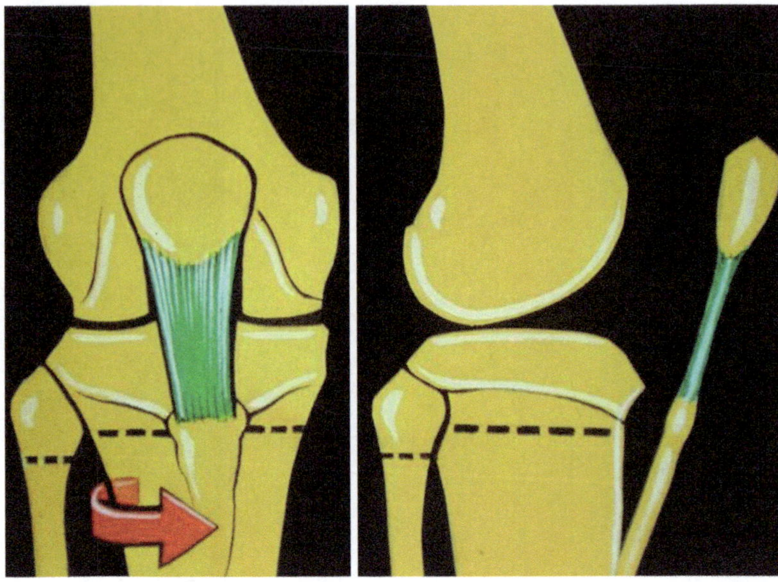

Fig. 24.12 Tibial rotational osteotomy with tibial tubercle osteotomy

Conclusions

When faced with a patellofemoral malfunction, it is important to check the articular geometry and the lower limb alignment. It is important to remember that small alterations in alignment can result in significant alterations in patellofemoral joint stresses. The clinical assessment of torsional malalignment may be difficult, and validated CT scan rotational studies are indicated. When patellofemoral symptoms or intrinsic patellofemoral pathology is the result of skeletal maltorsion, rotational osteotomy may be the only appropriate surgical treatment (Fig. 24.13). Surgical procedure may be considered aggressive, and the patient must be thoroughly informed about surgical morbidity and complications as neurovascular problems which must never be underestimated (Figs. 24.14, 24.15, and 24.16).

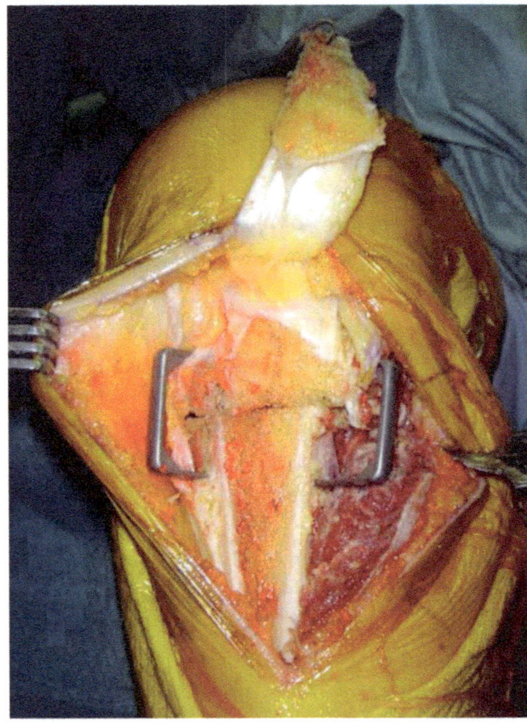

Fig. 24.13 Surgical image. Tibial rotational osteotomy. Fixation with two agrafes

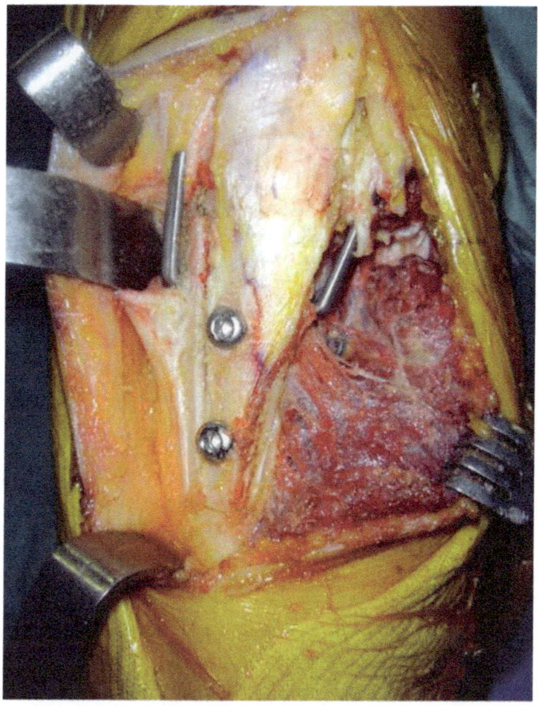

Fig. 24.14 Surgical image. Tibial rotational osteotomy with tibial tubercle osteotomy

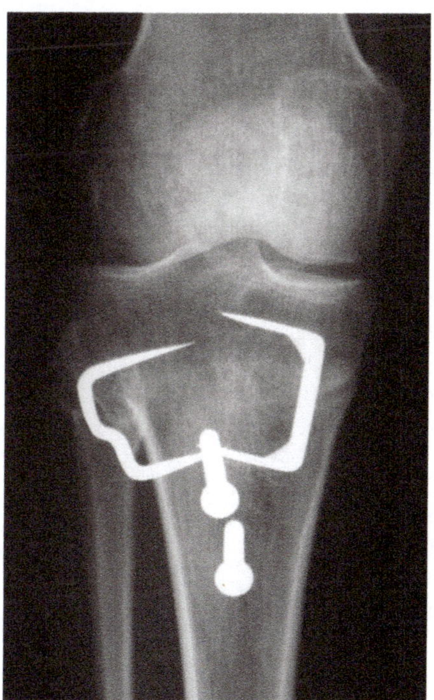

Fig. 24.15 Post-op radiography. Final result of tibial rotational osteotomy with tibial tubercle osteotomy

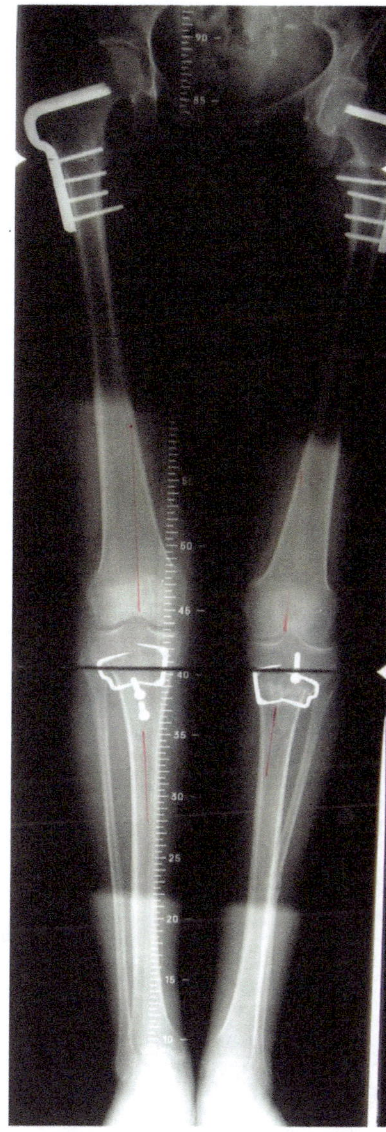

Fig. 24.16 Full-lenght limb radiography. Final result after bilateral femoral and tibial rotational osteotomy

References

1. Hvid J, Andersen LJ. The quadriceps angle and its relation to femoral torsion. Acta Orthop Scand. 1982; 53:577.
2. Bruce WD, Stevens PM. Surgical correction of miserable malalignment syndrome. J Pediatr Orthop. 2004; 24(4):1–5.
3. Brattstrom H. Shape of the intercondylar groove normally and in recurrent dislocation of the patella.

A clinical and x-ray anatomical investigation. Acta Orthop Scand. 1964;68(Suppl):1–147.
4. Kingsley PC, Olmstead KL. A study to determine the angle of anteversion of the neck of the femur. J Bone Joint Surg Am. 1948;30:745–51.
5. Delgado ED, Schoenecker PL, Rich MM, Capelli AM. Treatment of severe torsional malalignment syndrome. J Pediatr Orthop. 1996;16:484–8.
6. Dejour H, Walch G, Nave-Josserand L, Guier C. Factors of patellar instability: an anatomic radiographic study [invited lecture]. Knee Surg Sports Traumatol Arthrosc. 1994;2:19.
7. Stuberg W, Temme J, Kaplan P, Clarke A, Fucks R. Measurement of tibial torsion and thigh-foot angle using goniometry and computed tomography. Clin Orthop. 1991;272:208–12.
8. Lerat JL, Moen B, Bochu M, Galland O. Femoropatellar pathology and rotational or torsional abnormalities of the inferior limb: the use of CT scan. In: Muller W, Hackenbruch W, editors. Surgery and arthroscopy of the knee: second congress of the European society. Berlin: Springer; 1988.
9. Turner MS, Smillie IS. The effect of tibial torsion on the pathology of the knee. J Bone Joint Surg Br. 1981; 63:396.
10. Tamari K, Tinley P, Briffa K, Breidahl W. Validity and reliability of existing and modified clinical methods of measuring femoral and tibiofibular torsion in healthy subjects: use of different reference axes may improve reliability. Clin Anat. 2005;18: 46–55.
11. Aglietti P, Insall JN, Cerulli G. Patellar pain and incongruence. I: measurements of incongruence. Clin Orthop. 1983;176:217.
12. Staheli LT, Corbett M, Wyss C. Lower extremity rotational problems in children. Normal values to guide management. J Bone Joint Surg Am. 1985;67:39.
13. Carson WG, James SL, Larson RL, et al. Patellofemoral disorders: physical and radiographic evaluation. I: physical examination. Clin Orthop. 1984;185:165.
14. Murphy SB, Simon SR, Kijewski PK, et al. Femoral anteversion. J Bone Joint Surg. 1987;69A:1169–76.
15. Paulos L, Swanson SC, Stoddard GJ, Barber-Westing SD. Surgical correction of limb malalignment for instability of the patella: a comparison of 2 techniques. Am J Sports Med. 2009;37(7): 1288–300.
16. Staheli LT. Torsion: treatment indications. Clin Orthop. 1989;247:61–9.

Prosthetic Indications in Patellofemoral Osteoarthritis

25

Johan Bellemans

Contents

J. Bellemans
Knee and Sports Orthopaedics,
Ziekenhuis Oost-Limburg, Genk, Belgium
e-mail: johan.bellemans@uz.kuleuven.ac.be

A. Gobbi et al. (eds.), *The Patellofemoral Joint*,
DOI 10.1007/978-3-642-54965-6_25,

25.1 Introduction

Since the first successful attempt by McKeever in 1955 to replace the patellar surface using a Vitallium shell and since the results of the first artificial patellotrochlear replacements by Blazina et al. were published in 1979, the enthusiasm of surgeons towards artificial replacement of the patellofemoral joint has gone through ups and downs [1, 2].

The results of these implants were initially considered as unpredictable and inconsistent by most surgeons, contrary to what was observed for total knee replacements. Shortcomings in the available designs, difficulty in obtaining correct implant positioning, and failure to address correctly the underlying pathology were the main reasons for this lack of enthusiasm.

Recently, however, there has been a renewed interest in the use of patellofemoral arthroplasty, and there is a growing tendency to believe that artificial patellofemoral replacement has a well-defined place in the treatment of end-stage patellofemoral osteoarthritis.

The recent trend towards less invasive surgery as well as the revival of selective, unicompartmental resurfacing options has aroused the orthopaedic industry towards increasing the efforts in designing better and more anatomic patellofemoral prostheses.

In the meantime, a better understanding of patellofemoral physiology and pathology allowed surgeons to gain a better understanding on how and when patellofemoral arthroplasty should be

Table 25.1 Literature overview on isolated patellofemoral joint replacement

Series	Implant	Number of cases	Follow-up (years)	Good/excellent results (%)	Revision rate (%)
Blazina et al. (1979) [2]	Richards I/II	57	2	NA	35
Arciero and Toomey (1988) [3]	Richards II	25	5.3	85	12
Cartier et al. (1990) [4]	Richards II/III	72	4	85	10
Argenson et al. (1995) [5]	Autocentric	79	5.5	84	13
Krajca-Radcliffe et al. (1996) [6]	Richards I/II	16	5.8	88	6
De Cloedt et al. (1999) [7]	NA	45	6	NA	18
Tauro et al. (2001) [8]	Lubinus	62	7.5	45	28
de Winter et al. (2001) [9]	Richards II	26	11	62	19
Smith et al. (2002) [10]	Lubinus	45	4	69	19
Kooijman et al. (2003) [11]	Richards II	45	15.5	86	25
Board et al. (2004) [12]	Lubinus	17	1.5	53	12
Merchant et al. (2004) [13]	LCS	15	3.7	93	0
Lonner (2004) [14]	Lubinus	30	4	84	33
Lonner (2004) [14]	Avon/Nexgen	25	0.5	96	0
Argenson et al. (2005) [15]	Autocentric	66	16.2	NA	51
Ackroyd et al. (2005) [16]	Avon	306	2	NA	4
Cartier et al. (2005) [17]	Richards II/III	79	10	72	13
Leadbetter et al. (2006) [18]	Avon	30	2	83	7
Sisto and Sarin (2006) [19]	Kinamatch	25	6	100	0
Ackroyd et al. (2007) [20]	Avon	109	5.2	78	17
Gadeyne et al. (2008) [21]	Autocentric	43	6	67	24

performed in order to lead to consistent clinical results.

Like in any other operation, a successful clinical outcome depends on the correct patient selection and indication, as well as surgical technique and postoperative care. In this chapter we try to address the issue of patient selection and indication, based upon the evidence available in literature. Over the last few years, several reports have indeed been published on the results of patellofemoral arthroplasty as well as total knee arthroplasty for patellofemoral disease, and based upon these data, it is becoming increasingly clear what the exact place is of prosthetic patellofemoral surgery.

Review of the literature shows that all of the published studies on patellofemoral replacement are retrospective in nature and provide only level 3 or level 4 evidence [3–21] (Table 25.1).

No therapeutic level 1 or level 2 studies have indeed been performed in order to compare patellofemoral replacement to total knee replacement or any other treatment options for patellofemoral pathology.

25.2 Isolated Patellofemoral Arthroplasty

The typical indication for the use of a patellofemoral prosthesis has traditionally been the patient with disabling, isolated end-stage patellofemoral degeneration that has failed to respond to conservative or other surgical treatment options. Usually this means that the patient has full-thickness cartilage loss as documented by radiographic, arthroscopic, or other investigations.

In cases of subtotal cartilage damage without exposed bone, one should always consider alternative, more conservative surgical options first. Arthroscopic debridement may be helpful in cases of mechanical symptoms caused by unstable cartilage flaps. Microfracture, mosaicplasty, or even autologous chondrocyte transplantation may have a place in the younger patient with a fresh, posttraumatic lesion. Lateral retinacular release, soft tissue realignment of the extensor mechanism, and/or anteromedialization osteotomy of the tibial tubercle may all help to unload the damaged patellofemoral cartilage.

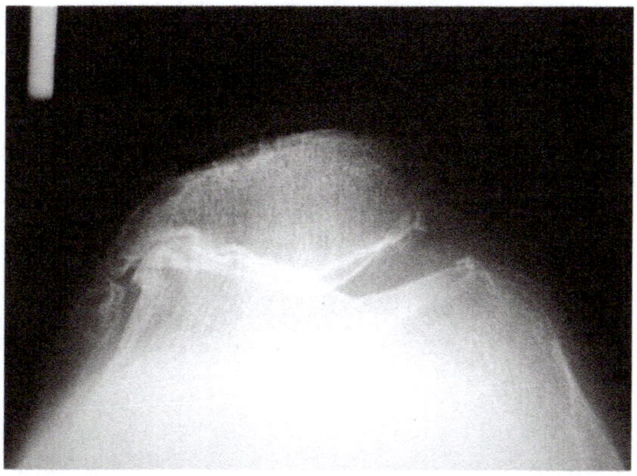

Fig. 25.1 End-stage patellofemoral osteoarthritis with full cartilage loss and lateral maltracking is a potential indication for isolated patellofemoral replacement but will require correction of the maltracking intra-operatively. Usually a limited lateral release or facetectomy will be sufficient to obtain this

In cases of erosive full-thickness damage, these options are however frequently inappropriate or insufficiently effective, requiring further and more drastic care. Patellectomy may be a theoretical option, but it is a mutilating operation, and history has taught us that the results are unpredictable with respect to both the subjective and functional outcome [22–24].

A more conservative approach with excision of just the eroded lateral facet, while leaving the patellar body in situ, may be a better alternative [22, 25–28].

Patellofemoral joint replacement effectively replaces the damaged cartilage layers and therefore provides a more logical solution for the predominant problem of the patient. This implies that concomitant issues such as underlying patellar malalignment or maltracking should be absent or corrected (Fig. 25.1). Likewise, there should be no evidence of other pathology in the knee such as tibiofemoral arthritis or an inflammatory arthropathy.

In view of this, Leadbetter et al. have recently outlined the optimal indications and contraindications for patellofemoral arthroplasty [18, 29]. Degenerative osteoarthritis limited to the patellofemoral joint and causing severe symptoms affecting daily activity is the primary indication, at least in case a lengthy period of nonoperative treatment was unsuccessful.

Posttraumatic osteoarthritis; extensive grade 3 chondrosis affecting the entire trochlea, the medial facet, or the proximal half of the patella; and failure of previous extensor unloading surgical procedures are additional indications according to these authors. In their opinion, contraindications to the procedure are the presence of tibiofemoral arthritis, systemic inflammatory arthropathy, patella infera, uncorrected patellofemoral malalignment, tibiofemoral malalignment, psychogenic pain, and loss of range of motion greater than 10° [18, 29].

Interestingly, factors that are known to be associated with the development of tibiofemoral pathology are indeed associated with inferior results after patellofemoral arthroplasty. Obesity, tibiofemoral malalignment, and limited range of motion fall in this category.

In most published series, the most frequent reason for revising a patellofemoral arthroplasty to a total knee replacement was the progression of the arthritic disease in the femorotibial compartments.

In a recent literature analysis, Leadbetter et al. have reported an overall average reoperation rate of 24 % after patellofemoral joint replacement [18]. Revision to total knee arthroplasty was necessary in 9 % (range 5–18 %) of the published cases, with progression of osteoarthritis in the remaining compartments as the most important cause. Uncorrected extensor malalignment with patellar maltracking or instability, knee joint stiffness, and patellar component loosening were the other reasons for conversion to total knee replacement.

Recent data available from international knee arthroplasty registries seem to confirm these findings. In the annual 2008 report of the Australian Hip and Knee Arthroplasty registry, 1,057 patellotrochlear replacements were reported, accounting for 0.5 % of all knee procedures [30].

Nine different designs were used, with the Avon, LCS, Lubinus, and RBK being the most frequent and accounting for 86 % of all procedures. Again, the revision rate was found to be relatively high compared to total or unicondylar knee arthroplasty, with 3.1 revisions per 100 observed component years and a 5-year cumulative percent revision of 13.8 % (versus 12.1 % at 7 years for unicondylar knee replacements and 4.3 % at 7 years for total knee replacements).

The main reason for revision of patellotrochlear replacements was progression of disease in 24 %, pain in 22 %, and loosening in 17 %. Interesting to note was that the outcome depended on age, with the 5-year cumulative percent revision declining with increasing age. Patients aged less than 55 years at surgery had a 5-year cumulative revision percent of 17 % versus 13 % for the age group 55–64, 12 % for age 65–74, and only 7 % for those over 75 years old. Males had a doubled risk of revision compared to females. Finally, revision rates were highly influenced by the type of prosthesis used [30].

In a recent German, nationwide survey, a total of 195 patellofemoral replacements were reported, accounting for 0.37 % of all knee replacements. Again, the main reason for failure was progression of tibiofemoral degeneration of the affected knee [31].

Careful patient selection is therefore crucial, and the clinical challenge is to select the patient with isolated patellofemoral full-thickness cartilage wear, absent or correctable malalignment, and absence of risk factors for developing tibiofemoral disease (Fig. 25.2). Such is not an easy task and requires careful clinical and technical investigation [18, 22, 24, 29, 32].

While interrogating and examining the patient, it should become clear that the pain is exclusively located in the anterior compartment and secondary to severe wear of the patellofemoral joint. Patellofemoral crepitus, retropatellar pain while squatting or while performing open chain extension against resistance, and pain during retropatellar palpation should be present. Femorotibial joint line tenderness or other signs of femorotibial or meniscal pathology should not be present. Also, other causes of anterior knee pain such as prepatellar bursitis, pes anserinus tendonitis, patellar tendonitis, or referred hip pain should be excluded. Patellar tracking should be closely examined, and maltracking should be corrected preferably before or at the latest during the patellofemoral replacement.

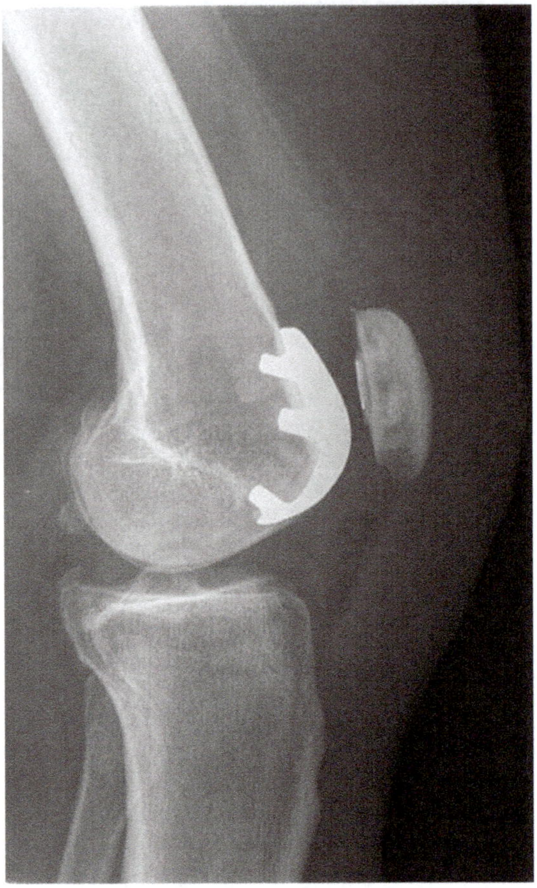

Fig. 25.2 Patellofemoral joint replacement in situ

Technical investigations should include standing AP and lateral knee radiographs in both extension and 30° flexion (Rosenberg or schuss view), in order to exclude tibiofemoral degeneration. On the lateral views, the presence of patella alta or baja can be noted. A patellar skyline (axial or Merchant) view should be taken to document cartilage loss as well as patellar tracking. Standing full-leg radiographs may be necessary to rule out tibiofemoral malalignment. CT scan or MRI may be helpful to further document cartilage status and to evaluate the tibiofemoral compartment.

Finally, patellofemoral arthritis can be the first, subtle indication of an otherwise subclinical inflammatory condition, and serum analysis may therefore be warranted in doubtful cases [24].

Based upon all these clinical and technical investigations, one should be able to determine

Table 25.2 Indication criteria for isolated patellofemoral arthroplasty

Isolated patellofemoral osteoarthritis (documented loss of patellofemoral joint space with osseous deformation)
Severe patellofemoral symptoms affecting activities of daily life
Nonresponsive to nonoperative treatment for at least 3–6 months
Absent patellofemoral malalignment (or corrected intra-operatively)
Absent tibiofemoral disease
Neutral tibiofemoral alignment
No obesity
No evidence of inflammatory arthritis

whether the patient fulfils the criteria for isolated patellofemoral replacement as shown in Table 25.2.

25.3 Total Knee Arthroplasty

Proponents of patellofemoral arthroplasty argue that despite the significant incidence of femorotibial degeneration necessitating revision to total knee arthroplasty, this argument does justify the systematic use of total knee replacement for the treatment of end-stage patellofemoral disease. In their point of view, total knee arthroplasty is an extreme and overly aggressive treatment for this indication (Table 25.3).

Despite this, several studies have been published indicating that total knee arthroplasty is an effective and reliable procedure for the treatment of isolated patellofemoral disease, with very little reoperation or revision rates, contrary to what is known for isolated patellofemoral replacements. Although no comparative studies exist, the mere fact that the revision and reoperation rate in published series is definitely lower for total knee replacement compared to published data on patellofemoral joint replacement is a strong argument in favour of TKA.

According to those in favour of TKA, the results after patellofemoral arthroplasty should at least become as good as those after TKA with respect to longevity and pain relief, in order to justify its use as a reasonable treatment option for isolated patellofemoral osteoarthritis.

Table 25.3 Contraindications for isolated patellofemoral arthroplasty

Presence of tibiofemoral disease
Inflammatory arthropathy
Uncorrected patellofemoral malalignment or instability
Tibiofemoral malalignment
Gross obesity
Fixed flexion contracture >10°
Evidence of psychosomatic component/chronic regional pain syndrome

Table 25.4 Total knee arthroplasty performed for isolated patellofemoral osteoarthritis: literature overview

Series	Number of cases	Follow-up (years)	KS score	Revision rate (%)
Dalury et al. (1995) [34]	33	5.2	96	0
Laskin and van Stejn (1999) [36]	53	7.4	96	0
Parvizi et al. (2001) [38]	31	5	89	3
Mont et al. (2002) [37]	33	6.8	93	0
Meding et al. (2007) [39]	33	6.2	88	0

Several authors have reported on the results after TKA for isolated patellofemoral osteoarthritis [33–39] (Table 25.4).

Laskin et al. have reported on 53 patients with an average follow-up of 7.4 years and noted a better subjective and functional outcome comparing this group with a matched series of patients with tricompartmental osteoarthritis [35]. Meding et al. retrospectively compared the outcomes of 33 TKAs with patellofemoral osteoarthritis with a matched group of primarily tibiofemoral osteoarthritis and noted similar results for both groups [36]. In their analysis of the literature, they pointed out that of the 167 TKAs performed for the treatment of isolated patellofemoral osteoarthritis, only one knee was revised and two knees underwent reoperation. Three of the studies reported no revisions or reoperations, and the highest revision rate was 3 % (one of 31 TKAs) (Table 25.4).

In view of these data, total knee arthroplasty can therefore be considered as an acceptable treatment option for the treatment of isolated end-stage patellofemoral disease and is the treatment of choice in case concomitant early tibiofemoral degeneration is present or in case risk factors for the development of such tibiofemoral degeneration exist.

Conclusion

The ideal indication for isolated patellofemoral joint replacement is the patient with end-stage patellofemoral osteoarthritis that has been nonresponsive to prolonged conservative treatment, causing him or her severe problems in activities of daily life. Underlying patellar maltracking should not be present or corrected during the procedure. Tibiofemoral degeneration or risk factors for developing tibiofemoral degeneration, such as obesity, tibiofemoral malalignment, or inflammatory arthropathy should not be present. In case they are, total knee replacement is the standard of choice. Progression of femorotibial degeneration is indeed the most frequent reason for failure of isolated patellofemoral replacements and occurs relatively common. Total knee replacement can avoid this and is justified under these conditions.

References

1. McKeever DC. Patellar prosthesis. J Bone Joint Surg Am. 1955;37:1074–84.
2. Blazina M, Fox J, Del Pizzo W, Broukhim B, Ivey F. Patellofemoral replacement. Clin Orthop Relat Res. 1979;144:98–102.
3. Arciero R, Toomey H. Patellofemoral arthroplasty. A three- to nine-year follow-up study. Clin Orthop Relat Res. 1988;236:60–71.
4. Cartier P, Sanouiller J, Greslamer R. Patellofemoral arthroplasty. 2–12 year follow-up study. J Arthroplasty. 1990;5:49–55.
5. Argenson JN, Guillaume J, Aubaniac J. Is there a place for patellofemoral arthroplasty? Clin Orthop Relat Res. 1995;321:162–7.
6. Krajca-Radcliffe J, Coker T. Patellofemoral arthroplasty. A 2- to 18 year follow-up study. Clin Orthop Relat Res. 1996;330:143–51.
7. De Cloedt P, Legaye J, Lokietek W. Les protheses femoro-patellaires. Etude retrospective de 45 cas successifs avec un recul de 3 a 12 ans. Acta Orthop Belg. 1999;65:170–5.
8. Tauro B, Ackroyd C, Newman J, Shah N. The Lubinus patellofemoral arthroplasty. A five to ten year prospective study. J Bone Joint Surg Br. 2001;83:696–701.
9. de Winter W, Feith R, van Loon C. The Richards type II patellofemoral arthroplasty: 26 cases followed for 1–20 years. Acta Orthop Scand. 2001;72:487–90.
10. Smith A, Peckett W, Butler-Manueal P, Venu K, d'Arcy J. Treatment of patello-femoral arthritis using the Lubinus patello-femoral arthroplasty: a retrospective review. Knee. 2002;9:27–30.
11. Kooijman HJ, Driessen AP, van Horn JR. Long-term results of patellofemoral arthroplasty. A report of 56 arthroplasties with 17 years of follow-up. J Bone Joint Surg Br. 2003;85:836–40.
12. Board TN, Mahmoud A, Ryan WG, Banks AJ. The Lubinus patellofemoral arthroplasty: a series of 17 cases. Arch Orthop Trauma Surg. 2004;124:285–7.
13. Merchant AC. Early results with a total patellofemoral joint replacement arthroplasty. J Arthroplasty. 2004;19:829–36.
14. Lonner JH. Patellofemoral arthroplasty. Pros, cons, and design considerations. Clin Orthop Relat Res. 2004;428:158–65.
15. Argenson JN, Flecher X, Parratte S, Aubaniac JM. Patellofemoral arthroplasty. An update. Clin Orthop Relat Res. 2005;440:50–3.
16. Ackroyd CE, Chir B. Development and early results of a new patellofemoral arthroplasty. Clin Orthop Relat Res. 2005;436:7–13.
17. Cartier P, Sanouiller JL, Khefacha A. Long-term results with the first patellofemoral prosthesis. Clin Orthop Relat Res. 2005;436:47–54.
18. Leadbetter WB, Seyler TM, Ragland PS, Mont MA. Indications, contraindications, and pitfalls of patellofemoral arthroplasty. J Bone Joint Surg. 2006;88(S4):122–37.
19. Sisto DJ, Sarin VK. Custom patellofemoral arthroplasty of the knee. J Bone Joint Surg Am. 2006;88:1475–80.
20. Ackroyd CE, Newman JH, Evans R, Eldridge JD, Joslin CC. The Avon patellofemoral arthroplasty: 5-year survivorship and functional results. J Bone Joint Surg Br. 2007;89:310–5.
21. Gadeyne S, Besse JL, Galand-Desme S, Lerat JL, Moyen B. Results of self-centering patellofemoral prosthesis: a retrospective study of 57 implants. Rev Chir Orthop Reparatrice Appar Mot. 2008;94:228–40.
22. Donell S, Glasgow M. Isolated patellofemoral osteoarthritis. Knee. 2007;14:169–76.
23. Lennox I, Cobb A, Knowles J, Bentley G. Knee function after patellectomy. A 12 to 48 year follow-up. J Bone Joint Surg Br. 1994;76:485–7.

24. Grelshamer R, Stein A. Current concepts review: patellofemoral arthritis. J Bone Joint Surg Am. 2006;88:1849–60.
25. Beltran J. Resection arthroplasty of the patella. J Bone Joint Surg Br. 1987;69:604–7.
26. Martens M, De Rycke J. Facetectomy of the patella in patellofemoral osteoarthritis. Acta Orthop Belg. 1990;56:563–7.
27. Wetzels T, Bellemans J. Patellofemoral osteoarthritis treated by partial lateral facetectomy: results at long-term follow up. Knee. 2012;19:411–5.
28. Yercan H, Selmi T, Neyret P. The treatment of patellofemoral osteoarthritis with partial lateral facetectomy. Clin Orthop Relat Res. 2005;436:14–9.
29. Weaver J, Wieder D, Derkash R. Patellofemoral arthritis resulting from malalignment. A long term evaluation of treatment options. Orthop Rev. 1991;20:1075–81.
30. Leadbetter WB, Rageland PS, Mont MA. The appropriate use of patellofemoral arthroplasty. An analysis of reported indications, contraindications and failures. Clin Orthop Relat Res. 2005;436:91–9.
31. Hip and knee Arthroplasty annual report. Australian Orthopaedic Association National Joint Replacement Registry. www.aoa.org.au. 2008. p. 117–168.
32. Becher C, Renke A, Heyse TJ, Schofer M, Tibesku CO, Fuchs-Winkelmann S. Patellofemoral arthroplasty – results of a nationwide survey in Germany and review of the literature. Z Orthop Unfall. 2008;146:773–81.
33. Lonner JH. Patellofemoral arthroplasty. J Am Acad Orthop Surg. 2007;15:495–506.
34. Dalury DF, Ewald FC, Christie MJ, Scott RD. Total knee arthroplasty in a group of patients less than 45 years of age. J Arthroplasty. 1995;10:598–602.
35. Dalury DF. Total knee replacement for patellofemoral disease. J Knee Surg. 2005;18:274–7.
36. Laskin RS, van Stejn M. Total knee replacement for patients with patellofemoral arthritis. Clin Orthop Relat Res. 1999;367:89–95.
37. Mont MA, Haas S, Mullick T, Hungerford DS. Total knee arthroplasty for patellofemoral arthritis. J Bone Joint Surg Am. 2002;84:1977–81.
38. Parvizi J, Stuart MJ, Pagnano MW, Hanssen AD. Total knee arthroplasty in patients with isolated patellofemoral arthritis. Clin Orthop Relat Res. 2001;392:147–52.
39. Meding JB, Wing JT, Keating EM, Ritter MA. Total knee arthroplasty for isolated patellofemoral arthritis in younger patients. Clin Orthop Relat Res. 2007;464:78–82.

Treatment of Patellofemoral Disorders in Skeletally Immature Athlete

26

Mahmut Nedim Doral, Egemen Turhan, Gürhan Dönmez, Defne Kaya, João Espregueira-Mendes, and Alberto Gobbi

Contents

26.1 Introduction

Patellofemoral problems are frequently encountered in adolescents and may affect daily activities and even limit athletic activities [1]. The two most common forms of these problems are patellofemoral instability and patellofemoral pain syndrome. Success in treatment program depends on a precise understanding of the origin or underlying cause of the patellofemoral problem. In this brief review, the current literature for treatment of immature and adolescent athletes with patellofemoral disorders were reviewed.

M.N. Doral (✉)
Department of Orthopaedics and Traumatology, Hacettepe University, Ankara, Turkey

Department of Sports Medicine, Hacettepe University, Ankara, Turkey
e-mail: mndoral@gmail.com

E. Turhan
Department of Orthopaedics and Traumatology, Hacettepe University, Ankara, Turkey

G. Dönmez
Department of Sports Medicine, Gaziantep Dr. Ersin Arslan State Hospital, Gaziantep, Turkey

D. Kaya
Department of Sports Medicine, Hacettepe University, Ankara, Turkey

J. Espregueira-Mendes
Clínica do Dragão – Espregueira-Mendes Sports Centre – FIFA Medical Centre of Excellence, F.C. Porto Stadium, 4350-415 Porto, Portugal

ICVS/3B's Research Group – Biomaterials, Biodegradables and Biomimetics, University of Minho, Headquarters of the European Institute of Excellence on Tissue Engineering and Regenerative Medicine, Caldas das Taipas, Guimarães, Portugal

Department of Orthopaedics, Minho University, Braga, Portugal
e-mail: jem@espregueira.com

A. Gobbi
Orthopaedic Arthroscopic Surgery International (O.A.S.I.) Bioresearch Foundation, Gobbi N.P.O., Milan, Italy
e-mail: gobbi@cartilagedoctor.it

A. Gobbi et al. (eds.), *The Patellofemoral Joint*,
DOI 10.1007/978-3-642-54965-6_26, © ISAKOS 2014

26.2 Patellofemoral Pain

Patellofemoral pain syndrome (PFPS) or anterior knee pain (AKP) is a common musculoskeletal disorder in adolescents. Previously various terminologies have been used to define PFPS: chondromalacia patellae, patellofemoral malalignment or maltracking, lateral hyperpressure syndrome, lateral patellar compression syndrome, retinaculitis, etc. [2]. Patellofemoral pain is more common among in female adolescents and young athletes, and it was reported that 26.6 % of the adolescent female athletes screened over three consecutive basketball seasons had AKP [3]. The incidence was found approximately the same in young dancers for PFPS in a large population (23.6 %) [4]. Symptoms of anterior knee pain likely persist after middle school-aged onset and reach peak prevalence during the high school years.

Although the reason of pain in PFPS is not exactly understood, synovial irritation caused by the cartilage degradation-related chemicals released into the joint, increased subchondral bone pressure, and excessive stress in peripatellar soft tissues are some of the proposed factors. The majority of patellofemoral pain in children and adolescent patients is caused by trauma and malalignment syndromes (or a combination of both) and can usually be managed successfully with nonoperative methods. The principle of management of the pain in PFPS is to modify pain-provoking activities until the symptoms completely disappear. Therefore, the most of the current literature focused on studies evaluating the reasons and risk factors for PFPS [5, 6].

Patellofemoral pain has often been attributed to abnormal hip and knee mechanics especially in females. Greater hip adduction, hip internal rotation, shank internal rotation, and weaker knee extension strength in female runners appears to be a risk factor for PFPS [7]. Finnoff et al. suggested that stronger pre-injury hip abductors (particularly in relation to their hip adductors) and weaker pre-injury hip external rotators (particularly in relation to their hip internal rotators) are associated with the development of PFP in high school running athletes [8]. Hip strengthening prior to functional exercises prove greater improvements than female athletes who perform quadriceps strengthening prior to the same functional exercises. In addition, patients with a greater quadriceps muscle size, lower eccentric knee strength, and less pain have a better short-term functional outcome after conservative treatment for PFPS [9]. Forefoot varus and excessive subtalar pronation can be also associated with patellofemoral pain. Thence, semiflexible orthoses or custom-made foot orthoses are significant in reducing symptoms of PFPS [10].

Although previous data indicate a relationship between higher relative body mass and overall knee injury, a study among middle school-aged female basketball players found no relationship between relative body composition or relative body mass to height to the propensity to develop PFP [11].

Almost 80 % of PFPS cases conservative measures are effective. Treatment protocol usually consists of quadriceps strengthening and stretching exercises and symptomatic pain-relieving agents. In patients not responding to the activity modification and exercise program, patellar taping, knee sleeves, foot orthotics, and physical therapy modalities (electrical stimulation, phonophoresis or iontophoresis) can be added to the conservative treatment program. Surgical treatment should be avoided in adolescents in whom a structural abnormality does not accompany with PFPS.

26.3 Patellofemoral Instability

Patellofemoral instability is common in the pediatric and adolescent population, yet prognosis after the first dislocation has been difficult to determine [12]. Acute patellar dislocation is a common and severe knee injury that occurs most often in adolescents, frequently associated with sporting and physical activities. It usually manifests with symptoms such as giving way, a sensation of lateral displacement of the patella accompanied by hemarthrosis, and tenderness over the medial epicondylar region to palpation. The etiology of acute patellar dislocation is complex and variable, with many components making different contributions in each individual, resulting in several distinct clinical presentations.

Unless associated with substantial articular cartilage damage, nonsurgical management is typically used to treat a first-time acute patellofemoral dislocation in a skeletally immature athlete. Traditionally, patellar instability has been treated with variable periods of immobilization, sporadic rehabilitation, and an expected full return to sports activity. Most athletes benefit from an initial nonoperative program that is aggressive, multidimensional, and responsive to early treatment outcomes [13]. In a comprehensive systematic review, it was stated that the current best evidence does not support the superiority of surgical intervention over conservative treatment for an acute patellar dislocation [14, 15]. Nonoperative treatment for first-time patellofemoral dislocation was reported as 62 % success rate [12]. However, skeletally immature patients with trochlear dysplasia had only a 31 % success rate with conservative management. Conversely, some authors claim better results and lower recurrence rates for acute patellar dislocations with surgical repair compared with conservative treatment [16].

In patients with patellar dislocation, osteochondral injury is often an indication for early surgical intervention. Osteochondral injury to the weight-bearing surface of the lateral femoral condyle may occur in a high percentage of patients following a lateral patellar dislocation and in a higher percentage of male adolescent athletes than girls [17]. Therefore, arthroscopic examination of the knee for possible osteochondral injuries is a reliable approach after patellar dislocation.

Recurrent lateral patellar dislocation is also a common knee injury in the skeletally immature adolescent. Habitual or recurrent dislocation of the patella in the skeletally immature patient is a particularly demanding problem since the etiology is frequently multifactorial. Patients presenting with a prior history of instability were more likely to be female and were older than first-time dislocation patients [18]. Risk was found highest among females 10–17 years old. Risk factors for recurrent dislocation may include various skeletal abnormalities, increased quadriceps angle, generalized ligamentous laxity, and family history. Body mass index and patella alta are not statistically associated with recurrent instability [12].

In the setting of recurrent instability, surgical reconstruction is usually recommended. However, because of the open physis, operative therapy in children and adolescents is challenging [19]. Technique of choice should be performed in a case-by-case basis, and no single method is efficient in every case. Anatomic and biomechanical studies have demonstrated that the medial patellofemoral ligament (MPFL) and the vastus medialis obliquus are the primary restraints to lateral translation and ultimately dislocation of the patella. Clinically, up to 94–100 % of patients suffer from MPFL rupture after first-time patellar dislocation [20]. The location of MPFL tears has been well documented in the adult population, with most occurring at the insertion of the ligament on the adductor tubercle. However, there is no consensus on the literature for MPFL injury pattern in skeletally immature patients [21–23].

The current goal of surgery is to restore the normal anatomy of the patellofemoral joint. Therefore, surgical treatment of patellofemoral instability in the skeletally immature athlete is evolving from nonanatomic extensor mechanism surgical procedures to anatomic restorative procedures based on reconstitution of the MPFL [13, 24]. No standard surgical procedure for MPFL reconstruction exists in skeletally immature patients with patellar instability. Consistent beneficial effects from surgical stabilization on clinical scores, postoperative stability, and radiographic assessment were proved. Moreover, there is no evidence for growth disturbance with surgical patellar stabilization in immature patients [14]. Recently Nelitz et al. reported satisfactory functional results with no recurrent dislocation after minimal invasive technique for anatomic reconstruction of MPFL that respects the distal femoral physis in skeletally immature patients [19]. The most important variables for clinical outcomes following anatomical MPFL reconstruction using an autologous gracilis tendon are trochlea geometry, medial patellofemoral ligament (MPFL), patella height, tibial tuberosity-trochlea groove distance (TT-TG), and the extensor muscles [25]. Since the importance of an anatomical reconstruction respecting the femoral insertion of the ligament has been proven, an insertion proximal of the physis has to be strictly avoided.

Numerous surgical treatments have been described for the treatment of recurrent patellar dislocation in adolescents, one of the most famous being Galeazzi's semitendinosus tenodesis as modified by Baker which produces good midterm clinical results [26]. The Galeazzi semitendinosus tenodesis is a soft-tissue reconstruction technique designed to stabilize the patella without altering the femoral or the bony structures about the knee. However, the dynamic CT showed that in those patients with high patellae, semitendinosus tenodesis alone is not enough to stabilize the patella [26]. Moreover, long-term data of Grannatt et al. suggested that it may not be as successful as previously reported. They reported that approximately 82 % of patients experienced recurrent subluxation or dislocation despite surgical intervention, which may be due in large part to various predisposing factors, and there was a 35 % rate of second surgeries [27]. However, given the variability in techniques and reported results of patellofemoral stabilizing procedures in skeletally immature patients, the Galeazzi procedure may still be a reasonable way to temporize the difficult problem of patellar instability until patients reach skeletal maturity, when bony realignment procedures can be more safely used.

Kumahashi et al. reported the MPFL reconstruction methods, using a double-stranded semitendinosus autograft and sparing the femoral physeal line in non-closure of the epiphyseal line, provide acceptable short-term results for the treatment of patellar instability [28]. The "sandwich" method was described as fixation of the patella between a double-stranded semitendinosus tendon through the posterior third of the femoral insertion of the medial collateral ligament as a pulley with a titanium interference screw in a single patellar tunnel. Yercan et al. reported a technique that preserves femoral and patellar insertion anatomy of MPFL using a free semitendinosus autograft together with tenodesis to the adductor magnus tendon without damaging open physis on the patellar attachment of MPFL, and no recurrent dislocation was observed in three patients (4 knees) at a mean follow-up time of 17.7 months [29]. Giordano et al. presented an original surgical procedure for reconstructing both the MPFL and medial patellotibial ligaments by semitendinosus tendon with gracilis autograft augmentation in skeletally immature patients with recurrent patellar dislocation which is effective and permits satisfactory patellar congruency documented by static and dynamic CT [30]. Wang et al. proposed a novel suture-tie technique of patellar side fixation in MPFL reconstruction for recurrent patellar dislocation and reported that this technique can restore patella stability without significant complication [31]. Ma et al. compared the medial retinaculum plasty and medial capsule reefing and found that medial retinaculum plasty is improving the subjective effects and decreasing the rate of patellar instability postoperatively in children and adolescent patients [32].

Many reports of patellofemoral instability treatment suffer the same flaws of inappropriate patient selection, poor injury definition, insufficient activity assessment, and, especially in skeletally immature patients, limited follow-up found in other orthopedic literature [13]. Late diagnosis of advanced disease results in a poor prognosis. Albuquerque et al. classified the patients according to the duration of preoperative symptoms, and they reported unsatisfactory results in the patients who had more than 10 years of complaint [33]. They claimed patients with long-lasting symptoms or more severe disease seem to achieve better results with combined proximal and distal realignment techniques. Also, trochlear dysplasia seems to be a major risk factor for failure of operative stabilization of recurrent patellofemoral instability in children and adolescents [8]. Patients with both immature physes and trochlear dysplasia had a recurrence rate of 69 % [12]. Joint hypermobility is not a contraindication to

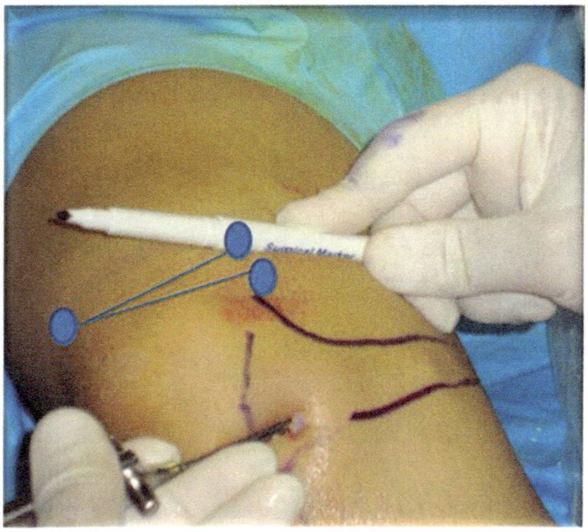

Fig. 26.1 Determination of topographic anatomy of MPFL

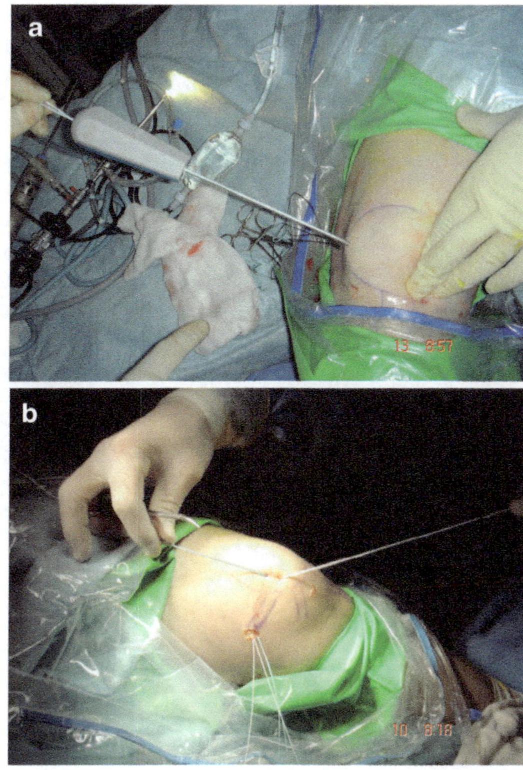

Fig. 26.2 (**a**, **b**) We use 2 number 3 or 5 suture anchors for this operation. One of the anchors is placed into the anatomic patellar attachment of MPFL, and the other one is placed into the femoral insertion from 2 stab incision. Margo medialis of patella is superficial so it is not difficult to identify the correct anatomic area of suture at patella, but this is not applicable at femoral side. We use imaging system in order to mark optimal anatomic area in femur through sagittal plane

MPFL reconstruction although caution is recommended in managing the expectations of patients with hypermobility before consideration of surgery [34]. Complete rupture of the MPFL during traumatic patellar dislocation may contribute to further patellar instability. Female patients with a history of atraumatic recurrent dislocation and all patients with history of previous surgery had a significantly worse outcome with MPFL reconstruction [35].

We improve a new minimal invasive technique for first-time patella dislocations [36]. First we draw topographic anatomy of MPFL on the skin over the medial side while the knee is in 30° flexion (Fig. 26.1).

Arthroscopic entrance is made from anterolateral portal. MPFL is not seen in this location because it is out of capsule. Sutures are directed by knowing its anatomic impression. In case of small amount of knee flexion, sutures start to be made preferably with #5 PDS (absorbable sutures or anchors, especially absorbable one is preferable) before the adductor tubercle level and by passing them thorough a position out of the capsule and under the skin (Fig. 26.2a, b).

Intra-articular section is entered by patella medial side and adductor tubercle, which is also the starting point, is reached again and three individual sutures are done at three levels from upside down and adductor is determined at the tubercle level while knots are at 30° flexion (Fig. 26.3a–c).

No forcing that pushes the patella to medial should be implemented. In this technique, direct procedures are done in case of acute dislocations, whereas same procedure is done with chronic ones by cutting medial capsule with radiofrequency. Lateral capsular relaxation absolutely should not be done if the outer tissues are not tensile. Physiotherapy is started in a short time after post-op compression. No devices are used. The main philosophy of this new technique is sutures are serving as an internal bracing to restore ideal length and environment for the biologic healing of the capsuloligamentous structures.

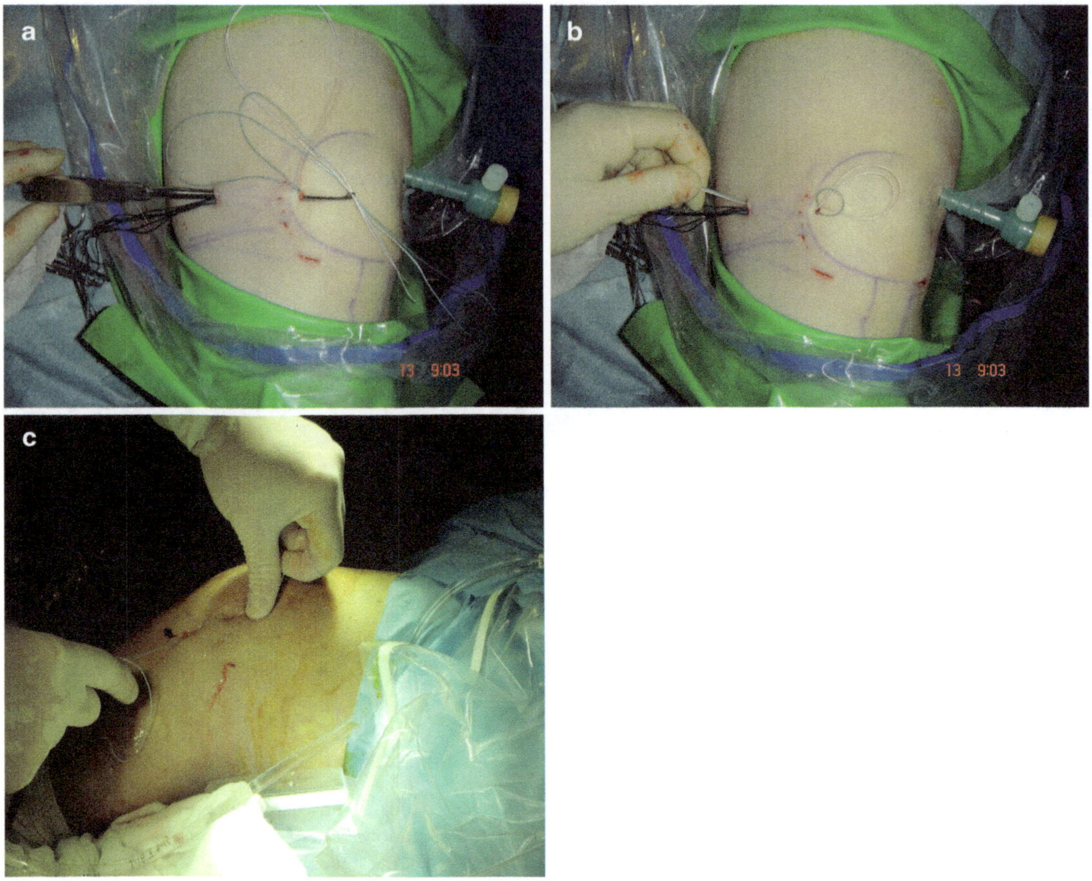

Fig. 26.3 (**a**) Carry the sutures of patella to the femoral side over the second layer by a suture passer. (**b**) It is very critical that the patellar relation with trochlea has to be controlled with arthroscope ın order to restore the ideal isometry and tension of the ligament. (**c**) Finally, we are tying the sutures at the femoral stab incision when knee is flexed approximately 30–40°

In conclusion, the treatment of patellofemoral instability in an adolescent athlete is a challenging clinical problem. The patient and parents should be informed about the possibility of ongoing instability episodes related to relative insufficiency of the distal realignment procedures and about the possible need for a subsequent operation after the bone maturation.

References

1. Frank JB, Jarit GJ, Bravman JT, Rosen JE. Lower extremity injuries in the skeletally immature athlete. J Am Acad Orthop Surg. 2007;15(6):356–66.
2. Aydogdu S. Management of patellofemoral problems in adolescents. In: Doral MN, editor. Sports injuries: prevention, diagnosis, treatment and rehabilitation. Part XIV: pediatric sports injuries. Springer verlag; 2011. p. 1017–22. ISBN978-3-642-15629-8.
3. Barber Foss KD, Myer GD, Chen SS, Hewett TE. Expected prevalence from the differential diagnosis of anterior knee pain in adolescent female athletes during preparticipation screening. J Athl Train. 2012;47(5):519–24.
4. Steinberg N, Siev-Ner I, Peleg S, Dar G, Masharawi Y, Zeev A, Hershkovitz I. Joint range of motion and patellofemoral pain in dancers. Int J Sports Med. 2012;33(7):561–6.
5. Kaya D, Doral MN. Is there any relationship between Q-angle and lower extremity malalignment? Is there any relationship between Q-angle and lower extremity malalignment? Acta Orthop Traumatol Turc. 2012; 46(6):416–9.
6. Kaya D, Atay OA, Callaghan MJ, Cil A, Cağlar O, Citaker S, Yuksel I, Doral MN. Hallux valgus in patients with patellofemoral pain syndrome. Knee

Surg Sports Traumatol Arthrosc. 2009;17(11): 1364–7.

7. Noehren B, Pohl MB, Sanchez Z, Cunningham T, Lattermann C. Proximal and distal kinematics in female runners with patellofemoral pain. Clin Biomech (Bristol, Avon). 2012;27(4):366–71.
8. Nelitz M, Theile M, Dornacher D, Wölfle J, Reichel H, Lippacher S. Analysis of failed surgery for patellar instability in children with open growth plates. Knee Surg Sports Traumatol Arthrosc. 2012;20(5):822–8.
9. Pattyn E, Mahieu N, Selfe J, Verdonk P, Steyaert A, Witvrouw E. What predicts functional outcome after treatment for patellofemoral pain? Med Sci Sports Exerc. 2012;44(10):1827–33.
10. Saxena A, Haddad J. The effect of foot orthoses on patellofemoral pain syndrome. J Am Podiatr Med Assoc. 2003;93(4):264–71.
11. Barber Foss KD, Hornsby M, Edwards NM, Myer GD, Hewett TE. Is body composition associated with an increased risk of developing anterior knee pain in adolescent female athletes? Phys Sportsmed. 2012;40(1):13–9.
12. Lewallen LW, McIntosh AL, Dahm DL. Predictors of recurrent instability after acute patellofemoral dislocation in pediatric and adolescent patients. Am J Sports Med. 2013;41(3):575–81.
13. Hinton RY, Sharma KM. Acute and recurrent patellar instability in the young athlete. Orthop Clin North Am. 2003;34(3):385–96.
14. Vavken P, Wimmer MD, Camathias C, Quidde J, Valderrabano V, Pagenstert G. Treating patella instability in skeletally immature patients. Arthroscopy. 2013;29(8):1410–22.
15. Apostolovic M, Vukomanovic B, Slavkovic N, Vuckovic V, Vukcevic M, Djuricic G, Kocev N. Acute patellar dislocation in adolescents: operative versus nonoperative treatment. Int Orthop. 2011;35(10): 1483–7.
16. Camanho GL, Viegas Ade C, Bitar AC, Demange MK, Hernandez AJ. Conservative versus surgical treatment for repair of the medial patellofemoral ligament in acute dislocations of the patella. Arthroscopy. 2009;25(6):620–5.
17. Beran MC, Samora WP, Klingele KE. Weight-bearing osteochondral lesions of the lateral femoral condyle following patellar dislocation in adolescent athletes. Orthopedics. 2012;35(7):e1033–7.
18. Fithian DC, Paxton EW, Stone ML, Silva P, Davis DK, Elias DA, White LM. Epidemiology and natural history of acute patellar dislocation. Am J Sports Med. 2004;32(5):1114–21.
19. Nelitz M, Dreyhaupt J, Reichel H, Woelfle J, Lippacher S. Anatomic reconstruction of the medial patellofemoral ligament in children and adolescents with open growth plates: surgical technique and clinical outcome. Am J Sports Med. 2013;41(1):58–63.
20. Panni AS, Vasso M, Cerciello S. Acute patellar dislocation. What to do? Knee Surg Sports Traumatol Arthrosc. 2013;21(2):275–8.
21. Felus J, Kowalczyk B. Age-related differences in medial patellofemoral ligament injury patterns in traumatic patellar dislocation: case series of 50 surgically treated children and adolescents. Am J Sports Med. 2012;40(10):2357–64.
22. Putney SA, Smith CS, Neal KM. The location of medial patellofemoral ligament injury in adolescents and children. J Pediatr Orthop. 2012;32(3):241–4.
23. Seeley M, Bowman KF, Walsh C, Sabb BJ, Vanderhave KL. Magnetic resonance imaging of acute patellar dislocation in children: patterns of injury and risk factors for recurrence. J Pediatr Orthop. 2012;32(2):145–55.
24. Hennrikus W, Pylawka T. Patellofemoral instability in skeletally immature athletes. Instr Course Lect. 2013;62:445–53.
25. Wagner D, Pfalzer F, Hingelbaum S, Huth J, Mauch F, Bauer G. The influence of risk factors on clinical outcomes following anatomical medial patellofemoral ligament (MPFL) reconstruction using the gracilis tendon. Knee Surg Sports Traumatol Arthrosc. 2013; 21(2):318–24.
26. Aulisa AG, Falciglia F, Giordano M, Savignoni P, Guzzanti V. Galeazzi's modified technique for recurrent patella dislocation in skeletally immature patients. J Orthop Sci. 2012;17(2):148–55.
27. Grannatt K, Heyworth BE, Ogunwole O, Micheli LJ, Kocher MS. Galeazzi semitendinosus tenodesis for patellofemoral instability in skeletally immature patients. J Pediatr Orthop. 2012;32(6):621–5.
28. Kumahashi N, Kuwata S, Tadenuma T, Kadowaki M, Uchio Y. A "sandwich" method of reconstruction of the medial patellofemoral ligament using a titanium interference screw for patellar instability in skeletally immature patients. Arch Orthop Trauma Surg. 2012; 132(8):1077–83.
29. Yercan HS, Erkan S, Okcu G, Ozalp RT. A novel technique for reconstruction of the medial patellofemoral ligament in skeletally immature patients. Arch Orthop Trauma Surg. 2011;131(8):1059–65.
30. Giordano M, Falciglia F, Aulisa AG, Guzzanti V. Patellar dislocation in skeletally immature patients: semitendinosous and gracilis augmentation for combined medial patellofemoral and medial patellotibial ligament reconstruction. Knee Surg Sports Traumatol Arthrosc. 2012;20(8):1594–8.
31. Wang JL, Li HP, Liu YJ, Wang N, Li ZL, Wang ZG, Wei M. Reconstruction of the medial patellofemoral ligament with a suture-tie technique of patellar side fixation. Chin Med J (Engl). 2012;125(11): 1884–8.
32. Ma LF, Wang CH, Chen BC, Zhang F, Zhou JW, Dong JT, Wang F. Medial patellar retinaculum plasty versus medial capsule reefing for patellar dislocation in children and adolescents. Arch Orthop Trauma Surg. 2012;132(12):1773–80.
33. Albuquerque RF, Pacheco AP, Hernandez AJ, Amatuzzi MM, Pécora JR, Kokron AE. Appraisal of surgical treatment of 47 cases of patellofemoral

instability. Rev Hosp Clin Fac Med Sao Paulo. 2002; 57(3):103–7.
34. Howells NR, Eldridge JD. Medial patellofemoral ligament reconstruction for patellar instability in patients with hypermobility: a case control study. J Bone Joint Surg Br. 2012;94(12):1655–9.
35. Howells NR, Barnett AJ, Ahearn N, Ansari A, Eldridge JD. Medial patellofemoral ligament reconstruction: a prospective outcome assessment of a large single centre series. J Bone Joint Surg Br. 2012; 94(9):1202–8.
36. Doral MN, Tetik O, Atay OA, Leblebicioğlu G, Aydoğ T, Akarcali I, Kaya D. Patellar instability: arthroscopic surgery, indications and techniques. Acta Orthop Traumatol Turc. 2004;38 Suppl 1:119–26. Review. Turkish.

Treatment of Patellofemoral Dislocation in Skeletally Immature

27

Masataka Deie and Mitsuo Ochi

Contents

M. Deie (✉)
Laboratory of Musculoskeletal Functional Research and Regeneration, Institute of Biomedical & Health Sciences, Hiroshima University, Hiroshima, Japan
e-mail: masatakadeie@yahoo.co.jp

M. Ochi
Department of Orthopaedic Surgery, Institute of Biomedical & Health Sciences, Hiroshima University, Hiroshima, Japan

27.1 Introduction

Patellar dislocation is relatively common, accounting for up to 16 % of acute knee trauma in young athletes with hemarthrosis [1]. The medial patellofemoral ligament (MPFL) is the primary medial soft-tissue restraint, preventing lateral displacement of the patella in the normal knee. MPFL also weighted 60 % to protect against lateral force [2]. When the patella is dislocated laterally, injury to the MPFL may predispose to recurrent instability; thus even young patients are recommended for surgical treatment. The current trend for treatment of patellar dislocation is the anatomical MPFL reconstruction; we have previously reported clinical results using this surgical procedure [3]. Including our procedure, many surgeons described MPFL reconstructions designed to fix the MPFL attachment of the femur side; however, this procedure presents difficulties in children including the original point of the MPFL femoral attachment being closed to the physis of the distal femoral. The senior surgeon (MO) performed original procedure for children from 1986 [4], and we reported results from our surgical procedures to treat patellar dislocation in children with 7.4 years' follow-up in 2003 [5]. In this paper we report additional results of our surgical procedure for immature patellar dislocation in patients treated since 2002.

A. Gobbi et al. (eds.), *The Patellofemoral Joint*,
DOI 10.1007/978-3-642-54965-6_27, © ISAKOS 2014

Table 27.1 Clinical details of the five children (six knees) who underwent MPFL reconstruction with a transferred semitendinosus tendon

Case	Gender	Age at surgery	Diagnosis	Follow-up terms (years)	Additional surgery	
1	F	9	Habitual	10	Distal realignment	After 3 years
2	F	6	Habitual	8		
2	F	7	Habitual	7		
3	F	9	Habitual	8	Distal realignment	After 5 years
4	F	4	Habitual	5		
5	F	11	Recurrent	2		
Mean		7.7		6.7		

27.2 Patients and Methods

We performed surgical treatments for 115 patellar dislocations from April 2002 to December 2012, including six immature patellar dislocated knees in five cases (Table 27.1). These immature patients were all females with a mean age at surgery of 7.7 years (4–11 years). They consisted of five habitual dislocation knees and one recurrent dislocation knee. The mean follow-up period was 6.7 years (range, 2–10 years).

27.3 Surgical Procedure

27.3.1 Lateral Release

For habitual patellar dislocation, we performed lateral release arthroscopically by using an electrodevice or plastic scissors. For the recurrent patellar dislocations, before surgery, the patellar stress radiography views were obtained at 45° knee flexion with 20 N stress from the medial to lateral direction and from the lateral to medial direction using a pushing apparatus (Imada, Toyohashi, Japan). If the patella showed only lateral instability, we chose to perform a lateral release.

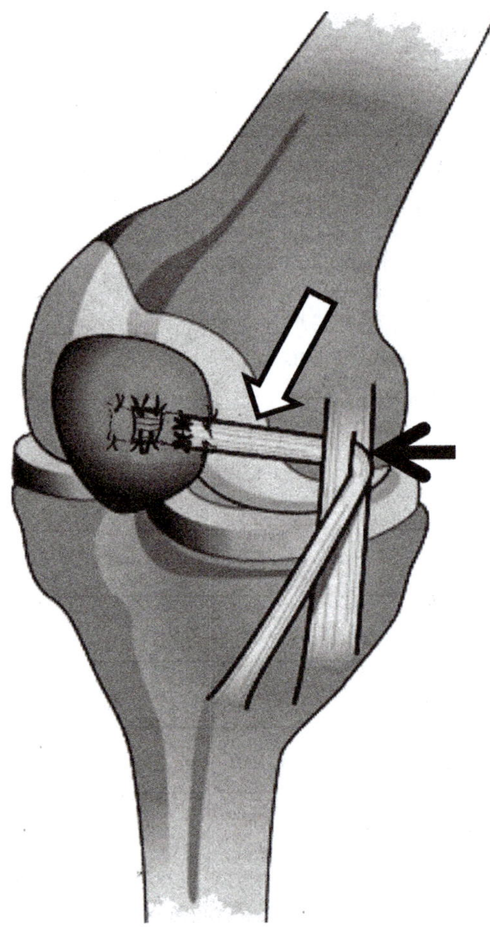

Fig. 27.1 Schema of MPFL reconstruction. The semitendinosus tendon (*white arrow*) is transferred to the patella using in the posterior one-third of the proximal attachment of MCL attachment (*black arrow*)

27.3.2 MPFL Reconstruction with the Transferred Semitendinosus Tendon

We performed this procedure as described previously [5]. Briefly, a 4-cm incision was made over the insertion point of the semitendinosus tendon (ST), which was then divided at its musculotendinous junction using an open tendon stripper,

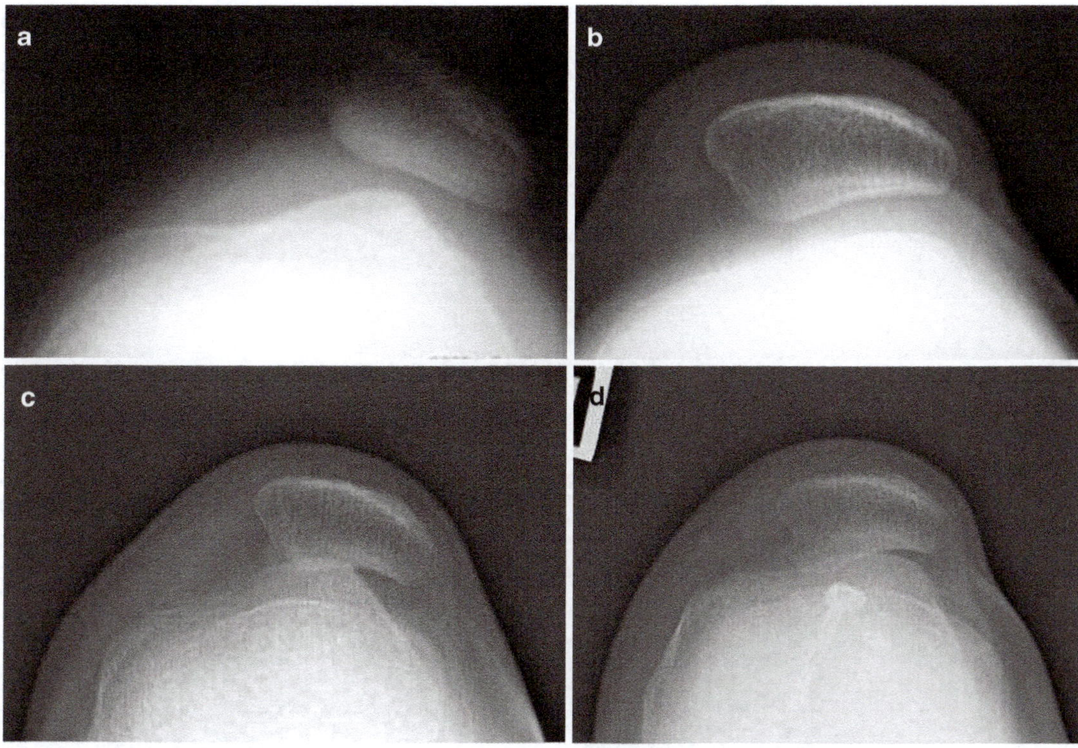

Fig. 27.2 Case 1: 9-year-old girl with left habitual dislocation. These X-ray images represent axial views. (**a**) The patella was laterally dislocated at 45° knee flexion before the MPFL reconstruction. (**b**) Left knee 1 year after MPFL reconstruction. The patient reported no patellar instability and could play sports. (**c**) Left knee 2 years after the MPFL reconstruction; at 11 years of age, the patient felt her left patellar laterally dislocate again. (**d**) Left knee at 7 years after the MPFL reconstruction and 4 years after the distal realignment; no further patella dislocation or apprehension signs were reported

leaving it attached distally. Through a 2-cm incision over the femoral attachment of the MCL, we made a 1-cm slit in the posterior one-third of the MCL to act as a pulley. Then a curved incision over the patella was applied and the ST was transferred to the patella through the pulley. The pulley point was slightly anterior to the original femoral attachment of the MPFL. The transferred tendon was then sutured to the surface of the patella with the knee flexed at 30° (Fig. 27.1). When the surgeon was suturing the transferred ST on the patellar surface, the surgical assistant held the patella to keep it in line with the lateral patella edge and lateral side of the femoral lateral condyle. Finally, the tracking of the patella and the transferred tendon from 0° to full flexion was assessed. Ideally, at this point, the tension should increase slightly when the knee is fully extended.

27.3.3 Postoperative Rehabilitation

After MPFL reconstruction surgery, the knee was fixed with a brace for 2 weeks, and then range of motion exercises were started. At 3 weeks after surgery, the patients were allowed to adopt a partial weight-bearing gait, and after 5–6 months, normal sports activity could be resumed.

27.4 Clinical Results

We surgically treated six patellar dislocations in immature patients by MPFL reconstruction and lateral release. In two of five habitual dislocations, re-dislocations occurred and treated again by the distal realignment surgery: one at 3 years after the MPFL reconstruction (Fig. 27.2a–d) and

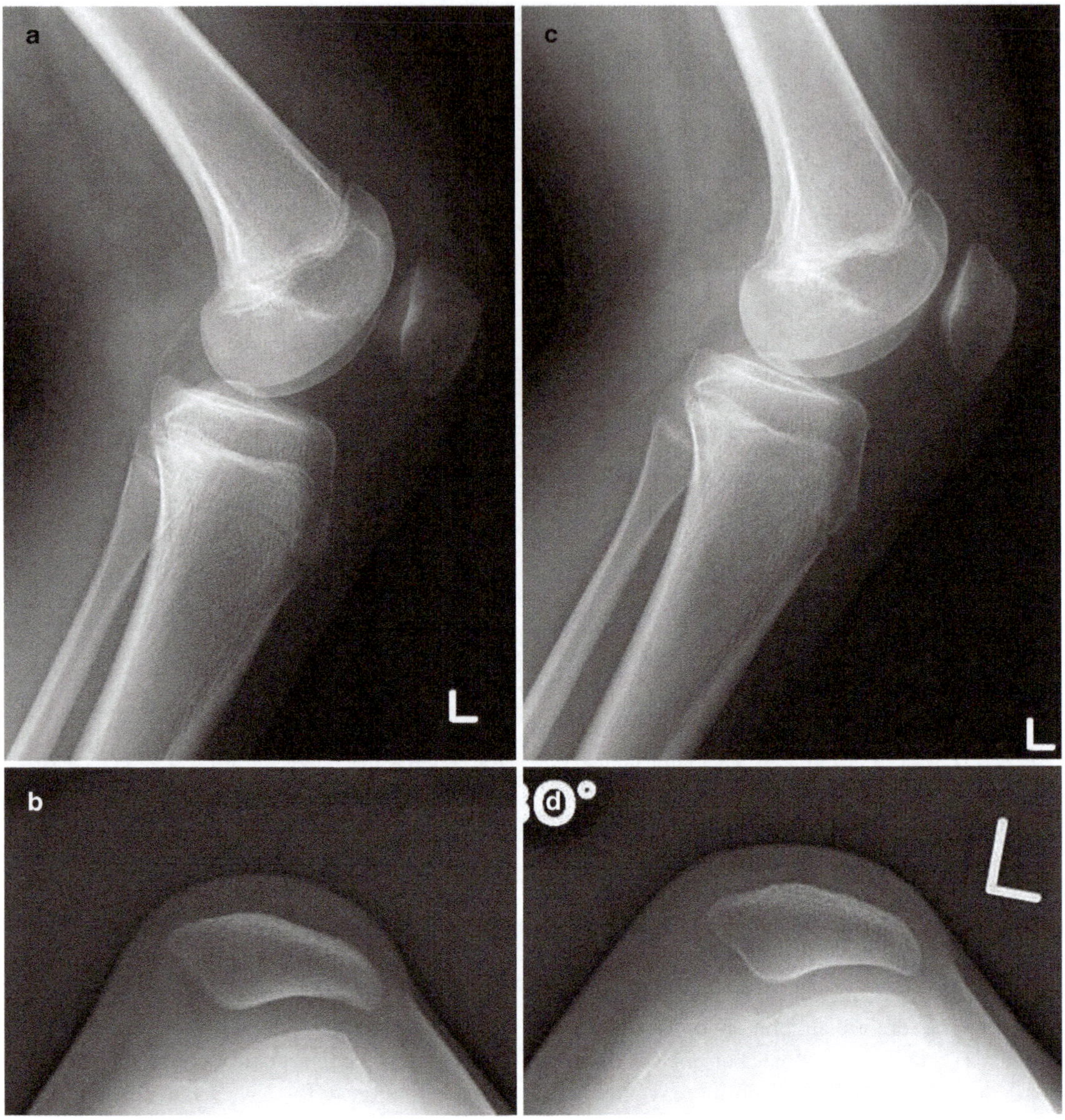

Fig. 27.3 Case 5:11-year-old girl with left recurrent patellar dislocation. These X-ray images represent lateral views (**a**, **c**) and axial views (**b**, **d**). The patella was laterally shifted and tilted at 30° knee flexion before MPFL reconstruction (**a**, **b**) and the patient showed positive apprehension signs. Left knee 2 years after the MPFL reconstruction (**c**, **d**). There was no patellar instability and she could be participated in playing any sports

the other at 5 years. The remaining four knees showed dislocation recurrence or reappearance of apprehension signs (Fig. 27.3a–d).

27.5 Discussion

Patellar dislocation is not rare to the immature children. Indeed, we have experienced both habitual and recurrent patellar dislocation in children. Currently, the first choice of surgical procedure for patellar dislocation is MPFL reconstruction, although such procedures present special difficulties in the immature patients because the femoral site of MPFL attachment is closed to the physis, increasing the risk of physis damage during MPFL reconstruction surgery. We reported our original MPFL procedure for immature children in 2003 [5]. The senior author (MO) developed this procedure based from the cadaver study [4]. The critical point in performing MPFL reconstructions in immature children is the pulley point sited at the posterior one-third of the MCL. Ochi et al. [4] found that the length changes from the posterior one – third of the MCL to the patellar insertion site of the MPFL are relatively less. Subsequent findings reported by Smirk et al. [6] on the length change from the femur site to the medial patellar site in cadaver also indicated that the pulley point used in our immature patients' MPFL reconstruction is relatively isometric.

In the current study, we also showed the transferred tendon grew at the same rate as the MPFL since no medial shift or tilt appeared with growth, consistent with our previous report [5]. Although two knees required distal realignment surgery for recurrent instability, we showed that our procedure is effective for treating immature children with habitual and recurrent patellar dislocation.

The study has some limitations. First, the transferred tendon does not mimic the native MPFL, which exists like a fan at the patella insertion, and second, the femoral position of the transferred tendon is not the natural attachment site. However, in this series, we experienced no major side effects such as difference of the leg length, growth disturbance, or valgus/varus knee deformity, while two habitual patellar dislocation patients required additional distal realignment operations due to recurrent patellar instability. Finally, we emphasize that this procedure could be applied to stabilize even habitual patellar dislocation knee safely in younger children.

References

1. Zaidi A, Babyn P, Astori I, White L, Doria A, Coleet W. MRI of traumatic patellar dislocation in children. Pediatr Radiol. 2006;36(36):1163–70.
2. Conlan T, Garth Jr WP, Lemons JE. Evaluation of the medial soft-tissue restraints of the extensor mechanism of the knee. J Bone Joint Surg Am. 1993;75: 682–93.
3. Deie M, Ochi M, Adachi N, Shibuya H, Nakamae A. Medial patellofemoral ligament reconstruction fixed with a cylindrical bone plug and a grafted semitendinosus tendon at the original femoral site for recurrent patellar dislocation. Am J Sports Med. 2011;39:140–5.
4. Ochi M, Okada Y, Manabe H, Mochiduki Y, Sasashige Y, Jitsuiki Z, Yamanaka T, Ikuta Y. The medial tenodesis for patellar dislocation – from the view points of the isometricity – (in Japanese). Hiza. 1989;15:45–50.
5. Deie M, Ochi M, Sumen Y, Yasumoto M, Kobayashi K, Kimura H. Reconstruction of the medial patellofemoral ligament for the treatment of habitual or recurrent dislocation of the patella in children. J Bone Joint Surg Br. 2003;85:887–90.
6. Smirk C, Morris H. The anatomy and reconstruction of the medial patellofemoral ligament. Knee. 2003;10: 221–7.

Matrix Autologous Chondrocyte Implantation of the Patella: From ACI to MACI to ICC

28

Stephen P. Abelow, Pedro Guillen, Isabel Guillen, Marta Guillen, Santiago Arauz de Robles, Elena Rodrigo Inigo, Tomas Fernandez Jaen, and Juan Manuel Lopez Alcorocho

Contents

28.1 Introduction

Patellofemoral cartilage lesions can often be difficult to treat. Because of the unique biomechanics of the patellofemoral joint, even subtle alterations of the mechanics secondary to instability, abnormal joint anatomy, or malalignment can lead to alteration in cartilage homeostasis with ultimate destruction of the articular cartilage. Treatment of large full-thickness articular cartilage lesions of the patella greater than 2.5–3.0 cm^2 that have failed previous conservative and operative treatment modalities are a considerable problem to deal with in young and middle-aged population (Figs. 28.1 and 28.2).

If encountered acutely or if a large fragment exists, often these lesions can be stabilized and internally fixed with metallic or bioabsorbable implants. Many of these lesions, however, progress and large osteochondral lesions of the patella that are too large to be treated by methods such as debridement and lavage, excision of the fragment and stimulative/reparative procedures such as marrow stimulating techniques (microfracture, drilling, abrasion chondroplasty), or restorative techniques such as osteochondral transfer procedures (OATS or mosaicplasty). A series of 85 patients with full-thickness chondral injuries of the knee treated with microfracture by Kreuz, Erggelet, and Steinwachs reported that lesions on the trochlea or on the patella had worse results (regardless of patient age) and deteriorated over time [1]. Regenerative processes such as osteochondral

S.P. Abelow (✉) • I. Guillen • M. Guillen
S.A. de Robles • E.R. Inigo • T.F. Jaen
J.M.L. Alcorocho
Unidad de Inverstigation Biomedica,
Clinica CEMTRO, Madrid, Spain
e-mail: sabelow@aol.com

P. Guillen
Unidad de Inverstigation Biomedica,
Clinica CEMTRO, Madrid, Spain

Universidad Catolica San Antonio de Murcia,
Murcia, Spain

A. Gobbi et al. (eds.), *The Patellofemoral Joint*,
DOI 10.1007/978-3-642-54965-6_28, © ISAKOS 2014

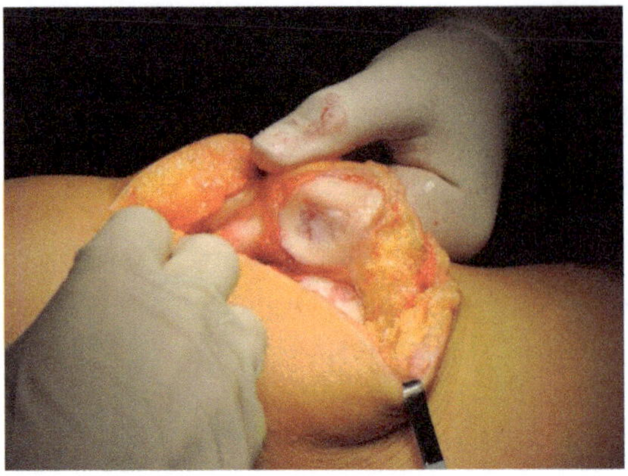

Fig. 28.1 Chondral defect of the patella

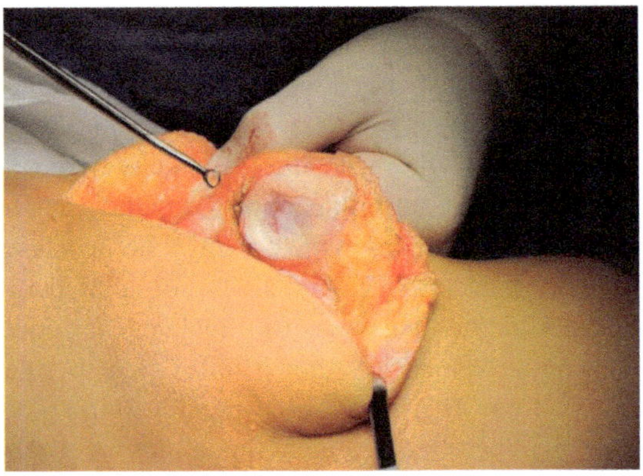

Fig. 28.2 Chondral defect of patella curetted

allograft or some form of autologous chondrocyte implantation may be able to treat these large lesions.

The indications for cartilage regenerative surgical techniques are symptomatic deep lesions classified by the International Cartilage Repair Society (ICRS) Grade 3: deep greater than 50 % cartilage depth and down to but not through subchondral bone and Grade 4: subchondral bone exposed (with lesions extending through the subchondral bone plate or deeper into the trabecular bone). There should be no uncorrected malalignment or instability and no significant arthritis. Any type of maltracking or malalignment of the patella must be corrected.

First-generation autologous chondrocyte implantation (ACI) is the implantation of in vitro cultured autologous chondrocytes using a periosteal tissue cover after expansion of the isolated chondrocytes.

Autologous chondrocyte implantation that was first reported by Brittberg, Peterson, and coworkers in Gothenburg, Sweden, in 1994 [2] has been successfully utilized in the knee. The procedure has yielded good-to-excellent results in greater than 77 % of deep chondral lesions (92 % in isolated femoral condyle lesions; 65 % in the patella) with more than 9 years' follow-up [3, 4].

ACI is a two-stage process. Articular cartilage chondrocytes are harvested by either arthroscopic or open techniques from a normal area of the knee. The chondrocytes are cultured in vitro for 3–5 weeks, expanded and reimplanted by arthrotomy or mini-arthrotomy. In the first-generation ACI, a periosteal graft was harvested and sutured in place over the chondral defect in a "watertight" manner (2–3 mm apart). The cultured autologous chondrocytes are then injected onto the defect under the periosteal patch and the arthrotomy incision is closed. This often requires a wide arthrotomy incision to be made to allow for proper suturing of the periosteal patch. The use of the periosteal patch has caused complications such as symptomatic periosteal hypertrophy, delamination of the defect, and intra-articular adhesions [3, 4].

A second-generation ACI utilizes a biocompatible bioabsorbable membrane to cover the chondral defect. A bilayer, absorbable, porcine collagen I/III membrane (Chondro-Gide, Geistlich Biomaterial, Wolhusen, Switzerland) has been used to avoid the problems of the periosteal patch. The membrane is degraded by enzymatic division (i.e., collagenase), and the resultant collagen fragments denature at 37 ° C to gelatin.

This collagen-covered autologous chondrocyte implantation (CACI) has been used extensively in Europe, Australia, and New Zealand instead of a periosteal patch. In a prospective study presented at the International Cartilage Repair Society in 2004, Steinwachs [5] described 163 patients treated for chondral defects of the knee with ACI using a periosteal patch of Chondro-Gide membrane instead of the periosteal patch. Seventy eight percent of patients in the periosteal group reported good or excellent, and 88 % of patients in the Chondro-Gide group reported good or excellent results. There was no case of membrane hypertrophy in the Chondro-Gide group (statistical significance was not discussed by the investigators) [5].

Matrix/membrane-induced autologous chondrocyte implantation (MACI) is a third-generation chondrocyte implantation process. Cultured autologous chondrocytes are impregnated onto a highly purified porcine collagen I/III membrane (Sanofi/Genzme Biosurgery). The MACI implant can be secured to the chondral defect by fibrin glue (with little or no suture necessary), suture, or bioabsorbable pins or tacks. No periosteal graft is needed. The procedure can be performed arthroscopically or by mini-arthrotomy [6].

Initially chondrocytes are harvested arthroscopically from a non-weight-bearing area of the ipsilateral knee (200–300 mg of healthy cartilage). The chondrocytes are then cultured, expanded in vitro (in 3–5 weeks), and then impregnated on an absorbable three-dimensional bilayered, purified porcine collagen I/III membrane. The bilayer structure has a smooth side that is nonporous acting as a natural barrier and faces the joint. Chondrocytes are seeded on the porous side of the matrix. The membrane is tear resistant and can be easily templated, trimmed, and cut to shape. The membrane is not self-adherent and can be "rolled-up" and handled with standard arthroscopic instrumentation, allowing for arthroscopic implantation of the membrane [6, 7] .The membrane is nonantigenic (telopeptides are split during the manufacturing process) and is bioabsorbable. The bioabsorbable membrane can be fixed to the patella cartilage defect with fibrin glue, pins, or suture.

Utilizing mini-arthrotomy or arthrotomy techniques, the cartilage defect of the patella is debrided and curetted with a sharp ring curette to remove the calcified fibrous cartilage layer without penetrating the subchondral bone (Avoid bleeding of the subchondral bone!) (Fig. 28.2). A stable cartilage rim with sharp vertical walls of healthy cartilage is created on the patella (Note: all "damaged" cartilage should be debrided back to a healthy stable border) (Figs. 28.3 and 28.4). Intralesional osteophytes, if any, should be removed. The chondral defect is measured and templated (Fig. 28.6). The MACI membrane is cut to the proper shape with a scalpel or scissors (Fig. 28.7). The membrane is then fixed with fibrin glue (Tisucol, Baxter, Spain). Suture is used for the patella (Figs. 28.9 and 28.10).

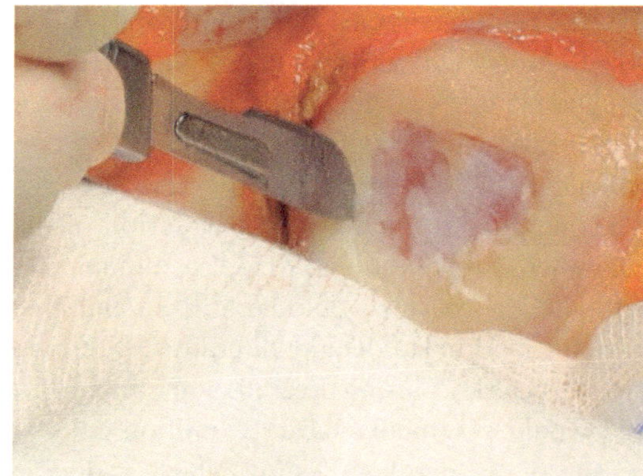

Fig. 28.3 Create a stable rim with stable walls

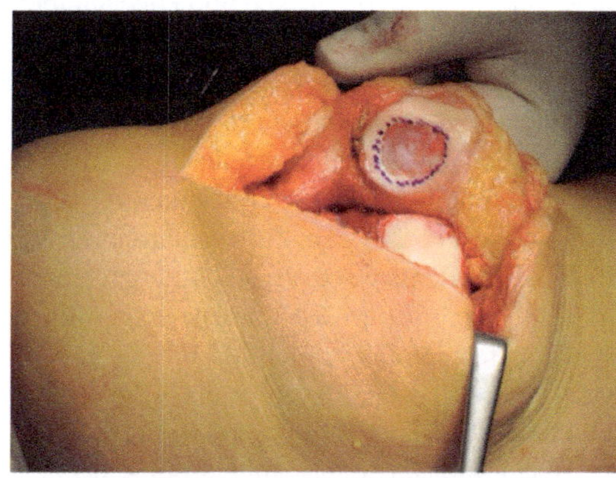

Fig. 28.4 Cartilage lesion patella with stable vertical wall

Postoperatively the patient is placed in a soft dressing and placed and on continuous passive motion (when available) for 8 weeks. The patient is kept to partial weight bearing activity for 8 weeks. Larger and more central lesions are kept partial weight bearing for 12 weeks.

The MACI membrane as currently supplied is now 3 × 5 cm and is seeded with one million chondrocytes per cm^2 for a total of 15 million chondrocytes (previously the membrane size was 4 × 5 cm^2 with a total of 20 million chondrocytes).

If one were to treat a 3 × 2 cm^2 lesion of the patella with the traditional MACI technique, six million chondrocytes would be utilized and nine million chondrocytes would "literally" be thrown away. The same lesion treated with traditional ACI would potentially have 12 million cells at the site of the cartilage lesion, which is double the amount of chondrocytes delivered to the same sized lesion treated with MACI. Since one side of the membrane is not porous, the MACI membranes really cannot be stacked upon one another.

At Clinica CEMTRO/Universidad Catolica San Antonio de Murcia (UCAM), the concept of "Cell Density" concerning chondrocytes was investigated. In a recent article in Cartilage, Foldger, Gomol, Lind, and colleagues reported, "In the absence of systematic evaluations of the effects of cell density and clinical outcome, many clinicians continue to use one or two million chondrocytes per cm^2, which, despite its lack of evidence and the fact that most in vitro studies point toward benefits of high intensities, has been associated with favorable clinical outcomes and nearly approximates the densities found in native adult articular cartilage" [8].

In an attempt to determine which type of cell (mesenchymal cell or chondrocyte) and the number of cells per square centimeter are "optimal," at clinica CEMTRO, we studied 15 female merino sheep with articular cartilage lesions treated with autologous chondrocytes or mesenchymal cells seeded onto a porcine collagen I/III membrane. Experimental groups were five million chondrocytes per cm^2, one million chondrocytes per cm^2, five million mesenchymal cells per cm^2, and microfracture. All samples were analyzed for cellular histology, type I collagen, type II collagen, and aggrecan. The expression of aggrecans was seen in all samples. The expression profile of Col II (marker of hyaline cartilage) showed the control group was greater than five million chondrocytes which was greater than one million chondrocytes, which was greater than five million mesenchymal cells, which was greater than microfracture. The expression profile of Col I was microfracture greater than five million mesenchymal cells greater than one million chondrocytes greater than five million chondrocytes. The results were statistically significant. The histology showed five million and one million chondrocytes to have a more hyaline-like cartilage structure than either the microfracture or implantation of five million mesenchymal cells. Increasing the density of chondrocytes improved the quality of the regenerated tissue [9].

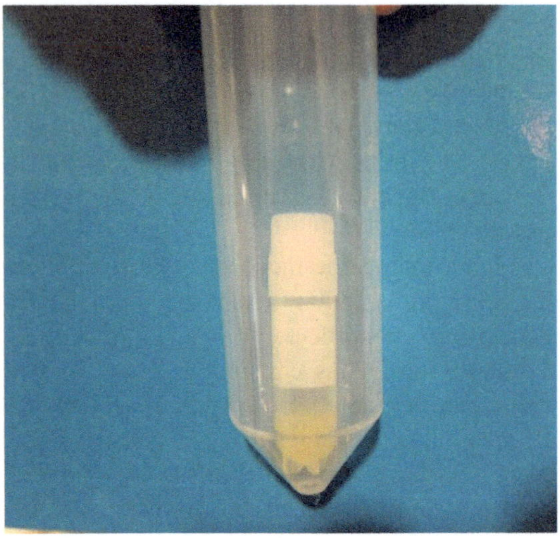

Fig. 28.5 Chondrocytes in cell culture

Based upon the fact that five million and one million chondrocytes demonstrated a better regenerative cartilage tissue, Clinica CEMTRO has developed a modification of the MACI procedure increasing the number of cells per cm^2 seeded on the collagen membrane (Instant CEMTROCELL-ICC, Madrid, Spain). After biopsy by arthroscopy, isolation of chondrocytes, and cell culture to 20 million cells, the cell suspension is transferred to the operating room (Fig. 28.5). The lesion is templated (Fig. 28.6). The Chondro-Gide membrane is cut to the size of the lesion (Figs. 28.7 and 28.8) and all of the cell

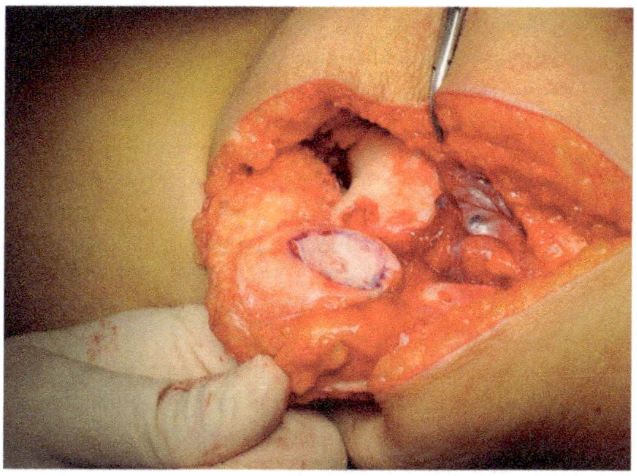

Fig. 28.6 Lesion of patella templated

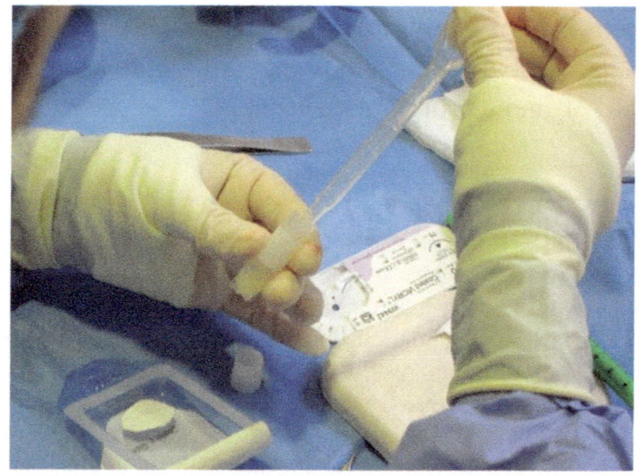

Fig. 28.9 ICC cultured chondrocytes placed on Chondro-Gide membrane

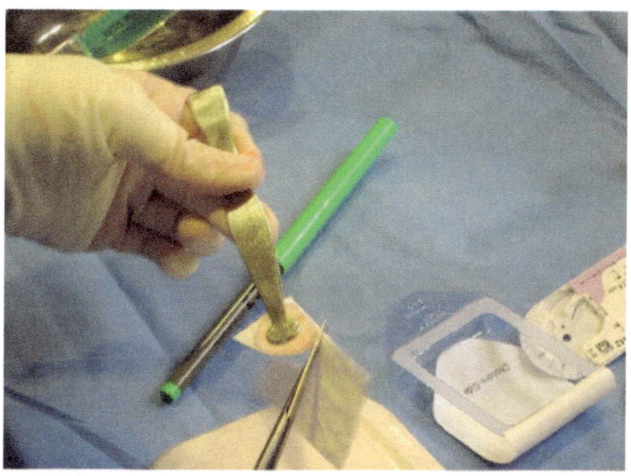

Fig. 28.7 Chondro-Gide membrane cut to size of lesion

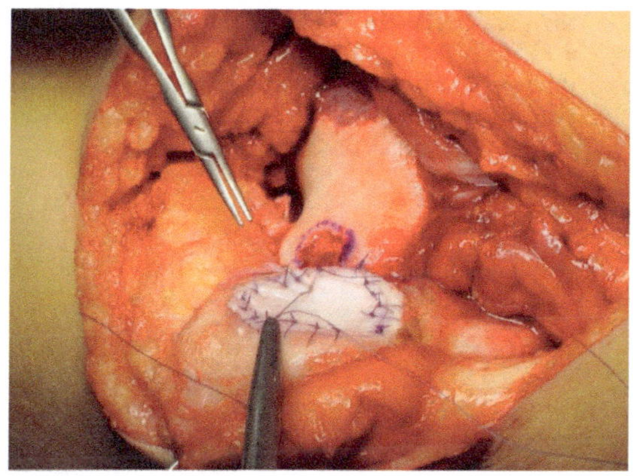

Fig. 28.10 ICC membrane sutured in place

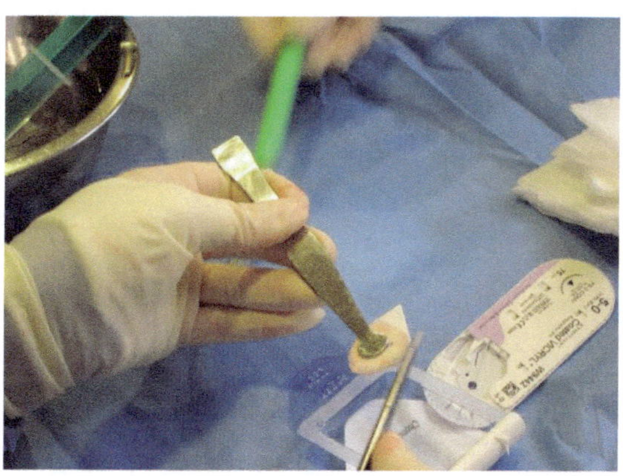

Fig. 28.8 Trimming the Chondro-Gide membrane

suspension is seeded on it (Fig. 28.9). The cells are seeded on the porcine collagen I/III membrane according to the method of Steinwachs [5]. The cultured chondrocytes are placed on the collagen membrane and after a 10-min period of time to allow for the absorption of chondrocytes, the membrane is implanted on the articular cartilage defect (e.g., a 2×3 cm^2 cartilage lesion would receive more than three million chondrocytes per cm^2) (Fig. 28.10).

Histological and genetic studies of ICC to date have shown a proliferation of collagen matrix, a population of viable mature chondrocytes, and immature population of chondrocytes with absence of expression of protein s-100, absence of atypical mitosis (absence of expression of P52), and a proliferative capacity.

References

1. Kure PC, Erggelet C, Steinbach's MR, et al. Is microfracture of chondral defects in the knee associated with different results in patients aged 40 years or younger? Arthroscopy. 2006;22:1180–6.
2. Brittberg M, Lindahal A, Nilson A, et al. Treatment of deep cartilage defects in the knee with autologous cartilage transplantation. N Engl J Med. 1994;331:889–95.
3. Peterson L. ACI surgical techniques and results at 2–10 years; in basic science clinical repair and reconstruction of articular cartilage defects: current status and prospects. Bologna: Timeo Editore; 2006. p. 325–32.
4. Peterson L, Minas T, Brittberg M, et al. Two to nine year outcome after autologous chondrocyte transplantation of the knee. Clin Orthop. 2000;374:212–34.
5. Steinwachs MR. ACI and resorbable collagen membrane: chondro-gide surgical technique and results, in basic science clinical repair and reconstruction of articular cartilage defects: current status and prospects. Bologna: Timeo Editore; 2006. p. 389–92.
6. Abelow SP, Guillen P, Ramos T. Arthroscopic techniques for matrix-induced autologous chondrocyte implantation for treatment of large chondral defects in the knee and ankle. Oper Tech Orthop. 2006;16(4):257–61.
7. Abelow SP, Guillen P, Fernandez T, Guillen I. Autologous chondrocyte implantation for the treatment of large chondral defects in the knee and ankle. Vancouver, BC: Athroscopy Association of North America Annual Meeting; 2005.
8. Foldager CB, Gomoll AH, Lind M. Spector: cell seeding densities in autologous chondrocyte implantation techniques for cartilage repair. Cartilage. 2012;3(2):108–17.
9. Guillen P, Rodriguez-Inigo E, Guillen I, Caballero R, Guillen M, Santos E, Abelow S, Lopez-Alcorocho JM. Histological and molecular study of cartilage repair with autologous implantation of chondrocytes or mesenchymal cells in the sheep animal model. Sitges: International Cartilage Repair Society World Congress; 2011.

Cartilage Lesions of the Patellofemoral Joint: Long-Term Results After Autologous Chondrocyte Implantation

29

Lars Peterson and Haris S. Vasiliadis

Contents

L. Peterson (✉)
Department of Orthopaedics,
University of Gothenburg, Gothenburg, Sweden
e-mail: peterson.lars@telia.com

H.S. Vasiliadis
Department of Orthopaedics,
University of Gothenburg, Gothenburg, Sweden

Department of Orthopaedics, School of Medicine,
University of Ioannina, Ioannina, Greece

A. Gobbi et al. (eds.), *The Patellofemoral Joint*,
DOI 10.1007/978-3-642-54965-6_29, © ISAKOS 2014

29.1 Introduction

The patellofemoral pain syndrome (anterior knee pain) is very common in the general population. It is often seen in young people who are very physically active in competitive and recreational sports and more often in teenage girls and young women. However, it is also very frequent in young people sustaining sedentary work (white collars or students). Patellar malalignment and instability with or without articular cartilage lesions in the patellofemoral joint are usually the source of pain. Articular cartilage injuries are very common in this patient group. Hielle et al. found that 17 % of patients having an arthroscopy had an articular cartilage injury located in the patella or trochlea [1]. Nomura et al. also found 35 patients with severe articular cartilage injuries of the patella in 37 patients with a first-time acute patellar dislocation [2].

Acute trauma or repetitive microtrauma may be the cause of cartilage lesions of the patellofemoral joint. Trauma with direct contusion of the patella is very common in many contact sports such as football, handball, ice-hockey, etc. Repetitive microtrauma with repeated loading/overloading is another common mechanism for cartilage erosions. Stress concentration between the articular surfaces results in progressive wear of the cartilage, subsequently causing kissing lesions.

Patellar malalignment and instability very often act as an undiagnosed background factor behind anterior knee pain, leading to or accelerating the cartilage lesions of the patella or trochlea [3].

The factors that may cause patellofemoral instability or contribute to it can be divided into three groups: skeletal, ligamentous, and muscular factors.

Skeletal factors include trochlear dysplasia, genu valgum, lateralized tibial tuberosity with increased Q angle, and patella alta. Besides the knee structures, other factors can also contribute including an increased femoral neck anteversion, excessive tibial torsion, tibia vara, and also flat foot [4].

Ligamentous factors include partial or total ruptures of the patella or knee joint stabilizers, ligament insufficiency after ruptures, or secondary lengthening [5].

Muscular factors include muscle imbalance especially between VMO and VL. VMO weakness or retarded VMO activation during the knee flexion has been accused. Besides, the pain inhibition will cause a prolonged reaction time of the VMO reducing the speed and strength of muscular function. That will also contribute to patellofemoral instability.

Treatment of the cartilage lesions of the patellofemoral joint alone may not have a satisfactory clinical outcome if other coexisting factors are ignored. Cartilage repair should be accompanied by the assessment of any potential malalignment or instability of the patella focusing on the correction of the background factors. The correction of those abnormalities is considered to be necessary for the cartilage treatment to be effective and durable [4].

29.2 Autologous Chondrocyte Implantation (ACI): Surgical Approach – Steps of Operation

Autologous chondrocyte implantation is a two-step procedure. The first step consists of the arthroscopic evaluation of the intra-articular pathology, confirmation of the cartilage lesion, and of indications for ACI. Biopsies are then retrieved from minor weight-bearing surfaces and are used as source of chondrocytes for cell culture.

29.2.1 Arthroscopy and Biopsy (1st Step of Procedure)

First, evaluate the intra-articular pathology including the characters of cartilage lesions. Assess the defect size, depth, location, and containment and also assess the opposing surfaces (for diagnosis of kissing lesions). Planning for the open procedure is undertaken, based on the arthroscopic findings along with the preoperative assessment (clinical and imaging). The planning should include the assessment of the lesions and potential need for correcting background factors (e.g., reconstruction of the extensor mechanism). Cartilage biopsies are taken from minor weight-bearing areas, usually from the upper medial trochlea, from the upper lateral trochlea, or from the lateral intercondylar notch. Our experience in approximately 1,600 biopsies (about 98 % taken from the upper medial trochlea) shows no donor site morbidity.

After the cartilage retrieval, any meniscus pathology should be corrected if needed.

29.2.2 In Vitro Culture of Chondrocytes

The harvested cartilage pieces are transferred from the operation theater to the cell-culture laboratory. The cartilage is placed in a sterile glass tube containing 0.9 % NaCl. Cell isolation procedure is initiated upon the arrival. The pieces are minced and washed twice in medium supplemented with antibiotics (usually gentamicin sulfate, amphotericin B, L-ascorbic acid, and glutamine). The minced cartilage is digested overnight (for 16–20 h) in a collagenase solution. The isolated cells are then washed and resuspended in flasks containing culture media with the addition of 10 % of the patient's own serum and antibiotics. The suspension is incubated in standardized conditions (7 % CO_2 in air at 37 °C). After about 1 week, the multiplied cells are trypsinized (released from the flask floor) and resuspended to more flasks. Finally, after about 2 weeks from the cartilage biopsy, the cultured cells are ready for the implantation according to the principles of first-generation ACI (cells are provided in suspension). Then, they can be trypsinized, isolated, and suspended in 0.3–0.4 mL of implantation medium, in a 1-ml syringe, in which they are transferred to the operation theater for the implantation. Another option is to be cryopreserved for future use, being on demand of the surgeon.

In second- and third-generation ACI, the cells are additionally injected (embedded) to a membrane (2nd generation) or cultured on and in a 3D scaffold (3rd generation). After a period of time necessary for the embedding or the culture, a membrane with the cells is ready to be contributed to the surgery for the implantation.

29.2.3 Autologous Chondrocyte Implantation and Assessment of Background Factors (2nd Step of Procedure)

Skin Incision: Approach

Use a short medial or lateral skin incision which should be adapted to the specific location of the cartilage defect, followed by a mini medial or lateral arthrotomy allowing tilting of the patella about 90° to be able to perform the surgery adequately.

Specifically in case of patellofemoral joint realignment procedures for cartilage lesions, a central skin incision is recommended allowing either medial or lateral parapatellar arthrotomy. When necessary adapt the incision accordingly to allow optimal access.

Use a central (midline) skin incision starting 1–2 cm proximal to basis patellae down and distal to the tibial tuberosity. After subcutaneous dissection start a medial parapatellar arthrotomy starting 1 cm proximal to the patella, between the rectus femoris and the vastus medialis obliquus, running 5–7 mm medial to the patellar insertion of the VMO and patellofemoral ligament. Continue down to the tibial condyle and incise the joint capsule and inspect the joint, especially the patella and trochlear groove. Evaluate the articular cartilage lesions along with the patellofemoral joint incongruity or any other malfunctions, and address the surgical plan for their treatment. Include a careful assessment of the Q-angle and the patellar tracking if the groove is dysplastic.

Correction of Background Factors

A specific surgical plan should be scheduled preoperatively according to the background factors that have been identified and need to be addressed. There is a specific order of all the stages that should be performed for the correction of the joint malalignment (also see Chap. 22). The order is shown below (Fig. 22.7):

(a) Lateral release.
(b) Tibial tuberosity transfer—unloading and correcting procedures (anteromedialization or anteriorization or distalization of tibial tuberosity). The screw fixation of tuberosity should be kept for after the implantation of cells.
(c) Trochlear grooveplasty (only in the cases with trochlear dysplasia).
(d) Autologous chondrocyte implantation (implantation of cells) (please see below).
(e) Screw fixation of tibial tuberosity (when a tibial tuberosity transfer is also performed).
(f) Medial plication.

Implantation of Cells

After the osteotomy of tibial tuberosity (in the cases where transfer is needed), proceed to the ACI (Fig. 29.1). Incise around the lesion and include all damaged cartilage, so to reach healthy cartilage around the lesion. Gently debride the damaged cartilage down to the subchondral bone without causing any bleeding.

For the first-generation ACI (with the use of periosteum), make a template of the defect and go to the proximal medial tibia for harvesting the periosteal flap. Make a skin incision in proximal medial tibia distal to the pes anserinus insertion, dissecting down to the bone. Care should be taken to remove all fat fibers and vessels covering the periosteum. Gently remove remaining fibers and the thin fascia covering the area, and also remove the thin fat layer directly overlying the periosteum making the harvested flap as thin as

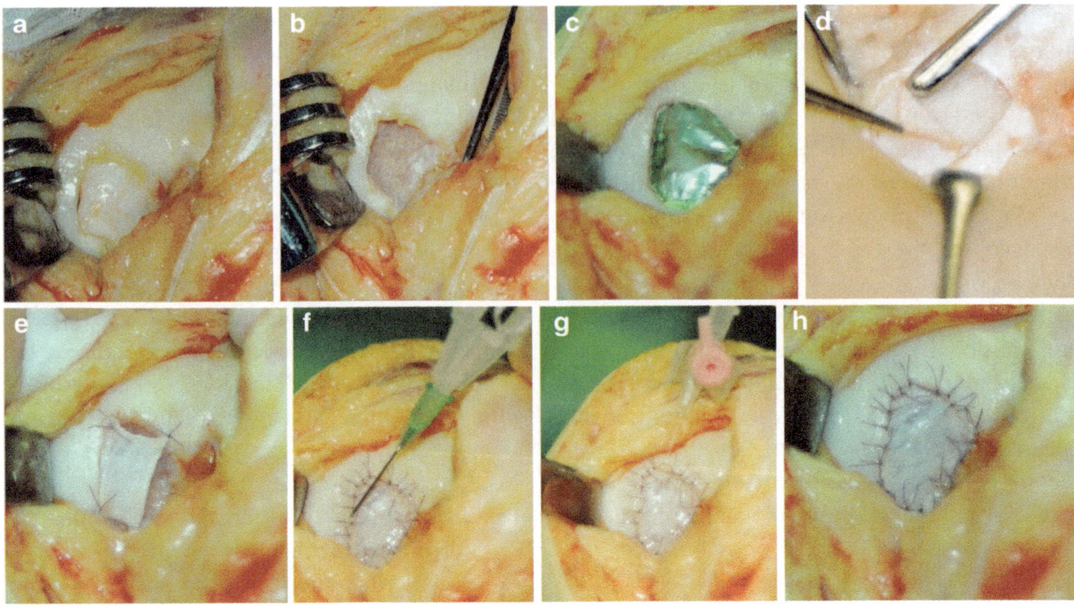

Fig. 29.1 Autologous chondrocyte implantation. Stages of the procedure: (**a**) cartilage lesion as first seen after an open arthrotomy, (**b**) the lesion area after debridement of the defect tissue, (**c**) a piece of foil is used to make a template of the debrided lesion area, (**d**) retrieval of a periosteal patch from the proximal tibia. Orientation of the patch is performed with the use of the template: (**e**) suturing of the patch on the defect area, (**f**) use of fibrin glue to water seal the area, (**g**) after water sealing injection of the suspension of chondrocytes beneath the periosteal patch, (**h**) the treated defect area before closure of the incision

possible. Care should be taken not to make it too thin or perforate it. Place the template on the periosteum and oversize the flap by 1–2 mm when incising the periosteum. Gently dissect the flap free from the cortical bone (periosteal patch).

A fabricated collagen membrane can be used instead of the periosteum. In this case, the procedure described above for the retrieval of periosteum is not needed. When a collagen membrane is used, the template of the cartilage lesion should be used as described above in order to cut the membrane accordingly to the lesion size and shape.

After the patch (periosteal patch or collagen membrane) is cut, place it on the cartilage lesion and suture it to the vertical edges of the defect by anchoring it with sutures in four corners. When periosteum is used, be sure that the cambium layer is facing the bone of the defect. Then, complete the suturing using 6-0 reabsorbable sutures with a 4–6 mm interval between them and try to grasp 5–6 mm into the normal cartilage (Figs. 29.1, 29.2, and 29.3). Leave a small area on the top of the defect (making a hole enough to be closed with one suture). Seal the intervals between the sutures with fibrin glue in order to make it watertight and then insert a plastic driver of a 20-gauge syringe through that hole. Test for leakage by gently injecting saline into the defect through the plastic syringe driver. When water sealing is achieved, aspirate the saline and inject the chondrocytes into the defect starting from distally and slowly withdrawing the syringe (Figs. 29.1 and 29.4). Close the injection site with a suture and some fibrin glue (Fig. 29.1). When using the collagen membrane injected with cells, place the rough surface to the bone and suture and seal as described above, but do not test for watertightness. Finally, inject any remaining cells under the membrane to ensure cell adhesion to the subchondral bone plate.

For 2nd- and 3rd-generation ACI, the cells are provided within a membrane (not in a suspension as in the 1st generation). In this case, just cut the

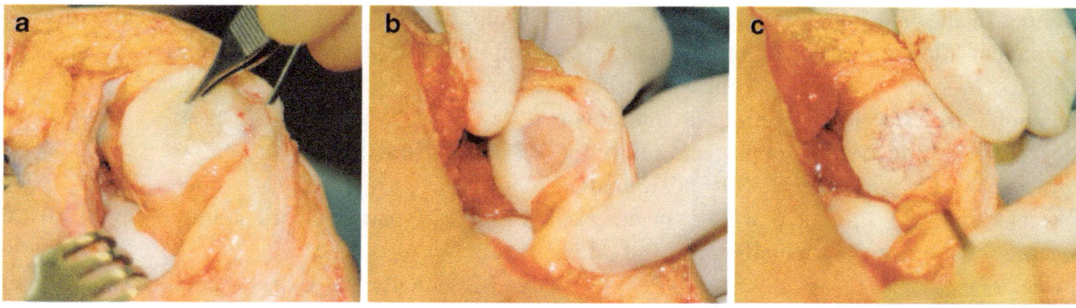

Fig. 29.2 (**a**) Large cartilage lesion of the patella. The defect is covered by necrotic cartilage tissue. It is fragile; thus, the subchondral bone is easily reached with a forceps, (**b**) the patellar lesion after the debridement and removal of the necrotic nonfunctional tissue, (**c**) the same lesion after the chondrocyte implantation

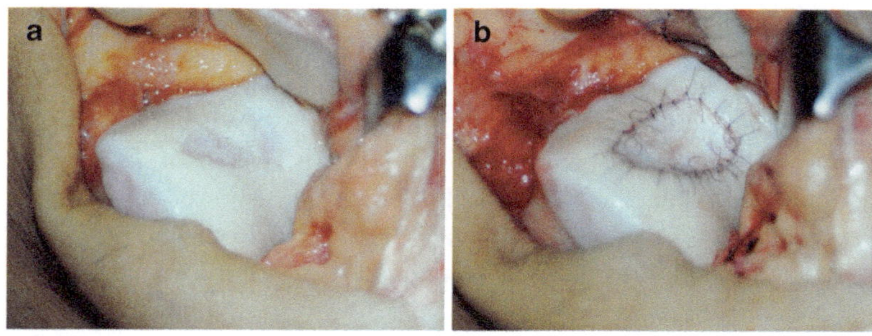

Fig. 29.3 (**a**) Trochlear lesion before debridement and (**b**) after the autologous chondrocyte implantation and suturing of the periosteum

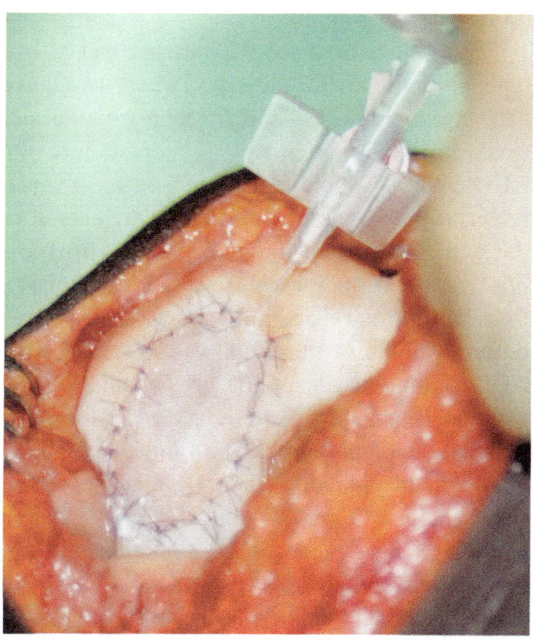

Fig. 29.4 After the suturing and waterproof sealing of the periosteal patch on the surrounding cartilage, the suspension of chondrocytes is inserted beneath the patch, through a small open area at the top of the graft

membrane accordingly to the cartilage lesion (with the use of a template as described above), and place it on the debrided lesion. BioGlue may be used to keep the membrane stuck on the subchondral bone. In bigger or uncontained lesions, we recommend sutures and sealing of the suture rim with fibrin glue to increase the stability of the graft.

29.3 Postoperative Treatment: Rehabilitation

In case of realignment procedures, use a brace allowing 0–90° of motion for the first 3 weeks, and then, open the brace to 0–120° for another 3 weeks. In case of isolated cartilage lesions, the use of a brace is not necessary. Antibiotics and thrombosis prophylaxis should be administered postoperatively, according to your routines. Use continuous passive motion machine (CPM), 0–30–40° to start with, after 8 h, for 48 h. Start quadriceps training and active range of motion training in the first postoperative day. Mobilize the patient allowing as much weight bearing as tolerated by pain. However, partial weight bearing should be advised when going up and down the stairs. Use crutches for support and safety for 3–6 weeks. Try to reach 60–90° flexion after 3 weeks and 100–120° at 6 weeks. In case of bracing, the brace is gradually removed after 6 weeks.

Physiotherapy is focused primarily on full extension and gradually on full flexion. Isometric quadriceps training and closed-chain strength training should be performed up to 4–6 months.

Functional training when full weight bearing is reached includes walking for increasing distances and bicycling with low resistance initially when at least 90° of flexion is reached on a stationary bike. Later on, outdoor biking with increasing distances is allowed. Swimming is also allowed when the wound is healed using freestyle leg work.

Return to sports is assessed on individual basis.

29.4 Long-Term Follow-Up of Patients Treated with ACI

29.4.1 Material

Since 1987 ACI has been performed in Gothenburg, Sweden, in more than 1,600 patients. Out of the first 442 patients operated with ACI, 244 included at least one patellar or trochlear lesion; 153 (35 %) had patella lesions, while 78 (18 %) had isolated patella lesions. Trochlear lesions were found in 91 patients (21 %), while in 18 of them (4.1 %), it was the only cartilage defect. Out of them, 42 patients (9.5 %) had kissing lesions of patellofemoral joint; in 28 of those other concomitant lesions of femoral condyles or tibia plateau were also found.

A 10–20-year follow-up study was conducted to show the long-term results of ACI. Two hundred and twenty-four patients who returned the questionnaires participated. Of the total number of patients, 80 were women and 144 were men. The average age was 33.3 years (range 15–61.6, SD 9.5) at the time of surgery while being 46.1 (range 25.8–74.2, SD 9.5) at the time of follow-up. The evaluation was performed 10–20 years after the implantation (average 12.8 years).

29.4.2 Results

From 224 patients with a follow-up of 10–20 years;

92 patients had either a patellar or trochlear lesion (39 isolated patella, eight isolated trochlea, 18 kissing). The mean age at the time of the ACI was 34.6 years (range 14–57) and at the follow-up was 12.6 years after the surgery (range 10–20). The mean size per lesion was 5.5 cm^2 (range 1–16), with a mean ratio of 1.7 lesions per patient.

Tegner-Wallgren score was 7, improved by 1 level compared with the preoperative values (Wilcoxon signed-rank test $p=0.01$). The mean Lysholm score was 68.1. The statistical analysis showed an improvement of 9 points in average from the preoperative values. However, this improvement was not found to be statistically significant (Wilcoxon signed-rank test $p=0.3$). Seventy-two percent of the patients reported that they were better or unchanged, while 93 % would do the operation again.

Thirty-eight of those 92 patients had also a type of realignment surgery. Twenty-two have had an extensor mechanism reconstruction and trochleaplasty (one had it 7 years before the ACI). Nine had only an extensor mechanism reconstruction (one of those had it 6 years after the ACI). One had medial plication and trochleaplasty and one lateral release and trochleaplasty. Five had a tibial osteotomy for varus or valgus deformity (one of those had it 4 years after the ACI). Twenty-seven of the 38 patients (71 %) responded that they were improved or the same compared with previous years, while 35 (92.1 %) would do the operation again. Lysholm score was 66.3 (range 17–100), while Tegner-Wallgren score was 8.1 (range 3–14). Patients with no realignment procedures had a final Lysholm score of 69.3 and Tegner-Wallgren score of 8.05. Patients with malalignment or instability that had a realignment procedure of any form had comparable outcomes to the cases which did not need any additional surgery ($p=0.5$ for Lysholm and $p=0.9$ for Tegner-Wallgren).

Patients with no kissing lesions appeared to have a better prognosis. Seventy-eight percent reported to be better or the same compared with previous years, and the mean Tegner-Wallgren score was 8.3. On the other hand, 44 % of the patients with kissing patellar and trochlear lesions appeared to be better or the same, with a mean Tegner-Wallgren score of 7.2 ($p=0.004$ and $p=0.04$, respectively). However, 94 % of the patients with kissing lesions would do the ACI again, similar to the patients with no kissing lesions (92 %).

Subdividing the patients according to the features of their lesions:

(i) We had, in total, 73 patients with patellar lesions (average 5.1 cm^2, range 1–12, SD 2.61). They had 1.7 lesions in average (from 1 to 4). Fifty-one (69.9 %) responded that they were improved or the same compared with the previous years, and 68 (93.2 %) would do the operation again. Lysholm score in the latest follow-up was 67.3 (range 17–100), and Tegner-Wallgren score was 7.9 [3–14].

(ii) Fifty-five patients sustained a patellar lesion without concomitant trochlear lesions. Eleven had also a medial femoral lesion and four had a lateral femoral lesion. Two patients had two patellar lesions. Forty-three (78.2 %) responded that they were better or the same compared with previous years, while 51 (92.73 %) would do the ACI again. Lysholm score in the 10–20-year follow-up was 69.5 (range 17–100), while Tegner-Wallgren score was 8.2 [3–14].

(iii) Thirty-nine patients (out of the 73) had an isolated patellar lesion (average 6.1 cm^2, range 1–12, SD 2.7). Thirty-one (79.5 %) responded that they were improved or the same compared with the previous years, while 35 (89.7 %) would do the operation again. Lysholm score was 66.4 (range 17–100), while Tegner-Wallgren score was 7.9 [3–14].

(iv) Thirty-seven patients had trochlear lesions (average 6.53 cm^2, range 1.2–20, SD 4.75). They had 2.3 lesions in average (from 1 to 4). Twenty-three (62.2 %) responded that they were improved or the same compared with the previous years, while 34 (92 %) would do the operation again. Lysholm score was 66.1 (range 17–100), while Tegner-Wallgren score was 7.9 [3–14].

(v) Nineteen of those patients sustained a trochlear lesions without concomitant patella lesions. Ten had also a medial femoral lesion and five had a lateral femoral lesion. One patient had two medial femoral lesions and one had a double trochlear lesion. Fifteen (79 %) responded that they were better or the same compared with previous years, while 17 (90 %) would do the ACI again. Lysholm score in the 10–20-year follow-up was 71.3 (range 43–100), while Tegner-Wallgren score was 8.7 [5–14].

(vi) Eight patients (out of the 37) had an isolated trochlear lesion (average 7.38 cm^2, range 2.8–15.8, SD 4.96). Eight (100 %) responded that they were improved or the same compared with the previous years, while all of them (100 %) would do the operation again. Lysholm score was 72.3 (range 50–91), while Tegner-Wallgren score was 8.9 [6–14].

(vii) Eighteen patients had kissing lesions in the patellofemoral joint (patellar lesion, average 3.7 cm^2, range 1–7.5, SD 1.9; trochlear lesion, 6.77 cm^2, range 1.3–20, SD 5). Eleven of those had also other lesions except from the trochlea and patella. Eight (44.4 %) responded that they were improved or the same compared with the previous years, while 17 (94.4 %) would do the operation again. Lysholm score was 60.9 (range 17–100), while Tegner-Wallgren score was 7.2 [3–10].

29.4.3 Complications

In 92 patients no general complications were seen. We had no cases with deep or superficial infections or deep venous thrombosis. Focal complications occurred in 52 patients (56 %), most of them being minor.

Five of the patients sustained a failure (5.4 %); three of the failures were due to the trochlear or patellar lesion (3.2 %). One of them had a revision of ACI on the trochlea, one had a revision of patellar lesion with carbon fibers by another surgeon, and one patient had finally a patellectomy. In two of the patients, the cause of failure was a medial femoral condyle lesion; a revision ACI was performed for these cases.

We had 18 screw extractions which were expected operations; thus, they were not considered a complication. Twenty-seven patients (29 %) sustained a periosteal hypertrophy. Most of them were focal or overlapping hypertrophy or fibrillation or flaps from the periosteum. About half of them were symptomatic. Nine had kissing patellofemoral lesions, and seven had concomitant procedures (three including one trochleaplasty, one osteotomy, and one posterolateral corner reconstruction). In all the cases, it was treated with debridement of the hypertrophic tissue.

Seven cases (five with patellar and two with kissing patellofemoral lesions) had an arthrofibrosis requiring arthroscopic mobilization shortly postoperatively. Four of them had also a realignment procedure, including trochleaplasty, and two had an ACL reconstruction during the ACI. In one case patellectomy was finally performed. Excluding this patient, which was considered as failure, no clinical impact was noticed on the long-term follow-up evaluation. Most of the arthrofibroses occurred early in the study. Any restriction in extension-flexion was later on carefully watched, and when present, active physiotherapy treatment was applied. This strategy reduced the final incidence of arthrofibroses.

Three patients sustained a partial delamination. The periosteal flaps were found delaminated and were removed by a arthroscopic debridement. One partial graft delamination was found on a concomitant lesion of a medial femoral condyle; the trochlear and patellar lesions of that patient were found intact. Two of the three cases had kissing lesions. There was a slight clinical impact for the two partial delamination patients, but not statistically significant ($p>0.05$). There were no cases with graft delamination on the patella or trochlea.

Two patients sustained a persistent bone marrow edema. Both of them were located on the medial femoral condyle, and both these patients had a tibial osteotomy during the ACI. One patient had four lesions (two on the medial femoral condyle, one on the lateral condyle, and one on the trochlea), and the other one had a large lesion of medial femoral condyle and patella. No bone marrow edema was seen in the subchondral bone of the patella or trochlea.

Realignment procedures were associated with less periosteal hypertrophies (16 % in patients with realignment procedures, 39 % in cases without realignment procedures, $p=0.01$). On the other hand, realignment procedures increased the incidence of serious complications (failures, arthrofibrosis, delamination, multiple surgeries). Thirteen percent of cases without realignment procedures and 29 % from those with realignment procedures sustained one of those complications ($p=0.05$). Although periosteal hypertrophy was the most common complication, no impact was found on the final clinical outcome (Tegner-Wallgren $p=0.4$, Lysholm $p=0.1$, operated again $p=0.3$, improvement $p=0.9$).

No association was found between the age at the time of the ACI or the size per lesion and any of the clinical outcomes.

29.4.4 Discussion

Articular cartilage injuries and anterior knee pain are difficult problems to adequately diagnose and treat. Especially in the young and middle-aged patients, it is important to analyze the history, clinical signs, and background factors of the condition. To successfully treat articular cartilage lesions in the patellofemoral joint is the most challenging task for the treating physician. The complexity of the mechanical function and the extraordinary loading acting over the joint creating compression and shearing forces has resulted in an increased thickness with impact resistance that is unique in human joints. These factors put

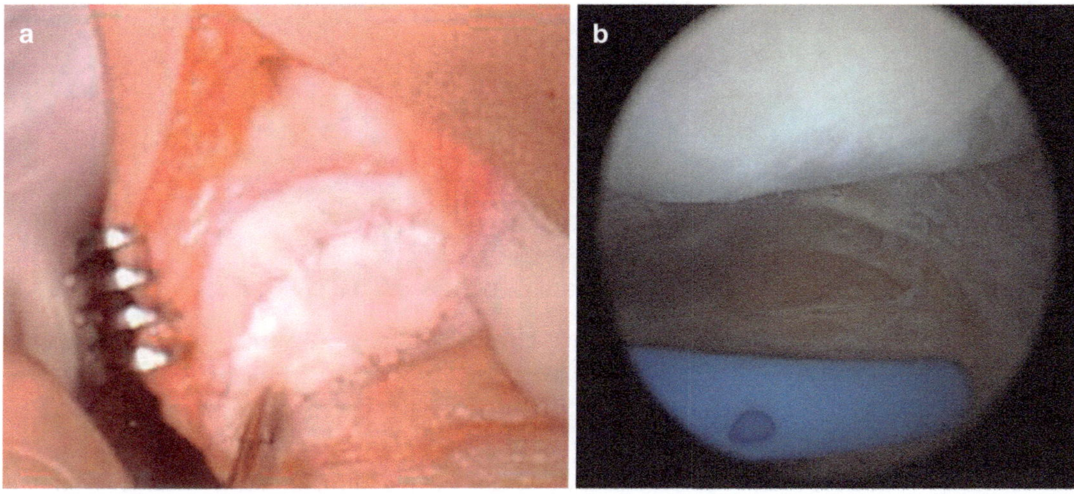

Fig. 29.5 (**a**) Patellar lesion after debridement and implantation, (**b**) second-look arthroscopy 5 years after surgery. Patient was asymptomatic

specific demands on the treatment for a successful outcome.

MRI and computerized tomography with and without quadriceps contraction have helped a lot to understand the mechanisms and anatomic abnormalities, giving information needed for an optimal treatment [6, 7]. Articular cartilage lesions are much more common than earlier recognized and if remained undiagnosed and untreated will lead to osteoarthritis over time. Curl et al. found that over 60 % of patients who underwent arthroscopy had articular cartilage lesions with the potential to deteriorate by enzymatic degradation and mechanic wear [8]. Hjelle et al. found about 20 % cartilage lesions on the patella and the trochlea in over 1,000 arthroscopies [1].

Earlier treatment options of chondral lesions mostly resulted in fibrous tissue coverage of the cartilage defect area. Spongialization, drilling, abrasion, or microfracturing of the subchondral bone plate or trabecular bone creates bleeding into the defect area, thus bringing mesenchymal stem cells and fibroblasts and causing vascularization. Finally, the cartilage lesion area is mainly covered by fibrous tissue or fibrocartilage, with inferior functional and mechanical properties compared with normal hyaline cartilage; that tissue fails to withstand the subjected forces during daily activities and sports and is subsequently leading to tissue wear down. Coverage of the defect area by a periosteal or a perichondral autologous flap has also been tried in the past, unfortunately also ending up in fibrocartilage or bone formation [9, 10].

Since 1987, autologous chondrocyte implantation has been used for the treatment of chondral lesions in the knee or other joints [11–14]. Isolated femoral lesions and osteochondritis dissecans have been followed for 5–11 years after ACI treatment, with good/excellent result in about 90 % and biopsies showing hyaline cartilage in 80 % of the patients. In the patella, the results were initially not satisfactory with only 28 % good/excellent results in 36-month follow-up [15]. Continued treatment with ACI combined with correction of background factors when indicated showed an improvement of the results. Contributing to this was also the modification of the rehabilitation program for trochlear or patellar lesions focusing on early weight bearing and to closed-chain exercises during the first 6 months postoperatively. This change of treatment showed improved results in later follow-ups (Fig. 29.5) [4, 11, 16–19].

Periosteal hypertrophy is the most common complication to the ACI technique and is treated, when symptomatic, with arthroscopic debridement. It seems that patella and trochlea are more prone to hypertrophy due to the high compression

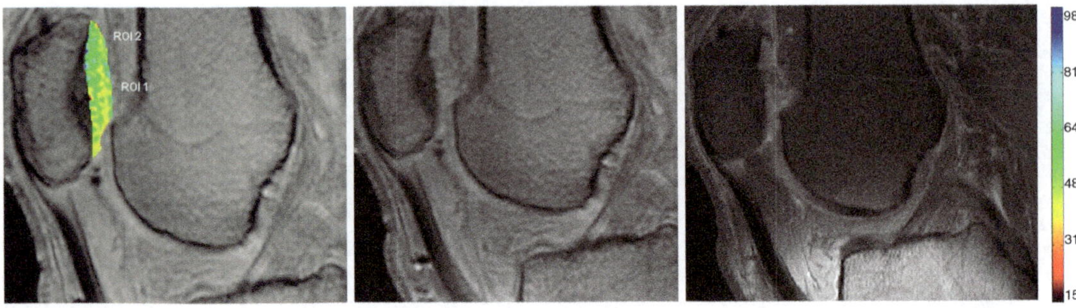

Fig. 29.6 Patellar lesion in a female, age of 29 at time of implantation. MRI with dGEMRIC technique 10.5 years after the surgery showing restoration of proteoglycan concentration in the defect area

and friction forces during weight-bearing activities. Especially in patella-trochlea kissing lesions, we see hypertrophy on both articulating surfaces [15, 20]. Our study provides evidence showing that decreasing the applied forces on the patellar or trochlear cartilage lesion, with an unloading procedure, may decrease the incidence of periosteal hypertrophies after an ACI. Meticulous dissection, sizing, and suturing of the periosteal patch could also reduce the frequency of this complication [21, 22]. However, despite the high rate of periosteal complications reported, it seems that these complications can be handled arthroscopically when symptomatic and would not affect the end result. There is also literature suggesting that the rate of hypertrophies is also decreased by using collagen membranes instead of periosteal patch [20].

From our last follow-up study of 10–20 years' evaluation, it seems that correcting the background factors with realignment, stabilizing, or unloading procedures is improving the results over time (Fig. 29.6). Even if they may contribute to some complications like arthrofibrosis and periosteal hypertrophy, an overall improvement in the result was shown, along with a long-term durability of good results of ACI.

29.4.5 Summary

Despite the initial controversy about the results and indication for ACI in patellofemoral lesions, it is clear that ACI provides a satisfactory result even for the difficult cases with concomitant patellar instability. Our study reveals preservation of the good results and of high level of patients' activities, even 10–20 years after the implantation.

Remarkable improvement of the results from the initial study to this last follow-up shows long-term durability in both isolated trochlear and patellar lesions and also in multiple and kissing lesions where an intervention could be considered as a salvage procedure. Over 90 % of the treated patients were satisfied with the ACI and would have the procedure again.

References

1. Hjelle K, Solheim E, Strand T, Muri R, Brittberg M. Articular cartilage defects in 1,000 knee arthroscopies. Arthroscpy. 2002;18(7):730–4.
2. Nomura E, Inoue M, Kurimura M. Chondral and osteochondral injuries associated with acute patellar dislocation. Arthroscopy. 2003;19(7):717–21.
3. Fulkerson JP. Diagnosis and treatment of patients with patellofemoral pain. Am J Sports Med. 2002;30(3):447–56.
4. Farr J. Autologous chondrocyte implantation and anteromedialization in the treatment of patellofemoral chondrosis. Orthop Clin North Am. 2008;39(3):329–35, vi.
5. Andrikoula S, Tokis A, Vasiliadis HS, Georgoulis A. The extensor mechanism of the knee joint: an anatomical study. Knee Surg Sports Traumatol Arthrosc. 2006;14(3):214–20.
6. Brossmann J, Muhle C, Schröder C, Melchert UH, Büll CC, Spielmann RP, et al. Patellar tracking patterns during active and passive knee extension:

evaluation with motion-triggered cine MR imaging. Radiology. 1993;187(1):205–12.
7. Vasiliadis HS, Danielson B, Ljungberg M, McKeon B, Lindahl A, Peterson L. Autologous chondrocyte implantation in cartilage lesions of the knee: long-term evaluation with magnetic resonance imaging and delayed gadolinium-enhanced magnetic resonance imaging technique. Am J Sports Med. 2010;38(5):943–9.
8. Curl WW, Krome J, Gordon ES, Rushing J, Smith BP, Poehling GG. Cartilage injuries: a review of 31,516 knee arthroscopies. Arthroscopy. 1997;13(4):456–60.
9. Homminga GN, Bulstra SK, Bouwmeester PS, van der Linden AJ. Perichondral grafting for cartilage lesions of the knee. J Bone Joint Surg Br. 1990;72(6):1003–7.
10. Alfredson H, Lorentzon R. Autologous periosteum transplantation for the treatment of full thickness patellar cartilage defects. Ortop Traumatol Rehabil. 2001;3(2):216–23.
11. Peterson L, Minas T, Brittberg M, Nilsson A, Sjögren-Jansson E, Lindahl A. Two- to 9-year outcome after autologous chondrocyte transplantation of the knee. Clin Orthop. 2000;374:212–34.
12. Micheli LJ, Browne JE, Erggelet C, Fu F, Mandelbaum B, Moseley JB, et al. Autologous chondrocyte implantation of the knee: multicenter experience and minimum 3-year follow-up. Clin J Sport Med. 2001;11(4):223–8.
13. Bentley G, Biant LC, Carrington RWJ, Akmal M, Goldberg A, Williams AM, et al. A prospective, randomised comparison of autologous chondrocyte implantation versus mosaicplasty for osteochondral defects in the knee. J Bone Joint Surg Br. 2003;85(2):223–30.
14. Browne JE, Anderson AF, Arciero R, Mandelbaum B, Moseley Jr JB, Micheli LJ, et al. Clinical outcome of autologous chondrocyte implantation at 5 years in US subjects. Clin Orthop. 2005;436:237–45.
15. Brittberg M, Lindahl A, Nilsson A, Ohlsson C, Isaksson O, Peterson L. Treatment of deep cartilage defects in the knee with autologous chondrocyte transplantation. N Engl J Med. 1994;331(14):889–95.
16. Peterson L, Brittberg M, Kiviranta I, Akerlund EL, Lindahl A. Autologous chondrocyte transplantation. Biomechanics and long-term durability. Am J Sports Med. 2002;30(1):2–12.
17. Peterson L, Vasiliadis HS, Brittberg M, Lindahl A. Autologous chondrocyte implantation: a long-term follow-up. Am J Sports Med. 2010;38(6):1117–24.
18. Mandelbaum B, Browne JE, Fu F, Micheli LJ, Moseley Jr JB, Erggelet C, et al. Treatment outcomes of autologous chondrocyte implantation for full-thickness articular cartilage defects of the trochlea. Am J Sports Med. 2007;35(6):915–21.
19. Vasiliadis HS, Wasiak J, Salanti G. Autologous chondrocyte implantation for the treatment of cartilage lesions of the knee: a systematic review of randomized studies. Knee Surg Sports Traumatol Arthrosc. 2010;18(12):1645–55.
20. Gooding CR, Bartlett W, Bentley G, Skinner JA, Carrington R, Flanagan A. A prospective, randomised study comparing two techniques of autologous chondrocyte implantation for osteochondral defects in the knee: periosteum covered versus type I/III collagen covered. Knee. 2006;13(3):203–10.
21. Jones DG, Peterson L. Autologous chondrocyte implantation. Instr Course Lect. 2007;56:429–45.
22. Peterson L. Autologous Cartilage Implantation: Cartilage Biopsy, Cartilage Implantation. In: Cushner FD, Norman Scott W, Scuderi GR, editors. Surgical Techniques for the Knee. New York: Thieme; 2005. p. 41–6.

Patellar Chondral Injury Treated with Autologous Osteochondral Transplantation

30

Diego Costa Astur, Gustavo Gonçalves Arliani, Nelson Astur, Camila Cohen Kaleka, and Moises Cohen

Contents

Cartilaginous injuries of the joint surface are challenging to treat due to a lesser regeneration rate and tissue regrown ability of the chondral cells. Articular cartilage is an avascular tissue composed mainly by extracellular matrix. It is a highly developed tissue that supports full contact and movement with low friction. Single lesions of the chondral tissue may develop with different etiologies and can affect all ages. When diagnosed in young and active patients, total knee joint replacement is not recommended due to a high probability of early wear of the implant, instability, and need for an early-age joint revision. For these patients, a different option of treatment is needed, and articular cartilage repair should be done with biological tissue [1].

The knee is frequently affected by cartilaginous injuries, seen in 63 % of knee arthroscopies [2]. A significant portion of those chondral defects is located in the patella or trochlea (17 %) [3]. Aroen et al. reported that 8 % of these lesions were located in the trochlea and 23 % in the patella [4]. Symptoms related to this injury usually are knee pain and joint swelling, and the most frequent complication is the development of early osteoarthrosis [5]. For this reason, repair of the articular cartilage lesions with a tissue with similar biomechanical and durability properties of natural cartilage is the main objective of all major joint preservation techniques [6].

The patella is a sesamoid bone with the thickest cartilaginous surface of the human body. This wide thickness increases the articular surface area as well as the femoropatellar joint load

D.C. Astur • G.G. Arliani • N. Astur • C.C. Kaleka
M. Cohen (✉)
Department of Orthopaedic, S. Paulo University, São Paulo, SP, Brazil
e-mail: moises@institutocohen.com.br

A. Gobbi et al. (eds.), *The Patellofemoral Joint*,
DOI 10.1007/978-3-642-54965-6_30, © ISAKOS 2014

distribution [7–9]. Daily activities generate high loads in this joint that can reach 6.5 times our body mass [10]. The normal sharing of these forces is affected by any full-thickness chondral injury in the patella, leading to the propagation of chondral tear, joint pain, and functional impairment of the knee [11]. For these reasons, osteochondral defects in the patella are difficult to treat [11].

Innumerous techniques and attempts to restore these injuries have been reported recently, with varying success rates [1, 11–20]. Nonsurgical treatment through rehabilitation and analgesia techniques is the first option to address pain control [18], but only surgical intervention will restore surface anatomy and cartilage repair.

Osteochondral autologous transplantation (AOT) was first reported and developed by Hangody in the early 1990s [21]. He described a technique that uses autologous osteochondral graft harvested within the peripheral non-load-bearing zone of the knee to fill cartilage defects. Good results have been published so far with this technique; however, one of its limitations is the small availability of donor areas, restricting its usefulness to smaller lesions [22].

AOT can restore the articular surface shape and stably filling the defect through a plug graft. Therefore, there will be less space for the development of fibrocartilage, which is a natural response of the body to repair cartilage defects [17]. AOT is a well-known surgical technique to treat cartilaginous femoral, trochlear, and condylar defects, but its application in patellar chondral injury is still a relative new technique [11, 14, 15, 20, 23].

30.1 Indications

Main indications for surgical repair of patellar chondral defects are [24]:

- Persistent clinical symptoms such as severe anterior knee pain, recurrent swelling, or clicking
- Radiological and arthroscopic proof of an osteochondral defect with an unstable fragment
- Patellofemoral osteoarthritis development
- Failed conservative treatment for at least 6 months

Large articular cartilage lesions in the patellofemoral compartment in younger patients are difficult to manage successfully. When conservative treatment fails to alleviate symptoms and functional limitations, surgery must be considered to treat both the cartilage lesion and any associated anatomic malalignment [25]. Surgical approaches to treat patellar cartilage injuries have not been reported as much as other site chondral lesions due to technical difficulties related to this therapy and few options available for repair. Recently, a wide range of operative procedures are applied for patellar chondral repair, from a simple arthroscopic debridement to more invasive techniques such as autologous osteochondral transplantation. Same stage procedures may be needed to repair an extensor mechanism malalignment or other conditions such as medial patellofemoral ligament insufficiency, excessive tightness of the lateral retinaculum, or miserable malalignment syndrome. The decision making process on which procedure to choose is difficult since there is no gold standard technique and no single technique for all types of lesion. This is especially true for patients younger than 50 years of age who have always had an active lifestyle and wish to restart athletic or recreational activities free of pain. AOT is the most popular technique for small- and medium-sized (1–3 cm^2) single chondral and osteochondral defects at the patella weight-bearing surface. There are no studies though relating this technique to optimal results for large symptomatic lesions ($\geq$4 cm^2). The upper age limit recommended for this technique is 50 years of age, since increasing age diminishes repair ability of the cartilaginous tissue. However, patients older than 50 years with a single patellofemoral lesion have been reported to be a candidate for this procedure.

30.2 Descriptive Surgical Technique

A standard arthroscopic joint evaluation is carried out confirming the injury diagnosis at the patella joint surface. Once the defect is located, it is characterized and the arthroscopy is taken out

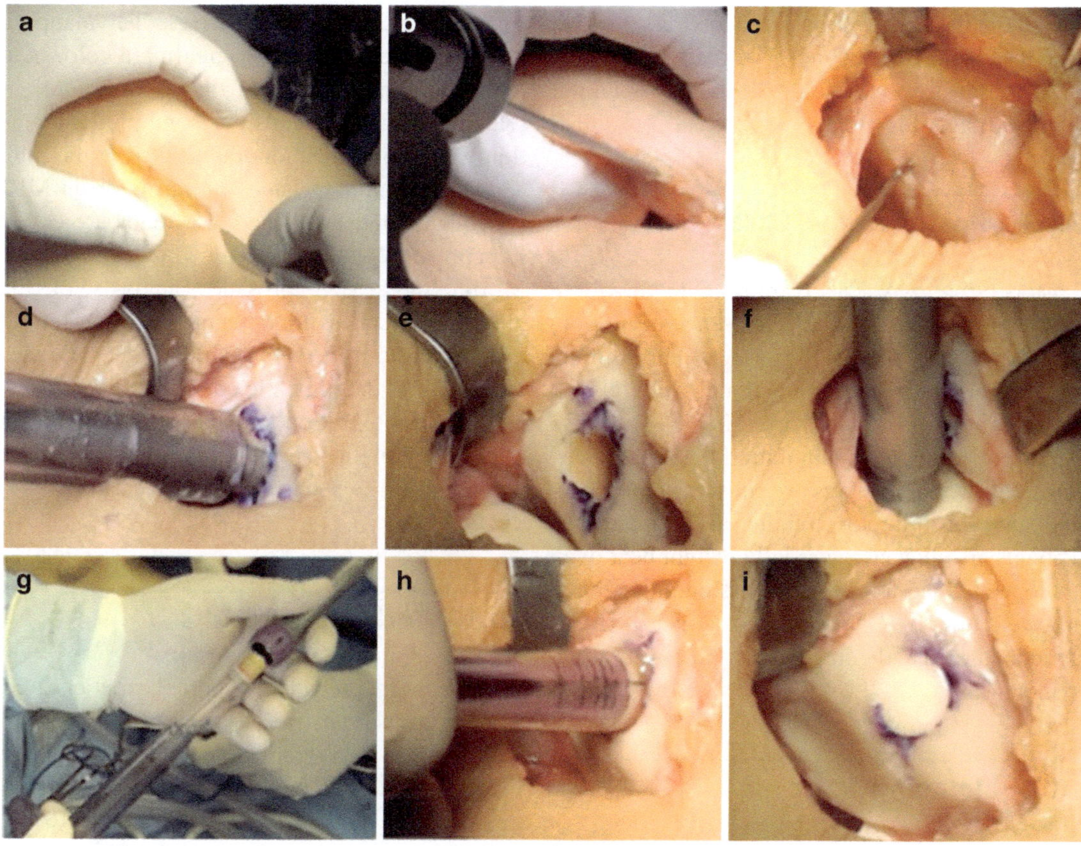

Fig. 30.1 Surgical technique for autologous osteochondral transplantation in a patellar chondral injury. (**a**) Incision and medial parapatellar arthrotomy. (**b**) K-wire fixation working as a joystick. (**c**) Identification of the osteochondral injury at the medial facet of the patella. (**d**) Debulking of the injured site with proper devices. (**e**) Receptor site aspect. (**f**) Osteochondral graft harvesting at the superior aspect of the medial femoral condyle, outside the load bearing. (**g**) Osteochondral plug graft aspect before its implantation. (**h**) Plug implantation at the receptor site. (**i**) Final aspect of the procedure showing adequate defect filling cartilaginous surface alignment

and switched for an open approach. A parapatellar 3-cm incision is made from the apex of the patella to its inferior edge. This approach can be medialized or lateralized according to which patellar facet is injured (Fig. 30.1a).

After subcutaneous dissection to expose the joint capsule, an arthrotomy is then carried out, and the patella is inverted for a perfect visualization of its joint surface. A K-wire is used as a joystick for a better exposure of the defect (Fig. 30.1b). At this point, the diameter of the chondral injury is measured with a metric guide; thus, the size of the osteochondral cylinder that will be harvested is defined at the donation site (Fig. 30.1c). The graft-harvesting device should always be a millimeter greater than the piercing device that will prepare the receiving site. At this point, the previously measured defect is drilled with a motor burr with the size of the designed diameter (Fig. 30.1d). The plug graft is usually 10–12 mm depth.

The receptor tunnel is then enlarged with an adequate device. After the receptor site is completed (Fig. 30.1e), the osteochondral graft is harvested with a flexed knee (Fig. 30.1f, g). The donor site is selected in a surrounding non-load-bearing zone, usually superior to the femoropatellar articular area. All drillings are performed perpendicular to the joint surface (Fig. 30.1h). The osteochondral cylinder is then implanted until its surface is leveled with the surrounding joint cartilage (Fig. 30.1i).

30.3 Postoperative Care

Postoperative physical therapy care involves early exercises to improve knee range of motion on a hard surface and under water. Gait training under deep water starts immediately. Exercising on a stationary bicycle starts after 3–4 weeks, according to patients' limitation and tolerance. This will lead to progressive muscle strengthening and sensorimotor training and stretching. Patients are allowed to partially weight bear over the operated leg for 2–3 weeks. Running is only allowed after 4–6 months of surgery. High-functional-demand sports are allowed after 6 months.

30.4 Imaging

Magnetic resonance image (MRI) is the gold standard study to diagnose and characterize a chondral lesion. Through this analysis, one can quantify the size of the lesion as well as its detachment grade. This will guide the treatment method to be chosen. For postoperative healing control purposes, patients with this type of injury should routinely undergo a standard MRI. Another MRI option technique is the T2 relaxation time mapping MRI study to evaluate osteochondral bone-plug integration. While routine MRI allows a subjective assessment of cartilage changes, quantitative T2 mapping provides objective data by creating a color map representation of the cartilage variations in relaxation time. We consider bone-plug integration when both the chondral surface of the plug and its surrounding have similar color on MRI (Grades III and IV). The color maps are coded to capture T2 values ranging from 20 to 70 ms. Morphologic MRI provides analysis of the chondral thickness maintenance and the reintegration tissue in between both surfaces. Grades III and IV at the plug site and around it indicate successful integration and also show a collagen network with the shape and overall structure similar to those seen in a regular cartilage [11, 26, 27].

30.5 Technique Results

In the authors' previously reported experience, 33 knees from patients (mean age 37.6 year) underwent an AOT for a symptomatic full-thickness cartilaginous injury on the patellar articular surface.

Twenty-seven knees had only a single 10×15 mm osteochondral graft implanted. Five knees required two graft cylinders due to a greater defect area.

The average Lysholm score [28] was 57.27 (±19.97) preoperatively and 80.76 (±12.26) postoperatively ($p<0.05$). The Fulkerson questionnaire [29] had a mean preoperative score of 54.24 (±18.89) and 80.42 (± 10.20) postoperatively ($p<0.05$). The average preoperative score for the Kujala questionnaire [30] was 54.76 (±17.61) points, while average postoperative score was 75.18 (±12.47) points ($p<0.05$) (Table 30.1).

Table 30.1 Preoperative (pre-op) and postoperative (post-op) functional results. There were statistically improved outcomes according to Lysholm, Fulkerson, and Kujala questionnaires

Variable	Average	SD	Median	Minimum	Maximum	*N*	*p*
Lysholm (pre-op)	57.27	19.97	58	9	98	33	<0.001
Lysholm (post-op)	80.76	12.26	85	51	99	33	
Fulkerson (pre-op)	54.24	18.89	56	2	93	33	<0.001
Fulkerson (post-op)	80.42	10.20	82	49	95	33	
Kujala (pre-op)	54.76	17.61	58	10	81	33	<0.001
Kujala (post-op)	75.18	12.47	77	49	97	33	

Table 30.2 Preoperative (pre-op) and postoperative (post-op) SF-36 results. There were statistically improved postoperative scores in 7 out of 8 criteria analyzed

Variable	Average	SD	Median	Minimum	Maximum	*N*	*p*
Physical function (pre-op)	45.91	13.31	50	25	75	33	0.006
Physical function (post-op)	63.64	29.11	70	10	95	33	
Role physical (pre-op)	43.94	35.37	50	0	100	33	0.001
Role physical (post-op)	73.48	32.44	75	0	100	33	
Bodily pain (pre-op)	51.73	20.98	51	21	100	33	<0.001
Bodily pain (post-op)	72.30	24.01	78	20	100	33	
General health (pre-op)	73.45	17.86	72	47	100	33	0.214
General health (post-op)	77.79	17.51	82	37	100	33	
Vitality (pre-op)	61.97	22.08	60	20	95	33	0.004
Vitality (post-op)	75.45	18.13	80	25	95	33	
Social function (pre-op)	61.70	15.91	62	38	100	33	0.017
Social function (post-op)	73.71	21.92	80	20	100	33	
Role emotional (pre-op)	44.18	36.60	37.5	0	100	33	0.016
Role emotional (post-op)	73.99	41.22	100	0	100	33	
Mental health (pre-op)	65.58	17.03	64	40	100	33	0.049
Mental health (post-op)	74.06	21.24	84	20	100	33	
Paired Wilcoxon test results							

There was a statistically significant difference ($p<0.05$) between pre- and postoperative analyses of the SF-36 for all its items (Table 30.2).

When the analysis between specific evaluations of the SF-36 subscales and all others knee questionnaires was performed, there was a correlation between the best scores for the Kujala with body pain, general health, and patients' social and physical functioning ($p<0.05$); Lysholm's with body pain and general health aspect ($p<0.05$); and Fulkerson's with body pain, general health, and vitality ($p<0.05$).

Six months after surgery, 83 % of the plugs had complete bony integration, which increased to 100 % 1 year postoperatively, according to conventional MRI (no articular surface incongruence was noted) supplemented with a T2 relaxation time mapping technique (Grades III and IV for the plugs' surface and its surroundings) (Fig. 30.2).

Complication rate was 9 % (3 patients), all due to arthrofibrosis of the knee. They were all successfully treated with an arthroscopic joint release of the knee. There were no intraoperative complications.

30.6 Future Perspectives

There is no one good evidence study in the literature comparing different techniques to treat cartilaginous injury of the patella. The existence of different injury types to indicate a specific technique such as microfractures or autologous chondrocyte transplantation makes it difficult to obtain a similar sample of patients for a comparative study. This may be the reason there is just a scarce number of randomized clinical trials available comparing surgical techniques for cartilaginous injury treatment. Thus, it is difficult to choose an optimal technique to treat most cases of patellar chondral injury. We agree that each technique needs its specific indication for good outcomes.

In the future, new studies with better quality are required (randomized controlled trials) in order to compare the different available types of treatment for chondral lesions in the patella.

From there, it will be possible to compare the current existing techniques, such as microfractures and autologous chondrocytes transplantation. Undoubtedly, the optimum technique will be able to restore the cartilage surface with the same characteristics that exist in the pre lesion stage.

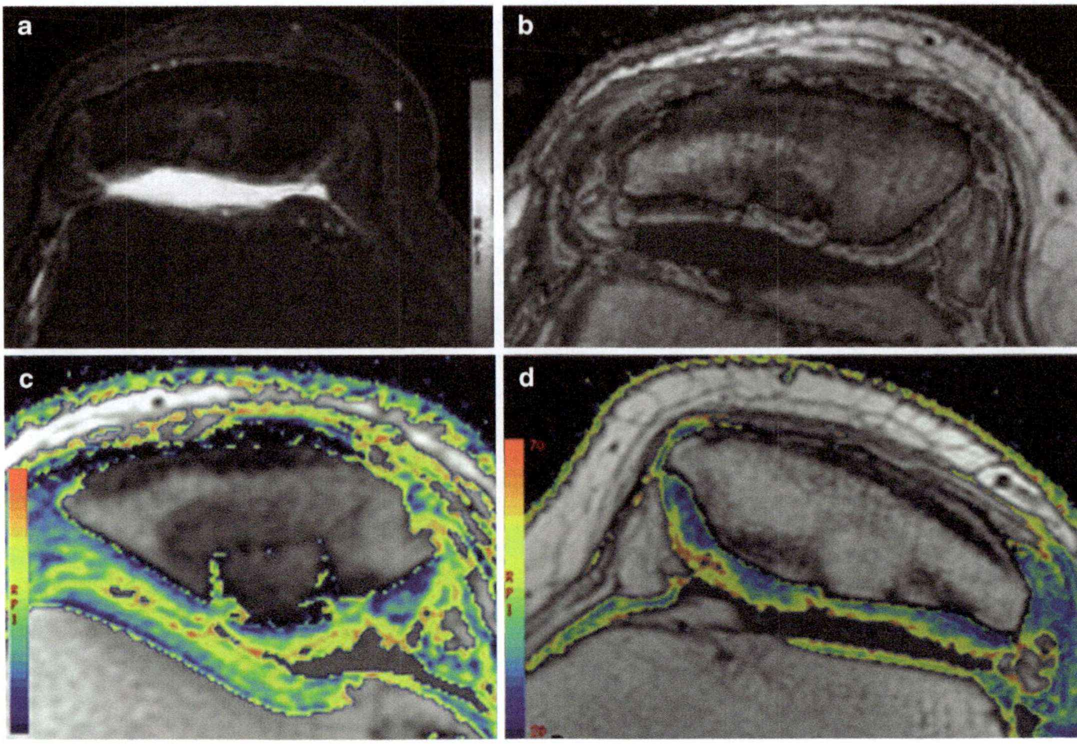

Fig. 30.2 Integrity of cartilage resurfacing in an axial T2 standard MRI in a 30-year-old patient with a patellar autologous osteochondral plug 6 months (**a**) and 12 months (**b**) after surgery and in an axial T2 relaxation time mapped MRI in a 32-year-old patient with a patellar autologous osteochondral plug 6 months without complete integration (**c**) and in a 20-year-old patient with a integrated patellar autologous osteochondral plug 12 months (**d**) after surgery. The color maps are coded to capture T2 values ranging from 20 to 70 ms, with blue and green reflecting shorter values and yellow and red reflecting longer T2 values. After 12 months, the repair cartilage maintains color stratification of normal cartilage

References

1. Marcacci M, Kon E, Delcogliano M, Filardo G, Busacca M, Zaffagnini S. Arthroscopic autologous osteochondral grafting for cartilage defects of the knee: prospective study results at a minimum 7-year follow-up. Am J Sports Med. 2007;35(12):2014–21.
2. Curl WW, Krome J, Gordon ES, Rushing J, Smith BP, Poehling GG. Cartilage injuries: a review of 31,516 knee arthroscopies. Arthroscopy. 1997;13(4):456–60.
3. Hjelle K, Solheim E, Strand T, Muri R, Brittberg M. Articular cartilage defects in 1,000 knee arthroscopies. Arthroscopy. 2002;18(7):730–4. Epub 2002/09/05.
4. Aroen A, Loken S, Heir S, Alvik E, Ekeland A, Granlund OG, et al. Articular cartilage lesions in 993 consecutive knee arthroscopies. Am J Sports Med. 2004;32(1):211–5. Epub 2004/02/03.
5. Buckwalter JA, Mankin HJ. Articular cartilage: degeneration and osteoarthritis, repair, regeneration, and transplantation. Instr Course Lect. 1998;47: 487–504.
6. Vanlauwe JJ, Claes T, Van Assche D, Bellemans J, Luyten FP. Characterized chondrocyte implantation in the patellofemoral joint: an up to 4-year follow-up of a prospective cohort of 38 patients. Am J Sports Med. 2012;40(8):1799–807. Epub 2012/07/13.
7. Cohen ZA, Roglic H, Grelsamer RP, Henry JH, Levine WN, Mow VC. Patellofemoral stresses during open and closed kinetic chain exercises. An analysis using computer simulation. Am J Sports Med. 2001;29(4):480–7.
8. Grelsamer RP, Weinstein CH. Applied biomechanics of the patella. Clin Orthop Relat Res. 2001;389: 9–14.
9. Lu AP, Hame SL. Autologous osteochondral transplantation for simple cyst in the patella. Arthroscopy. 2005;21(8):1008.
10. Huberti HH, Hayes WC. Patellofemoral contact pressures. The influence of q-angle and tendofemoral contact. J Bone Joint Surg Am. 1984;66(5):715–24.
11. Nho SJ, Foo LF, Green DM, Shindle MK, Warren RF, Wickiewicz TL. Magnetic resonance imaging and clinical evaluation of patellar resurfacing with press-

fit osteochondral autograft plugs. Am J Sports Med. 2008;36(6):1101–9.
12. Lane JG, Tontz Jr WL, Ball ST, Massie JB, Chen AC, Bae WC, et al. A morphologic, biochemical, and biomechanical assessment of short-term effects of osteochondral autograft plug transfer in an animal model. Arthroscopy. 2001;17(8):856–63.
13. Brittberg M, Winalski CS. Evaluation of cartilage injuries and repair. J Bone Joint Surg Am. 2003;85 Suppl 2:58–69.
14. Hangody L, Fules P. Autologous osteochondral mosaicplasty for the treatment of full-thickness defects of weight-bearing joints: ten years of experimental and clinical experience. J Bone Joint Surg Am. 2003;85 Suppl 2:25–32.
15. Atik OS, Uslu MM, Eksioglu F. Osteochondral multiple autograft transfer (OMAT) for the treatment of cartilage defects in the knee joint. Bull Hosp Jt Dis. 2005;63(1–2):37–40.
16. Miniaci A, Tytherleigh-Strong G. Fixation of unstable osteochondritis dissecans lesions of the knee using arthroscopic autogenous osteochondral grafting (mosaicplasty). Arthroscopy. 2007;23(8):845–51.
17. Hangody L, Rathonyi GK, Duska Z, Vasarhelyi G, Fules P, Modis L. Autologous osteochondral mosaicplasty. Surgical technique. J Bone Joint Surg Am. 2004;86 Suppl 1:65–72.
18. Chow JC, Hantes ME, Houle JB, Zalavras CG. Arthroscopic autogenous osteochondral transplantation for treating knee cartilage defects: a 2- to 5-year follow-up study. Arthroscopy. 2004;20(7):681–90.
19. Redman SN, Oldfield SF, Archer CW. Current strategies for articular cartilage repair. Eur Cell Mater. 2005;9:23–32.
20. Bentley G, Biant LC, Carrington RW, Akmal M, Goldberg A, Williams AM. A prospective, randomized comparison of autologous chondrocyte implantation versus mosaicplasty for osteochondral defects in the knee. J Bone Joint Surg Br. 2003;85(2):223–30.
21. Hangody L, Feczko P, Bartha L, Bodo G, Kish G. Mosaicplasty for the treatment of articular defects of the knee and ankle. Clin Orthop Relat Res. 2001;(391 Suppl):S328–36. Epub 2001/10/18.
22. Kish G, Hangody L. A prospective, randomised comparison of autologous chondrocyte implantation versus mosaicplasty for osteochondral defects in the knee. J Bone Joint Surg Br Vol. 2004;86(4):619, author reply –20. Epub 2004/06/04.
23. Lira Neto OA, Franciozi CE, Granata Júnior GS, de Queiroz AA, Carneiro Filho M, Navarro RD. Tratamento cirúrgico das lesões osteocondrais do joelho com mosaicoplastia. Rev Bras Ortop. 2010; 45(2):166–73.
24. Mouzopoulos G, Borbon C, Siebold R. Patellar chondral defects: a review of a challenging entity. Knee Surg Sports Traumatol Arthrosc. 2011;19:1990–2001.
25. Noyes F, Westin-Barber SD. Advanced patellofemoral cartilage lesions in patients younger than 50 years of age: is there an ideal operative option? Arthroscopy. 2013;29(8):1423–36.
26. Crema MD, Roemer FW, Marra MD, Burstein D, Gold GE, Eckstein F. Articular cartilage in the knee: current MR imaging techniques and applications in clinical practice and research. Radiographics. 2011;31(1):37–62.
27. Kijowski R, Blankenbaker DG, Munoz Del Rio A, Baer GS, Graf BK. Evolution of the articular cartilage of the knee joint: value of adding a T2 mapping sequence to a routine MR imaging protocol. Radiology. 2013;267(2):503–13.
28. Lysholm J, Gillquist J. Evaluation of the knee ligament surgery results with special emphasis on use of a scoring scale. Am J Sports Med. 1982;10:150–3.
29. Fulkerson JP, Shea KP. Disorders of patellofemoral alignment. J Bone Joint Surg Am. 1990;72:1424–9.
30. Kujala UM, Jaakkola LH, Koskinen SK, Taimela S, Hurme M, Nelimarkka O. Scoring of patellofemoral disorders. Arthroscopy. 1993;9:159–63.

New Techniques for Cartilage Repair of the Patella

31

Alberto Gobbi, Dnyanesh Lad, and Georgios Karnatzikos

Contents

A. Gobbi (✉) • D. Lad • G. Karnatzikos
Orthopaedic Arthroscopic Surgery International (O.A.S.I.) Bioresearch Foundation, Gobbi N.P.O., Milan, Italy
e-mail: gobbi@cartilagedoctor.it; giokarnes@gmail.com

A. Gobbi et al. (eds.), *The Patellofemoral Joint*,
DOI 10.1007/978-3-642-54965-6_31, © ISAKOS 2014

31.1 Introduction

Hyaline cartilage has long been recognized as having a limited capability to heal due to the fact that it is avascular, because of the presence of few specialized cells with a low mitotic activity and because of the lack of undifferentiated cells that can promote tissue repair. Once injured, cartilage gradually degenerates owing to both mechanical and biochemical factors leading to osteoarthritis (OA) [1], thereby mandating surgical intervention to achieve repair and to avoid subsequent cartilage degeneration.

Chondral or osteochondral lesions are frequently found during knee arthroscopy. In a study of 1,000 patients who underwent arthroscopy, the prevalence of osteochondral defects was 61 %, while 17 % of them were located in the patellofemoral joint (11 % patella, 6 % trochlea) [2]. Furthermore, patellofemoral maltracking and instability often acts as an undiagnosed background factor for articular cartilage lesions in the patella and trochlea [3].

31.2 Autologous Chondrocyte Implantation (ACI)

ACI was first introduced by Peterson [4] and represents a viable technique for cartilage full-thickness chondral lesion repair [5–7]. However, the apparent complexity of this technique, needing the sacrifice of periosteal tissue, the uncertain distribution of chondrocyte solution, and complications such as periosteal patch hypertrophy and arthrofibrosis prompted the scientific community to develop second-generation ACI [6, 8, 9]. However, essentially it remains a two-step procedure including an arthroscopic biopsy and subsequent implantation of the cultured chondrocytes. Apart from donor site morbidity, the risks of two surgical procedures, and the limited quantity of cartilage that could be harvested, the total cost of surgeries, scaffold, and in vitro culture still represent the major limitation of this technique [10, 11]. Also, patients who have undergone ACI for patellar lesions have demonstrated less satisfactory results compared to the treatment of femoral defects [12, 13].

31.3 Mesenchymal Stem Cells Implantation

A one-step surgical procedure is the order of the day. In this regard, the use of bone marrow aspirate concentrate (BMAC) cells, which contain multipotent mesenchymal stem cells (MSCs) and growth factors (GFs), can represent a possible alternative to regenerate cartilage tissue. It avoids the first surgery for cartilage biopsy and the subsequent in vitro chondrocyte cell cultivation, with a significant reduction of the cost and time of the treatment [14–17]. The rationale of this procedure is to paste the BMAC into the cartilage defect and protect the in-growth of the neotissue with a user-friendly scaffold impermeable to cells. This technique enhances cell-to-cell contact and provides a strong chondrogenic environment utilizing a collagen I/III matrix that endorses chondrogenic differentiation of MSC and cartilage regeneration.

MSCs secrete bioactive molecules that stimulate angiogenesis and mitosis of tissue-specific and intrinsic progenitors and reduce T-cell surveillance and inflammation [14, 18]; other authors have also recognized that the presence of other nucleated cells is able to restore the damaged tissue [19, 20]. This newly discovered capacity of MSCs to secrete bioactive factors that are both immunomodulatory and regenerative paves the way to strategies that mimic natural tissue repair. The easy availability coupled with the self-renewal capacity and multi-lineage differentiation potential of MSCs leading to generation of chondrogenic tissue offers a promising option in cartilage surgery [21–24]. These cells are characterized by their ability to adhere to plastic in standard culture conditions, expressing CD 105, CD 73, and CD 90 and lack the expression of CD 45, CD 34, CD14 or Cd11b, CD 79a or CD 19, and HLA-DR surface molecules [25]. Another crucial issue in the clinical application of MSCs for cartilage repair is their phenotypic stability. In fact, MSC-derived chondrogenic cells still possess a degree of plasticity and the tendency to proceed along the endochondral ossification route that can lead to calcification of the implant [26, 27]. The use of collagen-based membrane can possibly provide a suitable environment to maintain stable cell phenotype and cell stabilization into the defect.

31.4 Application in Clinical Practice: Surgical Technique

Under regional anesthesia, in a supine position and with sterile preparation and draping, diagnostic arthroscopy should be performed to evaluate the condition of the joint and to precisely locate and size the cartilage defects. Sixty milliliter of bone marrow is aspirated from the iliac crest using a dedicated aspiration kit and centrifuged using a commercially available system (BMAC Harvest Smart PreP2 System® – Harvest Technologies, Plymouth, MA) to concentrate the bone marrow cells 4–6 times (Fig. 31.1a, b). Using batroxobin enzyme (Plateltex®act-Plateltex SRO Bratislava, SK), the bone marrow concentrate is activated and a sticky clot material is generated. Through a mini arthrotomy, remove the calcified layer if present, while avoiding penetration of the subchondral bone. Damaged cartilage must be removed until a contained and shouldered defect remains, which is necessary to facilitate suturing the

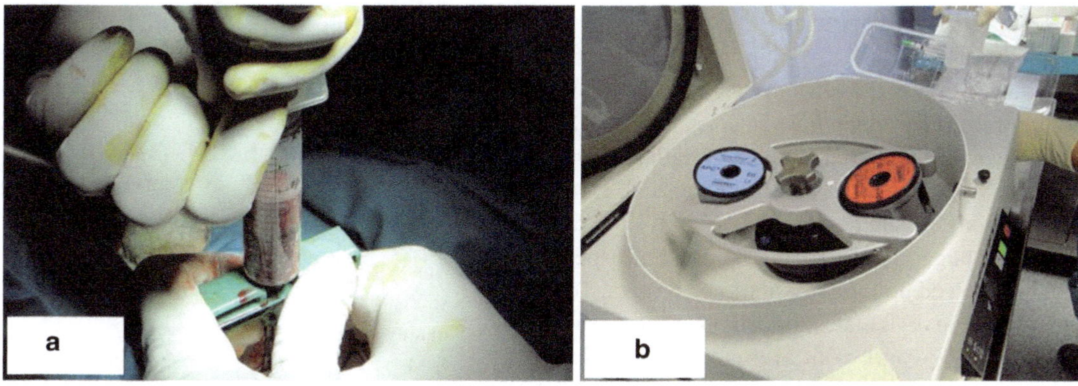

Fig. 31.1 (**a**, **b**) Aspiration of bone marrow from the iliac crest and its processing in the centrifuge

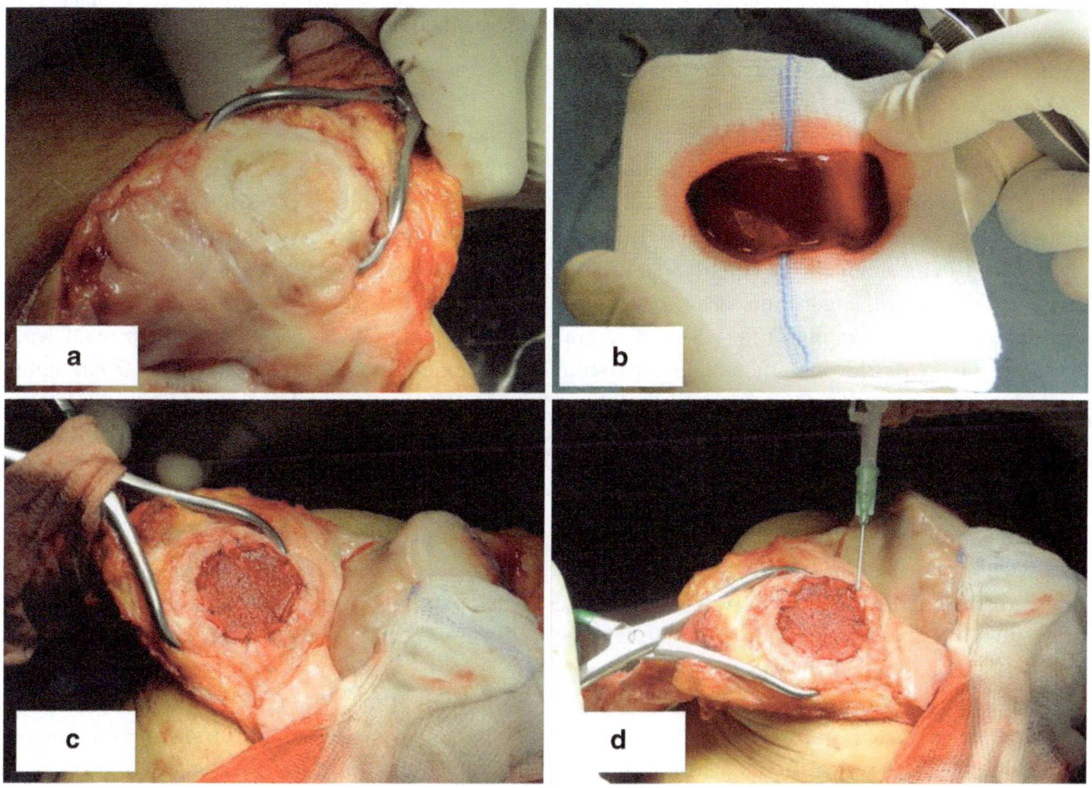

Fig. 31.2 (**a**) Prepared patellar chondral defect. (**b**) BMAC clot after activation. (**c**) Clot applied to defect, collagen membrane sutured with 6-0 PDS. (**d**) Application of fibrin glue to the periphery of the lesion

scaffold. Template the defect and then size the collagen membrane accordingly before pasting the prepared clot into the lesion. Cover the defect with a hyaluronan-based matrix (Hyalofast Anika Therapeutics, Abano T., Italy) in order to protect the MSCs. The membrane should be anchored to the surrounding cartilage using PDS 6-0 and sealed with fibrin glue (Tissucol, Baxter, Rome, Italy) (Fig. 31.2a–d) after which the knee should be moved through the full range of motion to assess the stability of the membrane.

31.5 Our Experience

In order to assess the efficacy of MSCs in patellofemoral (PF) chondral lesions, at OASI Bioresearch Foundation, Milan, 15 patients, presenting with chronic large full-thickness PF cartilage lesions, treated with BMAC and covered with a collagen type I/III matrix were prospectively followed up for a minimum of 3 years. Ten males and five females (mean age: 48 years; range 32–50 years), with average BMI of 24.8, who were nonprofessional athletes, were included. Cartilage lesions were diagnosed as grade 4 by ICRS classification on MRI and arthroscopy; patients with lesions in compartments other than the PF joint were excluded. Other exclusion criteria were tricompartmental OA, osteonecrosis of the knee, intra-articular steroid injections within 6 months prior to surgery, general systemic illness, and neurovascular diseases. Seven patients had lesions located at the patella (one of them had two patellar lesions), five at the trochlea, and three patients had kissing lesions. The average cartilage lesion size per patient was 7.3 cm^2, ranging from 1.5 to 18.75 cm^2 (total lesion area). Eleven patients had coexisting pathologies such as tibiofemoral axial malalignment, PF maltracking, anterior cruciate ligament (ACL) insufficiency, and meniscal tears, and these pathologies were treated before or during the same surgery. High tibial osteotomy (HTO) has been established as an effective treatment of the varus osteoarthritic knee in order to decrease the stress on the load-bearing cartilage in the medial compartment; however, only partial remodeling of the articular cartilage has been reported. In a recently published review, significantly greater survival at 5 years follow-up was seen after HTO with articular cartilage surgery than after isolated HTO [28]. Therefore, correction of tibiofemoral axis malalignment is indicated when articular cartilage restoration techniques are applied [29]. Similarly, patellofemoral maltracking, when present, should be addressed in order to reduce overload of the lateral patellofemoral joint and reduce the risk of future cartilage injuries [30].

Table 31.1 Clinical outcome

Score	Preoperative value	Final follow-up value
VAS	5	0.8
IKDC (subjective)	44.3	83.1
KOOS pain	66.4	94.2
KOOS symptoms	67.8	88.4
KOOS activity of daily living	69.7	94.8
KOOS sports	43.7	73.6
KOOS quality of life	38.5	76.3
LYSHOLM	62.1	91.7
MARX	3.7	9.5
TEGNER	2.4	5

VAS visual analogue scale, *IKDC* International Knee Documentation Committee, *KOOS* Knee Injury and Osteoarthritis Outcome Score

All patients followed the same rehabilitation protocol for at least 6 months, which is similar to rehabilitation after second-generation ACI, based on current knowledge of the graft healing biology and on functional criteria and therapy goal progression [15, 31, 32]. X-rays and MRI, visual analog scale (VAS) for pain, International Knee Documentation Committee (IKDC), Knee injury and Osteoarthritis Outcome Score (KOOS), Lysholm, Tegner, and Marx scores were collected at each follow-up.

All the patients showed significant improvement in all scores at 6-, 12-, 24-, and 36-month follow-up ($P<0.05$). No adverse reactions or postoperative complications were noted. IKDC objective score showed significant improvement in A and B subgroups from preoperative to final follow-up (Table 31.1).

Patients having single lesions and smaller lesions were found to have more improvement than patients with larger, multiple, and/or kissing lesions in all scores, (except for KOOS pain and KOOS symptom subgroups); however, the average KOOS values for these subgroups were comparable at final follow-up.

Posttreatment MRI showed complete filling of the defects, while no signs of hypertrophy were identified. Integration with adjacent cartilage was complete with restoration of the cartilage layer

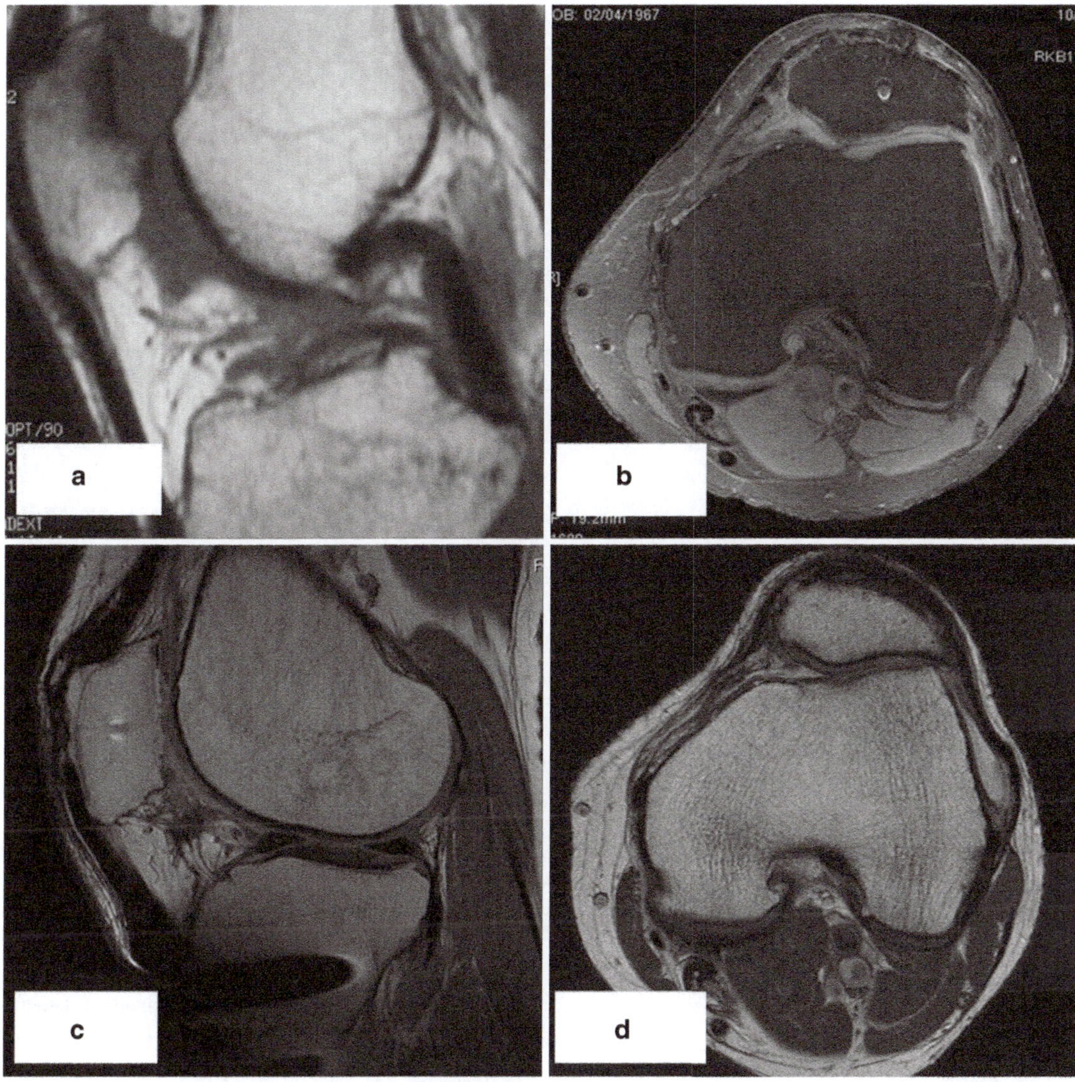

Fig. 31.3 (**a**, **b**) Sagittal and axial MRI scan through the knee showing patellar and trochlear chondral lesion. (**c**, **d**) Sagittal and axial MR images through the same knee, showing good articular continuity at final follow-up

and subchondral bone. We also did not identify edema, cysts, or sclerosis of subchondral bone (Fig. 31.3a–d).

Second-look arthroscopies in four knees revealed smooth, newly formed tissue with continuous intact to the healthy surrounding cartilage; no hypertrophy was identified. The stability of the implant appeared similar to the adjacent tissue when checked with a probe [33]. Macroscopic evaluation showed normal to nearly normal as classified by the ICRS visual scoring system. Histological examination of the biopsies showed the regeneration of new tissue with many hyaline-like cartilage features such as the presence of a noticeable proteoglycan component around the chondrons and also collagen type II content. There was a good organization of proteoglycans and collagen in the extracellular matrix, an intact superficial zone, and a well-defined tidemark (Fig. 31.4).

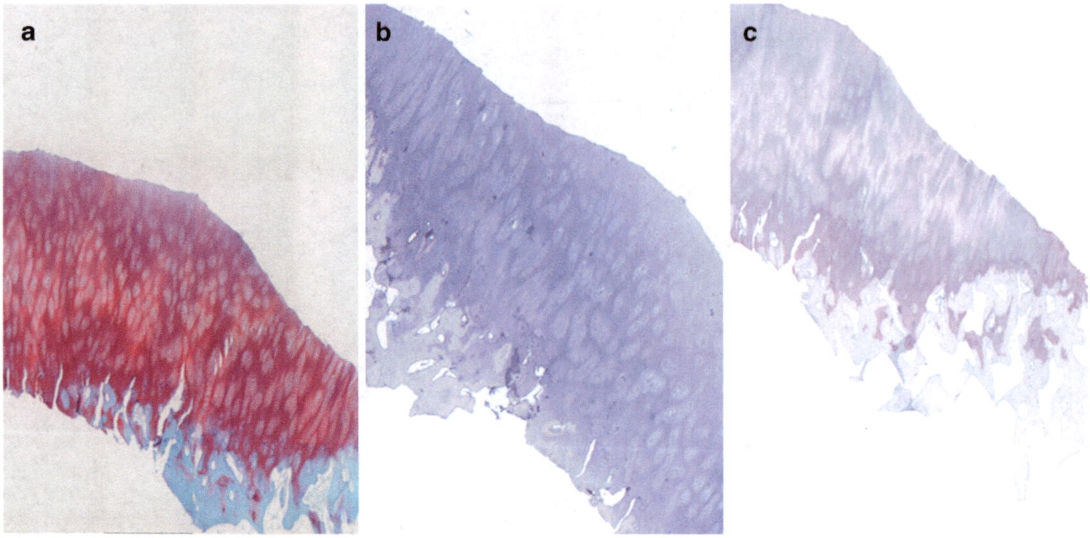

Fig. 31.4 Histological appearance of the tissue biopsy from the site of lesion obtained after12 months (original magnification, 40x). (**a**) The superficial layer is regular and the tidemark is well evident. The proteoglycan component is well represented and the cells show regular distribution along the extracellular matrix. The subchondral bone tissue is in a remodeling process. (**b**) Immunohistological analysis results of colagen type I are completely negative. (**c**) Collagen type II immunostaining is positive at the exracellular level

31.6 Advantages

1. One-step surgery.
2. The use of a collagen I/III-based matrix favors cell concentration in the defected area and allows early mobilization of the operated knee.
3. Lower cost if compared to the standard two-step ACI procedures.

Conclusion

The patellofemoral joint cartilage defects can be difficult to treat. The patella is not very amenable to treatment with microfracture given the mobility of the bone. The limitations of ACI have generated interest in one-step surgery with BMAC. Our experience with its use for patellofemoral cartilage lesions has been more than satisfactory. However, a larger number of patients and more randomized control trials would be essential to establish conclusively this relatively new option of treating patellofemoral cartilage lesions. Nonetheless, it is a very viable option for young patients who are not suited to undergo a metal resurfacing procedure and may buy important years of activity and patient satisfaction.

References

1. Mankin HJ. The response of articular cartilage to mechanical injury. J Bone Joint Surg Am. 1982;64(3): 460–6.
2. Hjelle K, Solheim E, Strand T, Muri R, Brittberg M. Articular cartilage defects in 1,000 knee arthroscopies. Arthroscopy. 2002;18(7):730–4.
3. Fulkerson JP. Diagnosis and treatment of patients with patellofemoral pain. Am J Sports Med. 2002;30(3):447–56. Review.
4. Brittberg M, Lindahl A, Nilsson A, Ohlsson C, Isaksson O, Peterson L. Treatment of deep cartilage defects in the knee with autologous chondrocyte transplantation. N Engl J Med. 1994;331(14): 889–95.
5. Behrens P, Bitter T, Kurz B, Russlies M. Matrix-associated autologous chondrocyte transplantation/ implantation (MACT/MACI): 5-year follow-up. Knee. 2006;13(3):194–202.
6. Peterson L, Brittberg M, Kiviranta I, Akerlund EL, Lindahl A. Autologous chondrocyte transplantation: biomechanics and long-term durability. Am J Sports Med. 2002;30(1):2–12.
7. Peterson L, Vasiliadis HS, Brittberg M, Lindahl A. Autologous chondrocyte implantation: a long-term follow-up. Am J Sports Med. 2010;38(6):1117–24.
8. Bekkers JE, Inklaar M, Saris DB. Treatment selection in articular cartilage lesions of the knee: a systematic review. Am J Sports Med. 2009;37 Suppl 1:148S–55.
9. Kon E, Gobbi A, Filardo G, Delcogliano M, Zaffagnini S, Marcacci M. Arthroscopic second-generation

autologous chondrocyte implantation compared with microfracture for chondral lesions of the knee: prospective nonrandomized study at 5 years. Am J Sports Med. 2009;37(1):33–41.
10. Henderson I, Francisco R, Oakes B, Cameron J. Autologous chondrocyte implantation for treatment of focal chondral defects of the knee: a clinical, arthroscopic, MRI and histologic evaluation at 2 years. Knee. 2005;12(3):209–16.
11. Minas T, Peterson L. Advanced techniques in autologous chondrocyte transplantation. Clin Sports Med. 1999;18(1):13–44.
12. Gobbi A, Kon E, Berruto M, Filardo G, Delcogliano M, Boldrini L, et al. Patellofemoral full-thickness chondral defects treated with second-generation autologous chondrocyte implantation: results at 5 years' follow-up. Am J Sports Med. 2009;37(6): 1083–92.
13. Farr 2nd J. Autologous chondrocyte implantation and anteromedialization in the treatment of patellofemoral chondrosis. Orthop Clin North Am. 2008;39(3):329–35, vi. Review.
14. Fortier LA, Potter H, Rickey E, Schnabel L, Foo L, Chong L, et al. Concentrated bone marrow aspirate improves full-thickness cartilage repair. Proceedings of the 55th ORS annual meeting. Las Vegas; 22–25 Feb 2009.
15. Gobbi A, Karnatzikos G, Scotti C, Mahajan V, Mazzucco L, Grigolo B. One-step cartilage repair with bone marrow aspirate concentrated cells and collagen matrix in full-thickness knee cartilage lesions: results at 2-year follow-up. Cartilage. 2011;2(3):286–99.
16. Sudo K, Kanno M, Miharada K, Ogawa S, Hiroyama T, Saijo K, Nakamura Y. Mesenchymal progenitors able to differentiate into osteogenic, chondrogenic, and/or adipogenic cells in vitro are present in most primary fibroblast-like cell populations. Stem Cells. 2007;25(7):1610–7.
17. Wakitani S, Imoto K, Yamamoto T, Saito M, Murata N, Yoneda M. Human autologous culture expanded bone marrow mesenchymal cell transplantation for repair of cartilage defects in osteoarthritic knees. Osteoarthritis Cartilage. 2002;10(3):199–206.
18. Fortier LA, Mohammed HO, Lust G, Nixon AJ. Insulin-like growth factor-I enhances cell-based repair of articular cartilage. J Bone Joint Surg Br. 2002;84(2): 276–88.
19. Caplan AI, Dennis JE. Mesenchymal stem cells as trophic mediators. J Cell Biochem. 2006;98: 1076–84.
20. Jones E, McGonagle D. Human bone marrow mesenchymal stem cells in vivo. Rheumatology (Oxford). 2008;47(2):26–31.
21. Cavallo C, Desando G, Cattini L, Cavallo M, Buda R, Giannini S, Facchini A, Grigolo B. Bone marrow concentrated cell transplantation: rationale for its use in the treatment of human osteochondral lesions. J Biol Regul Homeost Agents. 2013;27(1):165–75.
22. Gobbi A, Karnatzikos G, Nakamura N, Mahajan V. Next generation cartilage solutions. In: Doral MN, editor. Sports injuries – prevention, diagnosis, treatment and rehabilitation. Berlin/Heidelberg: Springer; 2012. p. 739–49.
23. Nakamura T, Sekiya I, Muneta T, Hatsushika D, et al. Arthroscopic, histological and MRI analyses of cartilage repair after a minimally invasive method of transplantation of allogeneic synovial mesenchymal stromal cells into cartilage defects in pigs. Cytotherapy. 2012;14(3):327–38.
24. Tateishi K, Ando W, Higuchi C, Hart DA, et al. Comparison of human serum with fetal bovine serum for expansion and differentiation of human synovial MSC: potential feasibility for clinical applications. Cell Transplant. 2008;17(5):549–57.
25. Dominici M, Le Blanc K, Mueller I, et al. Minimal criteria for defining multipotent mesenchymal stromal cells. The international society for cellular therapy position statement. Cytotherapy. 2006;8(4):315–7.
26. Pelttari K, Steck E, Richter W. The use of mesenchymal stem cells for chondrogenesis. Injury. 2008;39 Suppl 1:S58–65.
27. Scotti C, Tonnarelli B, Papadimitropoulos A, Scherberich A, Schaeren S, Schauerte A, et al. Recapitulation of endochondral bone formation using human adult mesenchymal stem cells as a paradigm for developmental engineering. Proc Natl Acad Sci U S A. 2010;107(16):7251–6.
28. Harris JD, McNeilan R, Siston RA, Flanigan DC. Survival and clinical outcome of isolated high tibial osteotomy and combined biological knee reconstruction. Knee. 2013;20(3):154–61.
29. Minzlaff P, et al. Osteochondral autologous transfer combined with valgus high tibial osteotomy long-term results and survivorship analysis. Am J Sports Med. 2013;41(10):2325–32.
30. Kramer DE, Kocher MS. Management of patellar and trochlear chondral injuries. Oper Tech Orthop. 2007; 17(4):234–43.
31. Gobbi A, Karnatzikos G, Mahajan V. Biologic arthroplasty. In: Brittberg M et al., editors. Cartilage repair: clinical guidelines. Guildford: DJO Publications; 2012. p. 269–79. ISBN 780956176004.
32. Gobbi A, Turgeon D, Karnatzikos G. Decision making in professional athletes with cartilage lesions. In: Brittberg M et al., editors. Cartilage repair: clinical guidelines. Guildford: DJO Publications; 2012. p. 185–208. ISBN 9780956176004.
33. Gobbi A, Karnatzikos G, Sankineani SR. One-step surgery with multipotent stem cells for the treatment of large full-thickness chondral defects of the knee. Am J Sports Med. 2014;42(3):648–57. doi:10.1177/0363546513518007. 23 Jan 2014.

Role of Mesenchymal Stem Cells in Patellofemoral Disorders

32

James H. Hui

Contents

32.1 Introduction

Patellofemoral disorders are a complex entity of conditions, resulting in cartilage lesions of the patella. These range from osteochondritis dissecans (OCD) in the young to patellofemoral osteoarthritis in the elderly. Chondromalacia patellae also result in anterior knee pain with resultant cartilage lesions on the patella. Early lesions may heal by cartilage or fibrous metaplasia [1]; however, as the disease progresses, it produces significant cartilage damage, which will result in worsening anterior knee pain refractory to conservative management.

The treatment of patella articular cartilage lesions is challenging due to the complexity of the patellofemoral joint and the limited capacity to heal. For OCD lesions, arthroscopic fixation with resorbable pins to stabilize the OCD has been advocated [2]. Marrow stimulation techniques, such as microfracture, to penetrate the subchondral bone have also been described with modest results [3]. The advent of cell-based therapy has brought significant clinical improvements to treatment of these conditions.

32.2 Cell-Based Therapy: Current Methods

Cell-based therapy consists broadly of two methods: autologous chondrocyte implantation (ACI) [4, 5] (Figs. 32.1, 32.2, and 32.3) and mesenchymal stem cell (MSCs) implantation.

J.H. Hui
Cartilage Repair Program, Tissue Engineering & Cell Therapy Lab, National University Hospital, Singapore, Singapore
e-mail: james_hui@nuhs.edu.sg

A. Gobbi et al. (eds.), *The Patellofemoral Joint*,
DOI 10.1007/978-3-642-54965-6_32, © ISAKOS 2014

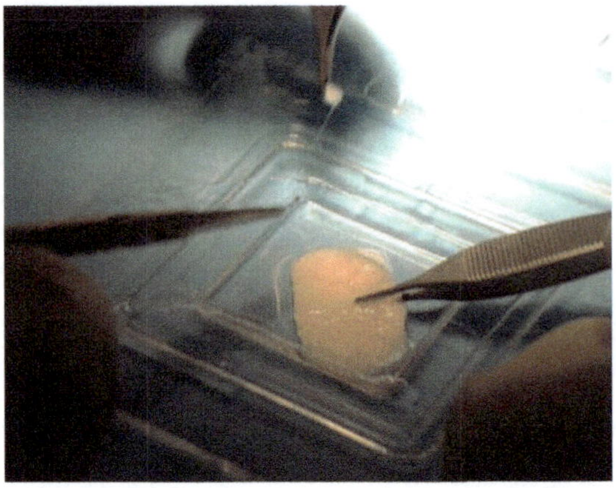

Fig. 32.1 Templated graft ready for implantation by Gobbi et al. [5]

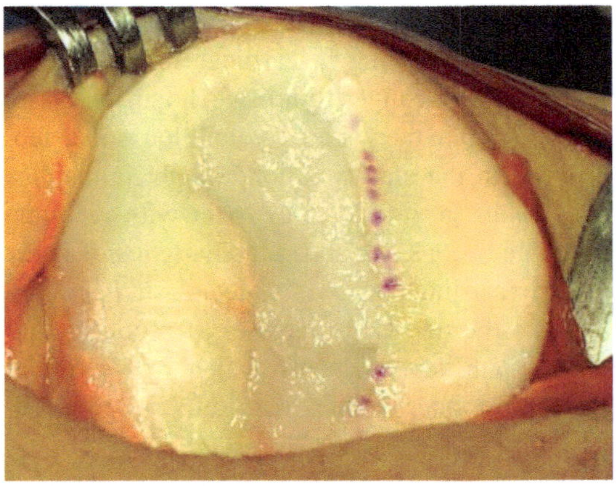

Fig. 32.2 Chondral defect after ACI by Gobbi et al. [5]

ACI has been clinically proven to be an effective tool; however, the treatment effect declines during the midterm follow-up. Gobbi et al. [5] reported improvements in the International Knee Documentation Committee (IKDC) subjective scores (46.09–77.06), Tegner-Lysholm (2.56–4.94) and EuroQol visual analogue scale (56.76–78.23) at 2 years using chondrocyte grafts but a decline of the IKDC subjective scores and Tegner-Lysholm scores in patients with multiple and patellar lesions from 2 to 5 years follow-up (Fig 32.4). There are some disadvantages to ACI, including general anaesthesia for both harvesting and implantation and donor site morbidity [6, 7].

When comparing ACI against MSCs, treatment of articular cartilage lesions with

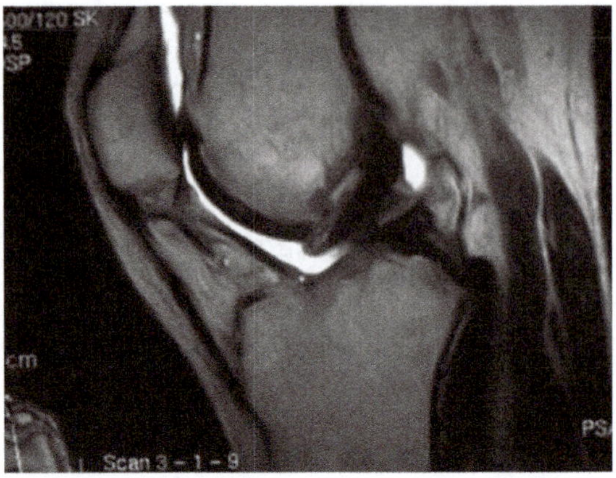

Fig. 32.3 Gobbi et al. [5] demonstrating good fill of patella and trochlear defects after 5 years post implantation

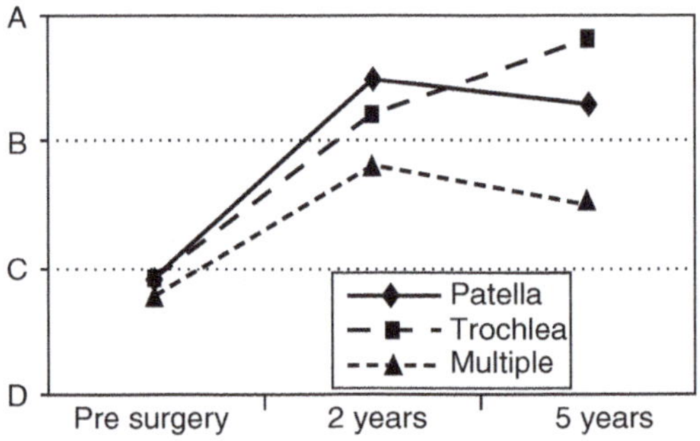

Fig. 32.4 Correlation between the site of the lesions and objective IKDC scores at 2 and 5 years' follow-up. By Gobbi et al. [5]

cultured bone marrow-derived MSCs (BMSCs) has been reported to be as effective as ACI in a clinical cohort study of 72 patients conducted by Nejadnik et al. [8] (Fig 32.5). Hence, BMSCs have been shown to be viable for cartilage repair.

The efficacy of cultured BMSCs specifically on the patellofemoral joint has seen significant clinical advancements in the last decade. Wakitani et al. investigated the clinical effectiveness of cultured BMSCs transplantation in treating cartilage defects via a case series of 9 defects in 5 knees of 3 patients (Fig. 32.6). He found that patients' clinical symptoms improved after 6 months and had remained well as long as 27 months post implantation. Histological grading of repaired cartilage shows promising fibrocartilaginous tissue [9].

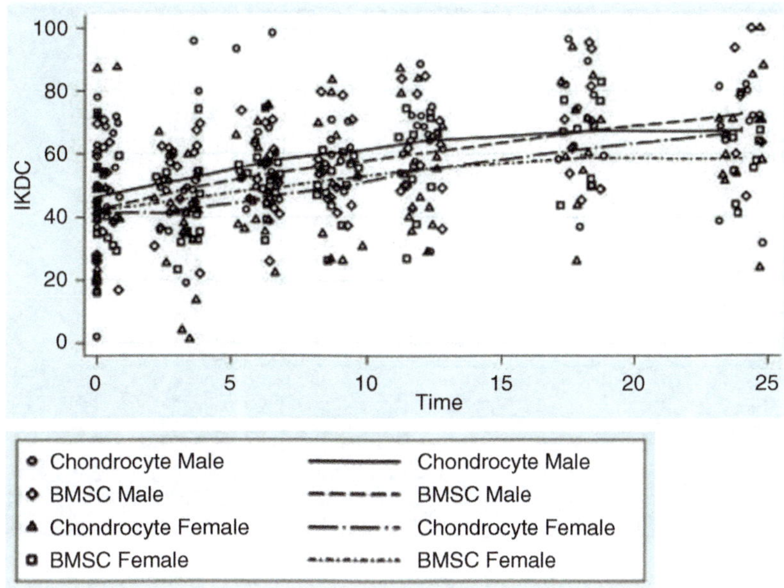

Fig. 32.5 IKDC scores preoperatively and at 3, 6, 9, 12, 18 and 24 months for both BMSC and chondrocyte groups in both genders by Nejadnik et al. [8]

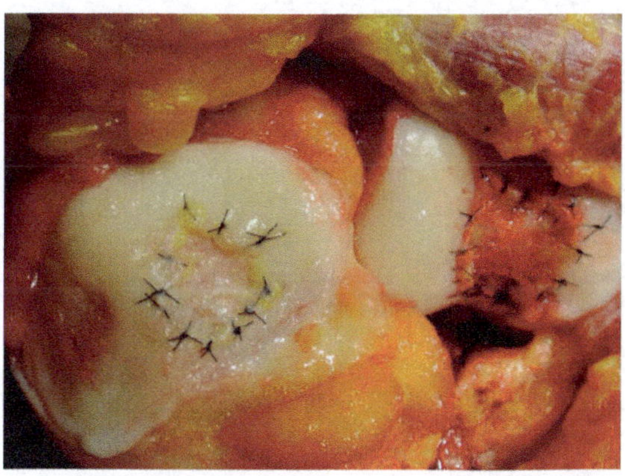

Fig. 32.6 Wakitani et al. demonstrating the use of BMSC transplantation in the patellofemoral joint [9]

32.3 Cell-Based Therapy: Paediatric and Young Adult Considerations

To assess the efficacy of cell-based therapy in children, Hui et al. conducted a clinical study looking at the efficacy of cell-based therapy in young patients who suffer from OCD. Twenty-three patients (from 12 to 21 years old) who underwent cell-based therapy with either ACI (20 patients) or cultured BMSCs implantation (3 patients) were retrospectively reviewed. Preoperative CT scans to assess patella subluxation, tilt and congruence angle to determine choice of treatment were carried out. Clinical scores (Tegner-Lysholm, Lysholm-Gillquist) were evaluated at 6, 12 and 24 months postoperatively. They found that clinical scores were better globally at the end of the 24-month follow up. Mean IKDC score, Tegner-Lysholm outcomes and Lysholm-Gillquist scale improved from 45, 2.5 and 50, respectively, at surgery to 75, 4 and 70, respectively. Cell-based therapy was thus generally safe in the use for children. A significant complication noticed in this age group was periosteal hypertrophy post implantation, which was detected in 2 patients (Fig. 32.7) [10].

32.4 Future Prospects: Intra-articular Mesenchymal Stem Cell Injections for Patella Cartilage Lesions

As ACI requires an additional operation for transplantation and coverage of the chondral defect, there has been a shift to using intra-articular injections of BMSCs instead. As BMSCs can be harvested at the same time as initial arthroscopic debridement (via iliac

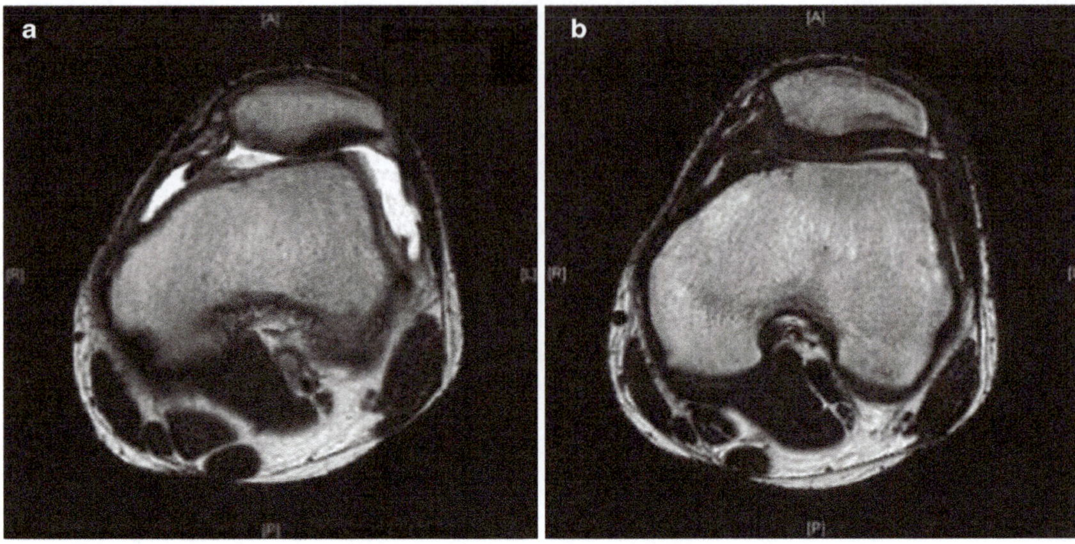

Fig. 32.7 Hui et al. demonstrate hypertrophy of periosteum in paediatric cases, (**a**) was MRI performed at surgery, and (**b**) was performed 4 months postoperatively [10]

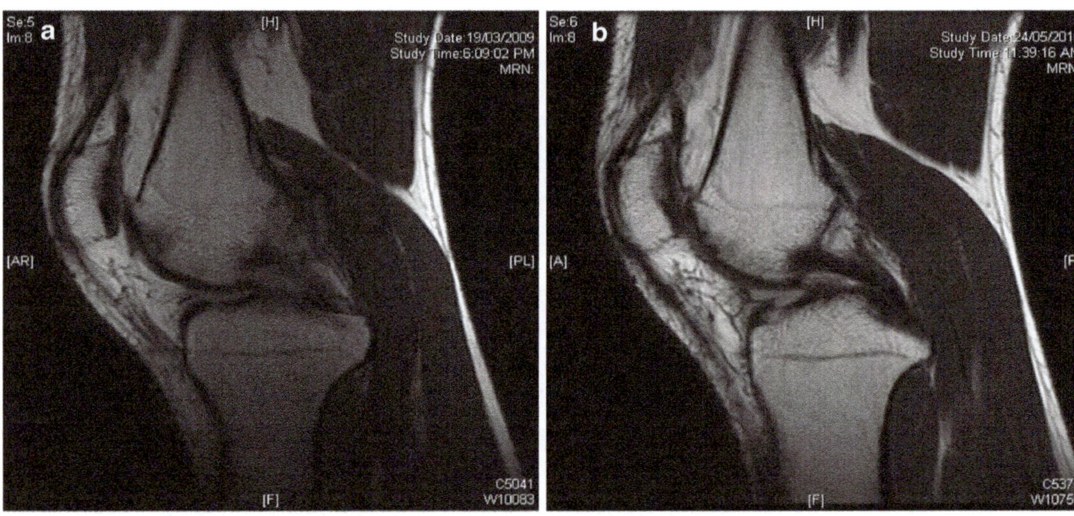

Fig. 32.8 (**a**) Lee et al. demonstrating the efficacy of intra-articular BMSC injections for cartilage repair. Preoperative, a full-thickness chondral defect with subchondral bone discontinuity and marrow oedema is seen [12]. (**b**) 1 year post intra-articular BMSC injection, the MRI shows neo-cartilage formation with good fill and significant reduction in the underlying marrow oedema [12]

crest methods), which the cultured BMSCs can then be introduced into the knee joint via intra-articular methods with local anaesthesia, this leads to 1 less knee surgery, reduced costs and minimized donor site morbidity.

Steadman et al. carried out a large animal study looking at the efficacy of intra-articular injections of MSCs in microfractured chondral defects in an equine model. They reported significant neo-cartilage in the microfractured defect, which was maintained up to 12 months [11].

Lee et al., who carried a prospective cohort study comparing the open technique of cartilage repair (including patellofemoral lesions) with BMSCs versus intra-articular injections of BMSCs, were able to demonstrate comparable efficacy compared to the open technique. The study was also able to validate the safety of this method, with no significant clinical complications reported during the 2-year follow-up (Fig. 32.8) [12].

To address the issue of localization of BMSCs to the specific area of the knee, such as the patellofemoral joint, Ochi et al. were able to utilize a pig model to demonstrate that magnetically labelled MSCs and magnetic delivery systems are capable of targeted MSC treatment to specific parts of the knee join [13]. We currently await the clinical trials to demonstrate the effect of patients.

Conclusion

The role of MSCs in patellofemoral disorders has been established in recent years, with efficacy comparable to ACI. Novel methods such as intra-articular injections of BMSCs have been efficacious in various clinical studies to promote cartilage repair, and these are currently being evaluated for use in specific patellofemoral disorders.

References

1. Bentley G. Articular cartilage changes in chondromalacia patellae. J Bone Joint Surg Br. 1985;67(5):769–74.
2. Matava MJ, Brown CD. Osteochondritis dissecans of the patella: arthroscopic fixation with bioabsorbable pins. Arthroscopy. 1997;13:124–8.
3. Mandelbaum BR, Browne JE, Fu F, Micheli L, Mosely Jr JB, Erggelet C, Minas T, Peterson L. Articular cartilage lesions of the knee. Am J Sports Med. 1998;26:853–61.
4. Macmull S, Jaiswal PK, Bentley G, Skinner JA, Carrington RW, Briggs TW. The role of autologous chondrocyte implantation in the treatment of symptomatic chondromalacia patellae. Int Orthop. 2012;36(7):1371–7. doi:10.1007/s00264-011-1465-6. Epub 2012 Jan 14.
5. Gobbi A, Kon E, Berruto M, Filardo G, Delcogliano M, Boldrini L, Bathan L, Marcacci M. Patellofemoral full-thickness chondral defects treated with second-generation autologous chondrocyte implantation: results at 5 years' follow-up. Am J Sports Med. 2009; 37:1083–92.
6. Bentley G, Biant LC, Carrington RW, Akmal M, Goldberg A, Williams AM, Skinner JA, Pringle J. A prospective, randomised comparison of autologous chondrocyte implantation versus mosaicplasty for osteochondral defects in the knee. J Bone Joint Surg Br. 2003;85:223–30.
7. Fu FH, Zurakowski D, Browne JE, Mandelbaum B, Erggelet C, Moseley Jr JB, Anderson AF, Micheli LJ. Autologous chondrocyte implantation versus debridement for treatment of full-thickness chondral defects of the knee: an observational cohort study with 3-year follow-up. Am J Sports Med. 2005; 33:1658–66.
8. Nejadnik H, Hui JH, Feng Choong EP, Tai BC, Lee EH. Autologous bone marrow-derived mesenchymal stem cells versus autologous chondrocyte implantation: an observational cohort study. Am J Sports Med. 2010;38:1110–6.
9. Wakitani S, Nawata M, Tensho K, Okabe T, Machida H, Ohgushi H. Repair of articular cartilage defects in the patello-femoral joint with autologous bone marrow mesenchymal cell transplantation: three case reports involving nine defects in five knees. J Tissue Eng Regen Med. 2007;1:74–9.
10. Teo BJ, Buhary K, Tai BC, Hui JH. Cell-based therapy improves function in adolescents and young adults with patellar osteochondritis dissecans. Clin Orthop Relat Res. 2013;471(4):1152–8. doi:10.1007/s11999-012-2338-z.
11. McIlwraith CW, Frisbie DD, Rodkey WG, Kisiday JD, Werpy NM, Kawcak CE, Steadman JR. Evaluation of intra-articular mesenchymal stem cells to augment healing of microfractured chondral defects. Arthroscopy. 2011;27(11):1552–61. doi:10.1016/j.arthro.2011.06.002. Epub 2011 Aug 20.
12. Lee KB, Wang VT, Chan YH, Hui JH. A novel, minimally-invasive technique of cartilage repair in the human knee using arthroscopic microfracture and injections of mesenchymal stem cells and hyaluronic acid–a prospective comparative study on safety and short-term efficacy. Ann Acad Med Singapore. 2012; 41(11):511–7.
13. Kamei G, Kobayashi T, Ohkawa S, Kongcharoensombat W, Adachi N, Takazawa K, Shibuya H, Deie M, Hattori K, Goldberg JL, Ochi M. Articular cartilage repair with magnetic mesenchymal stem cells. Am J Sports Med. 2013; 41(6):1255–64.

Platelet-Rich Plasma for the Treatment of Symptomatic Patellofemoral Cartilage Lesions

33

Georgios Karnatzikos, Sanyam Chaurasia, and Alberto Gobbi

Contents

33.1 Introduction

Articular cartilage has limited intrinsic healing potential due to the fact that it is avascular and the presence of few specialized cells with a low mitotic activity [1]. Trauma and/or chronic irritation may lead to progressive damage, joint degeneration, and early osteoarthritis (OA) [2]. Cartilage lesions on the patellofemoral joint present a frequent source of pain and dysfunction especially in active patients. Various associated pathologies such as objective or potential patellar instability, muscular dysfunction, and biomechanical disorders of the foot, knee, and hip might act as a cause of patellofemoral pain syndrome (PFPS), and they should be investigated and addressed when necessary.

Several conservative treatment options (such as oral and topical nonsteroidal anti-inflammatory drugs (NSAIDs), diacerein, and intra-articular corticosteroids and viscosupplementation) have been utilized and have yielded short-term efficacy and local or systemic side effects [3–6]. Platelet-rich plasma (PRP) represents a therapeutic application with promising preliminary clinical results [7–11].

33.2 Platelet-Rich Plasma

PRP can be defined as the volume of the plasma fraction from autologous blood with platelet concentration above baseline count (200,000 platelets/μl) [12]. Platelets contain

G. Karnatzikos (✉) • S. Chaurasia • A. Gobbi
Orthopaedic Arthroscopic Surgery International (O.A.S.I.) Bioresearch Foundation, Gobbi N.P.O., Milan, Italy
e-mail: giokarnes@gmail.com; gobbi@cartilagedoctor.it

A. Gobbi et al. (eds.), *The Patellofemoral Joint*,
DOI 10.1007/978-3-642-54965-6_33, © ISAKOS 2014

many important bioactive proteins and growth factors (GF); these factors when secreted regulate key processes involved in tissue repair, including cell proliferation, chemotaxis, migration, cellular differentiation, and extracellular matrix synthesis [13, 14]. The rationale for topical use of PRP is to stimulate the natural healing cascade and tissue regeneration by a "supraphysiological" release of platelet-derived factors directly at the site of treatment. Platelets α-granules contain a variety of GF including, in part, transforming growth factors (TGF-β1), platelet-derived growth factors (PDGF-BB), hepatocyte growth factor (HGF), basic fibroblast growth factors (b-FGF), epidermal growth factor (EGF), vascular endothelial growth factors (VEGF), and insulin-like growth factor 1 (IGF-I) [12]. GF mediate the biological processes necessary for tissue repair [15, 16]. Their mode of action is to bind to the extracellular domain of a target growth factor receptor that, in turn, activates the intracellular signal transduction pathways [17, 18]. In vitro studies in animal and human chondrocytes [14, 19] have demonstrated that PRP-secreted GF stimulate the proliferation and collagen synthesis. In clinical studies therapeutic application of PRP has shown promising results in the treatment of cartilage defects [8–11, 20, 21]; however, the clinical efficacy of PRP still remains under debate, and a standardized protocol has not yet been established.

We also investigated the possible positive effects of PRP intra-articular injections in active patients with symptomatic knee chondral defects.

33.3 Study Group

We prospectively followed up 50 patients (mean age 47.7) with symptomatic cartilage defects of the knee [11]. All patients (31 males and 19 females) were treated with 2 intra-articular injections (1 monthly) with autologous PRP and followed up for a minimum period of 1 year. Seventeen out of the 50 patients were presenting with lesions involving the patellofemoral joint: nine of the evaluated patients were presenting with multiple lesions involving patellofemoral joint, while eight patients were presenting with single lesions located on the patella or trochlea. Twenty-five patients had undergone a previous operative intervention for cartilage on the treated knee (shaving or microfracture) at least 1 year before PRP treatment. Patients' demographic data are provided in Table 33.1.

Inclusion criteria were age between 30 and 60 years, body mass index (BMI) < 30, normal complete blood count (BMC) and coagulation control, and participation in sports activities but not at professional level; patients with symptomatic osteoarthritic knees of grade 1–3 as per Kellgren-Lawrence classification [22] and cartilage lesions of grade 3 and 4 as per ICRS classification [23] evaluated by MRI and/or previous arthroscopy; and patients with stable knee, normal tibiofemoral alignment, or patellofemoral tracking. We excluded patients with blood and systemic metabolic diseases or immunodeficiency; pretreatment blood platelets value 25 % below the reference value; alcoholism, smoking, and drugs;

Table 33.1 Patients' demographic data

Patients	*n*	Age (average)	Gender (male/female)	Site (knee) right/left	Kellgren-Lawrence classification Grade		
					1	2	3
All	50	47.7 ± 2.52	31/19	20/30	11	19	20
With previous surgery	25	44.7 ± 2.01	14/11	7/18	3	11	11
Without previous surgery	25	50.4 ± 2.77	17/8	13/12	8	8	9
Cartilage shaving	12	44.4 ± 2.39	4/8	2/10	3	6	3
Microfracture	13	45.0 ± 1.68	5/8	5/8	0	5	8

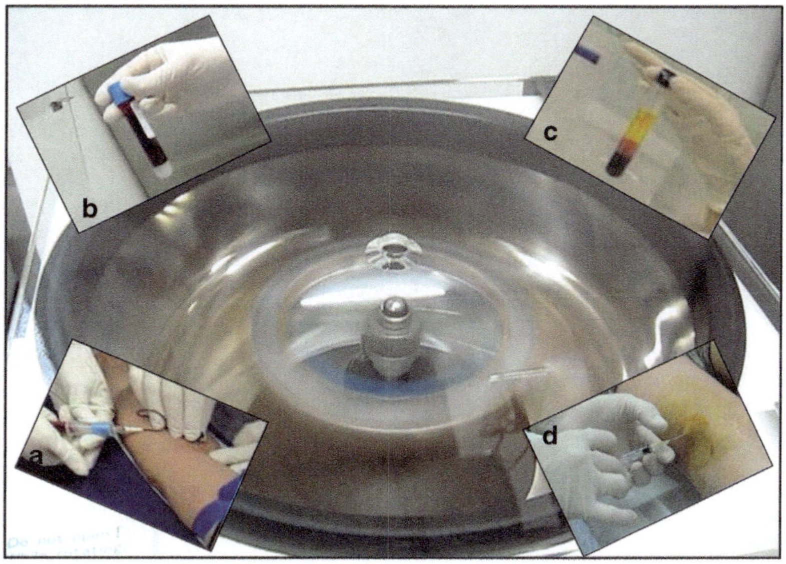

Fig. 33.1 Platelet-rich plasma preparation: (*a*) blood aspiration, (*b*) centrifugation of the blood sample for 9 min, (*c*) fraction of PRP after centrifugation (*yellow upper part* in tube), and (*d*) knee intra-articular infiltration

advanced and tricompartmental OA, rheumatoid or polyarticular arthritis, symptomatic hip and ankle OA, or symptomatic contralateral knee OA; and treatment with corticosteroids (<6 months) or medication that could interfere with platelet aggregation (<7 days).

The standard radiographic evaluation included a standing AP long-leg radiograph, including both hips and ankles, standing AP/lateral views of knees, skyline patellofemoral, and standing 45° bend knee views. Visual analogue scale (VAS) for pain (0 = no pain at all to 10 = worst pain), International Knee Documentation Committee subjective and objective score (IKDC), Knee Injury and Osteoarthritis Outcome Score (KOOS), and Tegner and Marx scores were collected at pretreatment evaluation and at 6- and 12-month follow-up.

33.4 Technique

All patients were treated with 2 intra-articular injections (1 month interval between injections) of autologous PRP (Regen® ACR-C, Regen Lab, Switzerland). After extraction of 8 ml of peripheral blood, the sample was centrifuged for 9 min at 3,500 rpm according to recommendation of the manufacturer. Subsequently we obtained 5 ml of PRP, and we proceeded to the intra-articular infiltration by a supra-patellar approach, under sterile aseptic conditions (Fig. 33.1a–d). After treatment patients were allowed weight bearing and were recommended the application of local ice for 24 h and restriction of vigorous activities of the knee, for at least 48 h.

33.5 Results

All patients showed significant improvement in all scores at 6 and 12 months follow-up ($p < 0.005$) and returned to previous activities including recreational sports (Table 33.2, Figs. 33.2 and 33.3) displaying that PRP injections could represent a valuable treatment in patients with symptomatic knee cartilage defects, including lesions located at the patellofemoral joint. No adverse reactions (like swelling or acute pain) or any major complication (like infection) were noted. There was no significant difference in improvement between operated patients and nonoperated patients. There was no significant difference in improvement between male and female patients.

Table 33.2 Summary of clinical outcome

					Post-HOC[c] test		
Variable[a]	Pretreatment	6 months	12 months	F^b test/p value	p 0–6 months	p 0–12 months	p 6–12 months
VAS	4.1±0.7	2.2±0.4	1.2±0.3	42.155/<.001	<.001	<.001	<.001
KOOS pain	73.6±4.3	81.9±4.3	88.7±2.9	32.333/<.001	<.001	<.001	<.001
KOOS symptoms	72.0±4.1	78.2±4.2	86.4±3.2	27.674/<.001	<.001	<.001	<.001
KOOS ADL	77.8±5.7	86.3±4.7	94.8±2.5	19.163/<.001	<.001	<.001	<.001
KOOS sport	42.3±7.3	50.6±7.6	63.8±6.7	22.176/<.001	<.001	<.001	<.001
KOOS QOL	41.3±5.3	52.5±5.2	68.0±5.6	43.305/<.001	<.001	<.001	<.001
IKDC subjective	48.2±3.5	65.2±2.6	75.4±3.4	82.900/<.001	<.001	<.001	<.001
IKDC objective	0A/16B/23C/11/D	16A/22B/10C/2D	29A/16B/5C/0D	<.001[d]	<.001	<.001	<.001
Marx	4.0±0.8	6.9±0.8	9.4±0.8	72.850/<.001	<.001	<.001	<.001
Tegner	2.9±0.4	3.9±0.4	4.8±0.5	18.942/<.001	<.001	<.001	<.001

[a]The variables are described as mean±standard error of the mean (SEM)

[b]*F*-test: General linear model – repeated measure test was performed to investigate within time improvement

[c]Post-hoc test with Bonferroni adjustment for multiple comparisons was performed to investigate the significance in improvement for each variable within time evaluation

[d]IKDC objective score Freedman test was performed

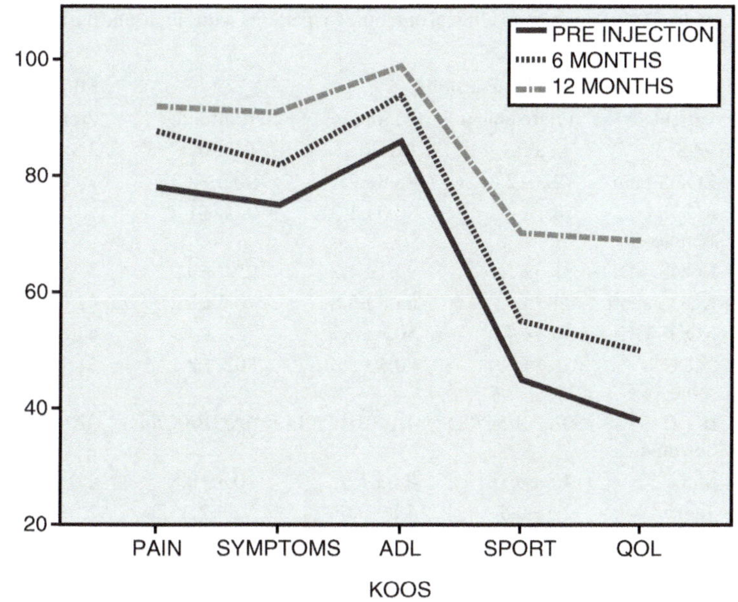

Fig. 33.2 Diagram showing KOOS improvement from preinjection evaluation to 6- and12-month follow-up in all patients ($p < 0.005$)

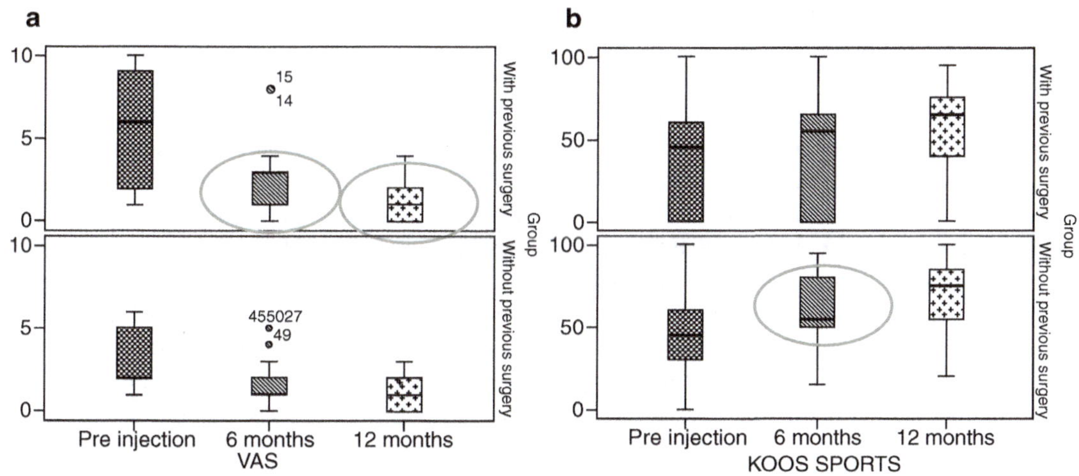

Fig. 33.3 Box plots showing the difference in improvement between operated and nonoperated patients in (**a**) VAS and (**b**) KOOS

An analysis of the results in patients presenting with single and multiple lesions on the patellofemoral joint showed significant improvement at 6- and 12-month follow-up; patients presenting with single lesions located either to patella or trochlea showed significant better improvement than patients presenting with multiple lesions (Table 33.3).

33.6 Discussion

Several studies have documented the effectiveness of PRP-derived growth factors in chondrogenesis [19, 24] and prevention of joint degeneration [25, 26] by controlling the synthesis and degradation of extracellular matrix proteins.

Table 33.3 Summary of clinical outcome for patients with single and multiple chondral lesions involving patellofemoral (PF) joint

	Single PF lesions			Multiple lesions including PF		
Variable[a]	Pretreatment	6 months	12 months	Pretreatment	6 months	12 months
VAS	3.2±0.3	1.3±0.2	1.0±0.2	3.5±0.4	1.6±0.3	1.4±0.3
KOOS pain	72.3±2.4	86.8±2.7	92.6±2.3	70.0±2.7	81.9±3.6	86.1±3.7
KOOS symptoms	74.1±2.8	86.0±1.3	92.8 ±1.7	70.2±2.0	80.8±4.3	83.1±4.0
KOOS ADL	81.6±2.1	93.1±1.2	95.9±0.9	77.8±3.9	87.6±2.1	91.8±3.5
KOOS sport	44.9±3.7	65.8±5.8	70.5±3.2	42.6±3.8	57.8±6.0	62.5±7.3
KOOS QOL	46.8±4.9	50.8±3.7	72.9±4.5	44.8±3.7	48.2±3.9	70.1±5.3
IKDC subjective	52.6±3.0	70.4±2.3	80.5±2.1	51.1±4.5	65.3±3.1	74.4±2.3
IKDC objective	0A/3B/5C/0D	4A/3B/1C/0D	5A/3B/0C/0D	0A/4B/4C/1D	3A/3B/3C/0D	5A/2B/2C/0D
Marx	4.8±0.7	8.6±1.6	10.8±0.8	4.0±0.4	6.6±0.4	9.8±1.0
Tegner	2.5±0.5	4.8±0.6	5.5±2.3	2.8±0.4	3.9±0.4	4.8±1.0

[a]The variables are described as mean ± standard error of the mean (SEM)

Nakagawa et al. [14] have reported the in vitro efficacy of autologous PRP in stimulating the proliferation and collagen synthesis of human chondrocytes, suggesting the use of this method in the treatment of cartilage defects. Kon et al. [7] have reported few interesting observations in their studies on PRP treatment in patients with chronic symptomatic degenerative condition of the knee: they demonstrated positive effects on the function and symptoms especially in patients with median age less than 60 years. However, in another study they also reported deterioration of results over 12–24 months of follow-up [9]. Other authors [8, 10, 27] also used intra-articular injections of PRP in patients with knee OA and presented good short-term results without provoking local or systemic adverse events. In accordance with the abovementioned results, all our patients showed significant improvement, which remained same at 1-year follow-up. Additionally, no deterioration of the results was noted at 1-year follow-up; a possible explanation might be that we included younger patients, while knees with advanced osteoarthritis were excluded. Furthermore, we didn't include patients presenting with associated pathologies such as knee instability and tibiofemoral and patellofemoral malalignment that can affect the clinical outcome and predispose to OA while increasing functional loads on the knee [28, 29]. Patients who had undergone previous cartilage shaving or microfracture also showed favorable results, indicating that PRP could be considered as an adjuvant in postoperative treatment of these patients. However, we should remark that platelet concentration varies widely in end-product PRP prepared by the different commercially available systems [12] and the impact on the efficacy of the PRP product is as yet relatively understudied; the differences in PRP products in terms of centrifugation protocols, platelets concentration, and presence of other cells – like leucocytes and erythrocytes – could be a possible reason for the different results in various clinical applications [30].

PRP represents a user-friendly therapeutic application, which is well tolerated and shows encouraging preliminary clinical results in active patients with symptomatic chondral defects of the knee involving patellofemoral joint. Therefore, PRP treatment would be beneficial in patients with patellofemoral pain and damage to the underlying cartilage, in terms of improving symptoms, function, and activity level. However, it is essential to investigate and assess, in advance, all the possible contributing factors such as overuse and overload of the patellofemoral joint, biomechanical problems, and muscular dysfunction.

References

1. Hunter W. Of the structure and disease of articulating cartilages. 1743. Clin Orthop Relat Res. 1995;317:3–6.
2. Mankin HJ. The response of articular cartilage to mechanical injury. J Bone Joint Surg Am. 1982; 64(3):460–6.
3. Bellamy N, Campbell J, Robinson V, Gee T, et al. Intraarticular corticosteroid for treatment of osteoarthritis of the knee. Cochrane Database Syst Rev. 2006;2, CD005328.
4. Habib GS. Systemic effects of intra-articular corticosteroids. Clin Rheumatol. 2009;28(7):749–56.
5. Habib GS, Saliba W, Nashashibi M. Local effects of intra-articular corticosteroids. Clin Rheumatol. 2010;29(4):347–56.
6. Zhang W, Nuki G, Moskowitz RW, Abramson S, et al. OARSI recommendations for the management of hip and knee osteoarthritis: part III: Changes in evidence following systematic cumulative update of research published through January 2009. Osteoarthritis Cartilage. 2010;18(4):476–99.
7. Kon E, Buda R, Filardo G, Di Martino A, et al. Platelet-rich plasma: intra-articular knee injections produced favorable results on degenerative cartilage lesions. Knee Surg Sports Traumatol Arthrosc. 2010;18(4):472–9.
8. Sampson S, Reed M, Silvers H, Meng M, Mandelbaum B. Injection of platelet-rich plasma in patients with primary and secondary knee osteoarthritis: a pilot study. Am J Phys Med Rehabil. 2010; 89(12):961–9.
9. Filardo G, Kon E, Buda R, Timoncini A, et al. Platelet-rich plasma intra-articular knee injections for the treatment of degenerative cartilage lesions and osteoarthritis. Knee Surg Sports Traumatol Arthrosc. 2011;19(4):528–35.
10. Wang-Saegusa A, Cugat R, Ares O, Seijas R, et al. Infiltration of plasma rich in growth factors for osteoarthritis of the knee short-term effects on function and quality of life. Arch Orthop Trauma Surg. 2011; 131(3):311–7.
11. Gobbi A, Karnatzikos G, Mahajan V, Malchira S. Platelet-rich plasma treatment in symptomatic patients with knee osteoarthritis: preliminary results in a group of active patients. Sports Health. 2012;4(2): 162–72.
12. Mazzucco L, Balbo V, Cattana E, Guaschino R, et al. Not every PRP-gel is born equal. Evaluation of growth factor availability for tissues through four PRP-gel preparations: FibrinetR, RegenPRP-KitR, PlateltexR and one manual procedure. Vox Sang. 2009;97(2): 110–8.
13. Bennett NT, Schultz GS. Growth factors and wound healing: biochemical properties of growth factors and their receptors. Am J Surg. 1993;165(6):728–37.
14. Nakagawa K, Sasho T, Arai M, et al. Effects of autologous platelet-rich plasma on the metabolism of human articular chondrocytes. Osteoarthr Cartil. 2007;15(Suppl 2):B134
15. Anitua E, Sanchez M, Nurden AT, et al. New insights into and novel applications for platelet rich and fibrin therapies. Trends Biotechnol. 2006;24(5):227–34.
16. Anitua E, Sanchez M, Orive G, Andia I. The potential impact of the preparation rich in growth factors (PRGF) in different medical fields. Biomaterials. 2007;28:4551–60.
17. Lieberman JR, Daluiski A, Einhorn TA. The role of growth factors in the repair of bone. Biology and clinical applications. J Bone Joint Surg Am. 2002;84-A(6):1032–44. Review.
18. Tabata Y. Tissue regeneration based on growth factor release. Tissue Eng. 2003;9 Suppl 1:S5–15. Review.
19. Akeda K, An HS, Okuma M. Platelet rich plasma stimulates porcine articular chondrocyte proliferation and matrix biosynthesis. Osteoarthritis Cartilage. 2006;14(12):1272–80.
20. Lopez-Vidriero E, Goulding KA, Simon DA, Sanchez M, et al. The use of platelet-rich plasma in arthroscopy and sports medicine: optimizing the healing environment. Arthroscopy. 2010;26(2):269–78.
21. Sánchez M, Anitua E, Orive G, Mujika I, Andia I. Platelet-rich therapies in the treatment of orthopaedic sport injuries. Sports Med. 2009;39(5):345–54.
22. Kellgren JH, Lawrence JS. Radiological assessment of osteo-arthrosis. Ann Rheum Dis. 1957;16(4):494–502.
23. Brittberg M, Aglietti P, Gambardella R, et al. International Cartilage Repair Society (ICRS) classification. The ICRS clinical cartilage injury evaluation system. Presented at the International Knee Society Meeting, Sweden; 2000.
24. Wu W, Chen F, Liu Y, Ma Q, et al. Autologous injectable tissue-engineered cartilage by using platelet-rich plasma: experimental study in a rabbit model. J Oral Maxillofac Surg. 2007;65(10):1951–7.
25. Frisbie D, Kawcak C, Werpy N, et al. Clinical biochemical and histological effects of intraarticular administration of autologous conditioned serum in horses with experimentally induced osteoarthritis. Am J Vet Res. 2007;68(3):290–6.
26. Saito M, Takahashi KA, Arai Y, Inoue A, et al. Intraarticular administration of platelet-rich plasma with biodegradable gelatin hydrogel microspheres prevents osteoarthritis progression in the rabbit knee. Clin Exp Rheumatol. 2009;27(2):201–7.
27. Cugat R, Carrillo JM, Serra I, et al. Articular cartilage defects reconstruction by plasma rich growth factors. In Zanasi S et al. (eds.)., Basic science, clinical repair and reconstruction of articular cartilage defects: current status and prospects. Bologna, Italy. TIMEO. 2006:801–7.
28. Block JA, Shakoor N. Lower limb osteoarthritis: biomechanical alterations and implications for therapy. Curr Opin Rheumatol. 2010;22(5):544–50.
29. Gomoll A, Farr J. Future developments in cartilage repair. In: Cole B, Gomoll A, editors. Biologic joint reconstruction. Thorofare: SLACK Incorporated.
30. Harrison S, Vavken P, Kevy S, Jacobson M, et al. Platelet activation by collagen provides sustained release of anabolic cytokines. Am J Sports Med. 2011;39(4):729–34.

Pulsed Electromagnetic Fields for the Treatment of Symptomatic Patellofemoral Cartilage Lesions of the Knee

34

Alberto Gobbi, Georgios Karnatzikos, and Sanyam Chaurasia

Contents

34.1 Introduction

Articular cartilage injury in patellofemoral joint presents a clinical challenge and has been identified as a cause of pain, dysfunction and progressive joint degeneration. Any concomitant pathology such as objective or potential patellar instability, muscular dysfunction and biomechanical disorders of the foot and hip can act as a cause of patellofemoral pain syndrome (PFPS) and they should be investigated and addressed when necessary. Cartilage has a poor intrinsic healing potential [1]. Therefore, when left untreated cartilage lesions can progress and lead to chondromalacia and early osteoarthritis (OA) [2]. Research today is moving toward preventive interventions and cost-effective treatments in order to find a way to improve clinical outcomes and retard the progression of OA thereby delaying the need for total joint replacement.

Recently, the use of pulsed electromagnetic fields (PEMFs) has received attention for the treatment of osteoarthritis and symptomatic focal cartilage lesions of the knee. In vitro and in vivo studies have demonstrated that PEMFs have the ability to influence cartilage metabolism through pro-anabolic and anti-catabolic activities [3–8]. PEMFs were introduced in the clinical setting in the 1970s. Although it has been proven to be a successful method in fracture healing (non-union and delayed union) [9], the effects on knee OA are equivocal, with several investigations reporting conflicting results [10–15]. Trock et al. [15] in a randomized clinical trial that included 86

A. Gobbi (✉) • G. Karnatzikos • S. Chaurasia
Orthopaedic Arthroscopic Surgery International (O.A.S.I.) Bioresearch Foundation, Gobbi N.P.O., Milan, Italy
e-mail: gobbi@cartilagedoctor.it; giokarnes@gmail.com

A. Gobbi et al. (eds.), *The Patellofemoral Joint*,
DOI 10.1007/978-3-642-54965-6_34, © ISAKOS 2014

patients treated with PEMFs versus placebo for knee OA reported significant improvements in symptoms and ADL in the PEMFs group. On the other hand, Thamsborg et al. [14] conducted a randomized clinical trial (83 patients) and did not demonstrate significant differences in the outcome scores in the group treated with PEMFs compared to a placebo group. However, a recent meta-analysis that included nine randomized clinical trials with a total of 483 patients concluded that evidence of a beneficial effect of PEMFs on functional outcomes in patients with knee OA exists [16]. PEMFs have also been applied in patients undergone knee arthroscopy for cartilage lesions. Zorzi et al. [17] in a randomized clinic trial evaluating the outcomes of arthroscopic chondroabrasion or perforation followed by treatment with PEMF showed that the treatment with PEMFs aided patient recovery after arthroscopic surgery, reducing the use of NSAIDs. The use of PEMFs was associated to improved functional outcomes with a long-term effect.

We also investigated the efficacy of the treatment with PEMFs in a group of active individuals presenting with symptomatic cartilage lesions of the patellofemoral joint [18].

34.2 Study Group

Twenty-two subjects (11 males and 11 females) with symptomatic cartilage lesions of the knee were treated with PEMFs according to the protocol presented below and prospectively followed up. Eighteen patients had single lesions while 4 had multiple; among them 7 patients were presenting with single patellar lesions. Patient demographics and localization of cartilage lesions are reported in Table 34.1.

Table 34.1 Patients' demographics and localization of cartilage lesions

Variables	Data
Number of patients	22 (11 male/11 female)
Mean age	48.1 ± 2.6 (range: 30–60)
Mean follow-up (years)	2.1
Age ≥ 45 years	12
Age < 45 years	10
Single lesions	18 (7 PAT, 7 MFC, 1 MTP, 1 LFC, 2 LTP)
Multiple lesions	4 (3 LFC/LTP, 1MFC/MTP)

Abbreviations: *PAT* patella, *MFC* medial femoral condyle, *LFC* lateral femoral condyle, *LTP* lateral tibial plateau, *MTP* medial tibial plateau

Fig. 34.1 I-ONE PEMFs generator

Patients were included according to the following criteria: age between 30 and 60 years; grade 2 or 3 cartilage lesion according to the ICRS classification [19], evaluated by MRI and previous arthroscopy; symptomatic with functional limitations; sports participation (recreational); and minimum 2 year follow-up. Exclusion criteria were malalignment of the lower limbs (varus-valgus greater than 8° from physiological); knee instability or patellofemoral maltracking; previous knee surgery (including arthroscopy performed within 6 months prior to treatment); previous intra-articular injections with corticosteroid, PRP, or hyaluronic acid (within 6 months prior to the study); inflammatory arthritis; smoking habits (>20 cigarettes/day); severe cardiovascular disease; and body mass index >30.

Visual analog scale (VAS) for pain, International Knee Documentation Committee (IKDC) objective, Tegner, Knee Injury and Osteoarthritis Outcome Score (KOOS) scores were recorded before treatment, then at 1- and 2-year follow-up. Primary outcomes of the study were pain relief, improvement of symptoms, and improvement of activity level.

34.2.1 Treatment Protocol

All patients underwent biophysical treatment with PEMFs (I-ONE therapy, IGEA S.p.A., Carpi, Italy), see Fig. 34.1. The protocol included a 4-h treatment per day, for a total of 45 days. The treatment could also be divided into two applications of 2 h each, at different times of the day. The maximum intensity of magnetic field was 1.5 mT and frequency 75 Hz.

34.3 Results

Patients showed a significant improvement in all scores at 1-year follow-up ($p=0.008$). At 2-year follow-up, results deteriorated but were still superior to pretreatment levels ($p=0.02$) (Figs. 34.2, 34.3 and 34.4). The mean values obtained in KOOS, VAS and Tegner scores before treatment, at 1 and 2 year follow-up are presented in Table 34.2. The analysis of IKDC objective score demonstrated improvement at 1-year follow-up, while a decline was seen at 2 years follow-up. Pretreatment IKDC objective resulted B in 2 patients, C in 18, and D in 2 (A=normal; B=nearly normal; C=abnormal; D=severely abnormal). After 1-year follow-up, the IKDC objective score was A in 7 patients, B in 13, and C in 2, while at 2 years the IKDC objective resulted A in 5 patients, B in 7 patients,

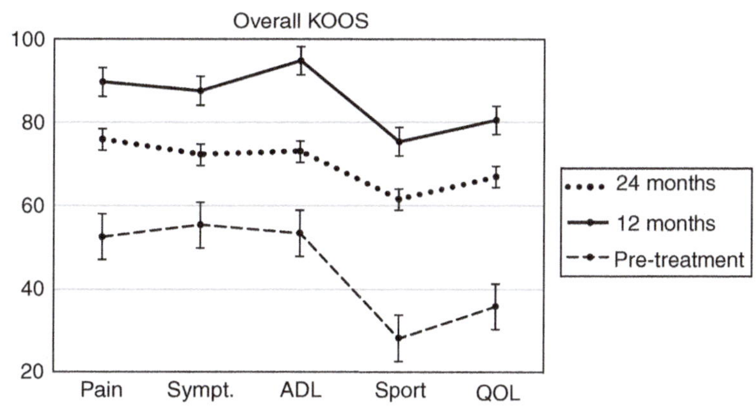

Fig. 34.2 Trend of KOOS score improvement from pretreatment to 1- and 2-year follow-up

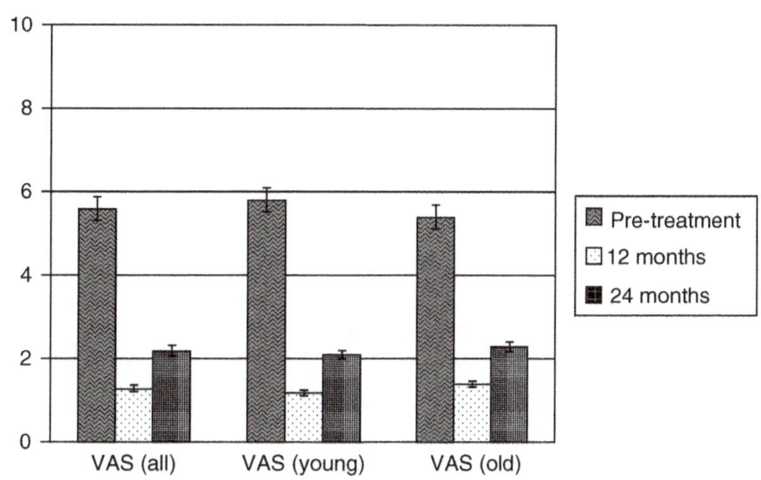

Fig. 34.3 VAS scale before treatment, at 1- and 2-year follow-up: overall results and results in the subgroups of patients under 45 years old and over 45 years old

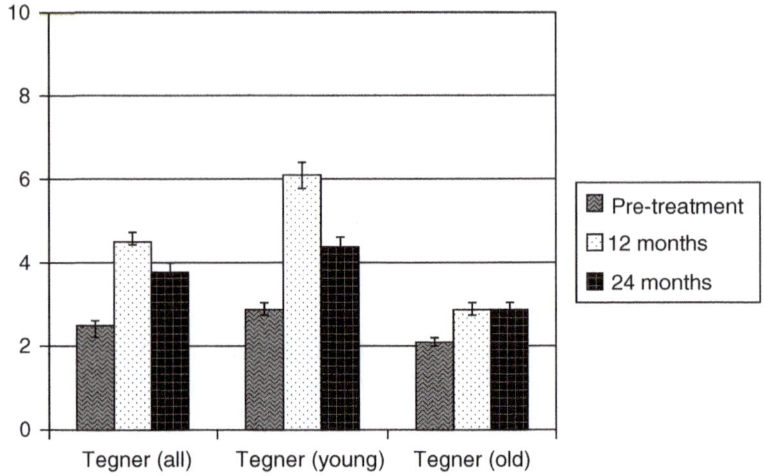

Fig. 34.4 Tegner score before treatment, 1- and 2-year follow-up: overall results and results in the subgroups of patients under 45 years old and over 45 years old

Table 34.2 Clinical outcome: overall results

Variable	Pretreatment	1 year	2 years
KOOS pain	52.4±4.9	89.7±4.3	75.9±3.6
KOOS symptoms	55.2±5.0	87.5±3.5	72.2±3.7
KOOS ADL	53.3±5.6	94.8±2.9	72.9±3.9
KOOS sport	28.0±5.9	75.4±6.2	61.4±5.5
KOOS QOL	35.6±4.5	80.5±4.7	66.8 ±6.1
VAS	5.6±0.3	1.3±0.4	2.2±0.6
Tegner	2.5±0.5	4.5±0.5	3.8±0.5

Note: The variables are expressed as mean ± SEM (standard error of the mean)
Abbreviations: *VAS* visual analog scale, *KOOS* Knee injury and Osteoarthritis Outcome Score, *ADL* activities of daily living, *QOL* quality of life

C in 9 patients, and D in 1 patient. An analysis of the results in patients under 45 years old revealed better outcomes in this subgroup compared to patients over 45 years of age (Figs. 34.3 and 34.4). Patients presenting with single lesions on the patella or the medial femoral condyle (MFC) showed significant improvement at 1-year follow-up; however, results deteriorated at 2 years. Patients presenting with MFC lesions showed better improvement than patients presenting with patellar lesions (Table 34.3).

No adverse reactions or side effects were seen. All the patients were compliant with the prescribed treatment protocol.

34.4 Discussion

All patients showed statistically significant improvement in pain, symptoms, quality of life, and activity level after PEMFs treatment. The scores significantly declined at 2-year follow-up, but the values were still higher than pretreatment levels. At final follow-up, the level of satisfaction was high and 85 % of patients stated that they would like to repeat the treatment in case to preserve the clinical improvement. These findings have a clinical relevance suggesting the need to repeat the treatment after 1 year. Interestingly, we observed superior outcomes in younger patients. Subjects younger than 45 years of age displayed higher scores in all of the scales that were evaluated and resumed regular sport participation.

We believe the posttreatment improvement is related to a reduction of the anti-inflammatory and chondroprotective action over time. The mechanism of action is also known and has been demonstrated in several in vitro and in vivo studies [3–8, 20, 21]. The anti-inflammatory effect of PEMFs is associated with the modulation of adenosine A_{2A} receptors through upregulation, as demonstrated in both bovine and human chondrocytes and synovial fibroblast [7, 21]. The modulation of these receptors having anti-inflammatory activity is considered to be one of

Table 34.3 Comparison of outcomes in patients presenting with single patellar and MFC lesions

	Pretreatment		1 year		2 years	
Variable	Patella	MFC	Patella	MFC	Patella	MFC
KOOS pain	51.0±4.3	55.0±6.0	82.6±7.1	95.9±1.2	70.1±3.7	80.2±3.4
KOOS symptoms	53.0±3.9	53.7±6.1	82.0±5.2	90.6±5.4	68.3±4.1	75.8±3.2
KOOS ADL	52.7±6.3	52.9±5.8	90.1±4.6	94.4±2.4	68.9±3.1	76.2±2.5
KOOS sport	24.3±6.8	23.6±2.8	71.4±9.2	76.7±4.7	60.1±6.2	65.3±2.7
KOOS QOL	33.1±6.1	35.0±3.9	75.1±7.0	83.1±3.4	62.5±4.3	69.3±4.2
VAS	5.9±0.5	4.7±0.6	1.7±0.5	0.6±0.3	2.4±0.4	1.8±0.6
Tegner	2.4±0.2	2.1±0.6	4.0±0.4	4.9±0.4	3.6±0.5	4.0±0.6

Note: The variables are expressed as mean ± SEM (standard error of the mean)
Abbreviations: *MFC* medial femoral condyle, *VAS* visual analog scale, *KOOS* Knee injury and Osteoarthritis Outcome Score, *ADL* activities of daily living, *QOL* quality of life

the mechanisms by which the PEMF counteracts the effect of pro-inflammatory cytokines in explants of cartilage and synovial fibroblasts and prevents the progression of OA [3, 7, 8]. On the other hand, PEMFs through the synergy with insulin-like growth factor 1 (IGF-1) exerts a pro-anabolic activity enhancing chondrogenic differentiation and synthesis of extracellular matrix component, as shown in both human and bovine models [5, 6, 22]. On a macroscopic level, in vivo studies conducted on Dunkin Hartley guinea pigs showed that PEMFs were able to reduce tissue fibrillation, preserve cartilage thickness, and prevent the sclerosis of the subchondral bone in lateral and medial compartment of the knee [8, 23].

This investigation showed that the use of PEMFs in a group of athletically active individuals with symptomatic cartilage lesions of the knee, including patients with isolated patellar lesions, led to significant improvement in symptoms, pain, function, and activity level at 1-year follow-up. We observed a significant decline in the outcome scores at 2 years follow-up. PEMFs represent a valid alternative to other conservative treatments, with the advantage of being free of side effects and well accepted by the patients.

References

1. Hunter W. Of the structure and disease of articulating cartilages. 1743. Clin Orthop Relat Res. 1995;317:3–6.
2. Mankin HJ. The response of articular cartilage to mechanical injury. J Bone Joint Surg Am. 1982;64: 460–6.
3. Ciombor DM, Aaron RK, Wang S, Simon B. Modification of osteoarthritis by pulsed electromagnetic field- a morphological study. Osteoarthritis Cartilage. 2003;11:455–62.
4. De Mattei M, Caruso A, Pezzetti F, Pellati A, Stabellini G, Sollazzo V, Traina GC. Effects of pulsed electromagnetic fields on human articular chondrocyte proliferation. Connect Tissue Res. 2001;42:269–79.
5. De Mattei M, Pasello M, Pellati A, Stabellini G, Massari L, Gemmati D, Caruso A. Effects of electromagnetic fields on proteoglycan metabolism of bovine articular cartilage explants. Connect Tissue Res. 2003;44:154–9.
6. De Mattei M, Pellati A, Pasello M, Ongaro A, Setti S, Massari L, Gemmati D, Caruso A. Effects of physical stimulation with electromagnetic field and insulin growth factor-I treatment on proteoglycan synthesis of bovine articular cartilage. Osteoarthritis Cartilage. 2004;12:793–800.
7. De Mattei M, Varani K, Masieri FF, Pellati A, Ongaro A, Fini M, Cadossi R, Vincenzi F, Borea PA, Caruso A. Adenosine analogs and electromagnetic fields inhibit prostaglandin E(2) release in bovine synovial fibroblasts. Osteoarthritis Cartilage. 2009;17:252–62.
8. Fini M, Giavaresi G, Torricelli P, Cavani F, Setti S, Canè V, Giardino R. Pulsed electromagnetic fields reduce knee osteoarthritis lesion progression in the aged Dunkin Hartley guinea pig. J Orthop Res. 2005;23:899–908.
9. Sharrard WJ. A double-blind trial of pulsed electromagnetic fields for delayed union of tibial fractures. J Bone Joint Surg Br. 1990;72:347–55.
10. Callaghan MJ, Whittaker PE, Grimes S, Smith L. An evaluation of pulsed short wave on knee osteoarthritis using radioleucoscintigraphy: a randomised, double blind, controlled trial. Joint Bone Spine. 2005;72: 150–5.
11. Laufer Y, Zilberman R, Porat R, Nahir AM. Effect of pulsed short-wave diathermy on pain and function of subjects with osteoarthritis of the knee: a placebo-controlled double-blind clinical trial. Clin Rehabil. 2005;19:255–63.

12. Nicolakis P, Kollmitzer J, Crevenna R, Bittner C, Erdogmus CB, Nicolakis J. Pulsed magnetic field therapy for osteoarthritis of the knee – a double-blind sham-controlled trial. Wien Klin Wochenschr. 2002; 114:678–84.
13. Pipitone N, Scott DL. Magnetic pulse treatment for knee osteoarthritis: a randomised, double-blind, placebo-controlled study. Curr Med Res Opin. 2001;17:190–6.
14. Thamsborg G, Florescu A, Oturai P, Fallentin E, Tritsaris K, Dissing G. Treatment of knee osteoarthritis with pulsed electromagnetic fields: a randomized, double-blind, placebo-controlled study. Osteoarthritis Cartilage. 2005;13:575–81.
15. Trock DH, Bollet AJ, Markoll R. The effect of pulsed electromagnetic fields in the treatment of osteoarthritis of the knee and cervical spine. Report of randomized, double blind, placebo controlled trials. J Rheumatol. 1994;21:1903–11.
16. Vavken P, Arrich F, Schuhfried O, Dorotka R. Effectiveness of pulsed electromagnetic field therapy in the management of osteoarthritis of the knee: a meta-analysis of randomized controlled trials. J Rehabil Med. 2009;41:406–11.
17. Zorzi C, Dall'oca C, Cadossi R, Setti S. Effects of pulsed electromagnetic fields on patients' recovery after arthroscopic surgery: prospective, randomized and double-blind study. Knee Surg Sports Traumatol Arthrosc. 2007;15:830–4.
18. Gobbi A, Lad D, Petrera M, Karnatzikos G. Symptomatic early osteoarthritis treated with Pulsed Electromagnetic Fields: 2 years follow-up. Cartilage. 2014;5(2):76–83. doi:10.1177/1947603513515904
19. Brittberg M, Aglietti P, Gambardella R, et al. International Cartilage Repair Society (ICRS) classification. The ICRS clinical cartilage injury evaluation system. Presented at the International Knee Society Meeting, Sweden; 2000.
20. Massari L, Benazzo F, De Mattei M, Setti S, Fini M, CRES Study Group. Effects of electrical physical stimuli on articular cartilage. J Bone Joint Surg Am. 2007;89 Suppl 3:152–61.
21. Varani K, De Mattei M, Vincenzi F, Gessi S, Merighi S, Pellati A, Ongaro A, Caruso A, Cadossi R, Borea PA. Characterization of adenosine receptors in bovine chondrocytes and fibroblast-like synoviocytes exposed to low frequency low energy pulsed electromagnetic fields. Osteoarthritis Cartilage. 2008;16: 292–304.
22. Ongaro A, Pellati A, Masieri FF, Caruso A, Setti S, Cadossi R, Biscione R, Massari L, Fini M, De Mattei M. Chondroprotective effects of pulsed electromagnetic fields on human cartilage explants. Bioelectromagnetics. 2011;32:543–51.
23. Fini M, Torricelli P, Giavaresi G, Aldini NN, Cavani F, Setti S, Nicolini A, Carpi A, Giardino R. Effect of pulsed electromagnetic field stimulation on knee cartilage, subchondral and epiphyseal trabecular bone of aged Dunkin Hartley guinea pigs. Biomed Pharmacother. 2008;62:709–15.

Patellofemoral Injuries in Soccer Players

35

Ramon Cugat, Gilbert Steinbacher, Pedro Alvarez, and Montse Garcia-Balletbo

Contents

R. Cugat (✉) • G. Steinbacher • P. Alvarez
M. Garcia-Balletbo
Mutualitat Catalana de Futbolistes-Real Federacion Española de Futbol, Fundacion Garcia Cugat, Barcelona, Spain
e-mail: ramon.cugat@sportrauma.com

35.1 Anterior Knee Pain

Patellofemoral pain syndrome (PFPS) is a vague, diffuse term which is employed when there is pain or dysfunction in the anterior aspect of the knee with unknown origin or cause. This anterior pain is only a symptom and it is also known as anterior knee pain syndrome (AKPS).

Chondromalacia patellae and patellar chondropathy also occur with anterior knee pain but are different from AKPS and PFPS as they include cartilage lesions.

The literature lists 56 factors that cause AKPS which are tied to the diagnosis [1].

35.1.1 Injury History

Pain may be described as:

(a) Peripatellar or retropatellar that changes with sport and going up or down the stairs.
(b) Associated with functional weakness; this impedes playing soccer.
(c) Accompanied by creeking/crackling noises at the femoropatellar joint. Sometimes noises are present without pain.
(d) At the distal pole of the patella due to overloading which creates insertion tendinopathy. Sometimes this pain, during growth periods, is in the anterior tibial tuberosity (ATT), and this condition may develop into Osgood-Schlatter disease.

A. Gobbi et al. (eds.), *The Patellofemoral Joint*,
DOI 10.1007/978-3-642-54965-6_35, © ISAKOS 2014

35.1.2 Clinical Examination

The anamnesis is essential for a good diagnosis, as sometimes it is complicated and difficult, although in the soccer player, as with all athletes, pathology due to overloading should be kept in mind.

1. Orthostatic Examination

   The exam is to be performed with the patient in underwear and barefoot: One should examine at the iliac crests to rule out lower extremity dysmetria.

   Stand in front of the player with eyes focused on the knees, to see if any rotational alteration exists: femur, tibia, patella, etc.

   Check for foot arches, flat feet, valgus deformity, forefoot or rearfoot rotation, etc.
2. Dynamic Examination

   Invite the patient to walk and observe if the gait cycle is normal, if there are any alterations like lameness, etc.

   Then with the patient supine, both knees are explored and compared: Start with inspection, palpation, and the quadriceps trophism circumferential measurement, comparing the two thighs to see if there is a degree of atrophy, which is usually seen in most patellofemoral pain cases. With any knee problem that involves some quadriceps atrophy, pain at the inferomedial fibers close to the patellar tendon is evident quite frequently. If pain is not eliminated, rebalancing of the quadriceps becomes difficult in soccer players.

35.1.3 Complementary Examinations

(a) Radiological studies help in two ways: They strengthen the diagnosis and rule out certain diseases.

The radiographs requested are:

- AP of both knees
- Axial view of both patellas at 30°
- Profile views of both knees

  The images taken in the axial radiographs show the types of patellar and trochlear disorders and can help diagnose trochlear dysplasia, calcification in the medial patella, signs of chronic medial patellar retinaculum tear, bipartite patella, etc.

  The images taken in the profile radiographs show the positioning of the patella: normal, high, or low.

(b) CT scan helps to examine the bones and bone axis.

(c) MRI is useful for studying the cartilage, subchondral bone, and soft tissue (Fig. 35.1).

(d) Ultrasonography is useful for the study of soft tissues, tendons, ligaments, muscles, and the vascular system. It allows for static and dynamic examination.

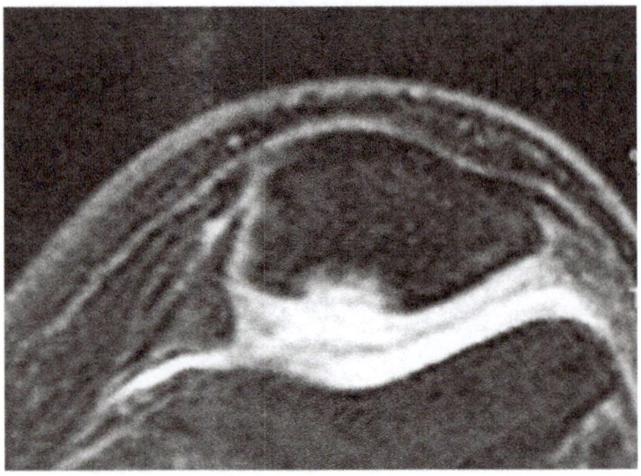

Fig. 35.1 Patellar osteochondral defect

35.1.4 Treatment

Conservative

1. Physiotherapy, ultrasound, electrotherapy.
2. Intra-articular injection of growth factors.
3. Rehabilitation: All lower extremity muscle groups should be worked out, seeking a balance between groups and insisting on stretching exercises and strengthening of the quadriceps.
4. Readaptation on the soccer pitch with total and integral preparation allowing for return to sport.

If the cause of the pain is known, then of course this knowledge should be acted upon (e.g., osteochondral injury to the trochlea or patella).

Upon returning to the sport, the athlete should continue with maintenance exercises as one should be in shape to play the sport!

35.2 Patellar Dislocation

In soccer, there are three types of patellar or femoropatellar dislocations:

1. *Primary traumatic patellar dislocation* without apparent anatomic alterations. It occurs in few cases. Conservative treatment in amateurs and surgical treatment in professionals.
2. *Primary traumatic patellar dislocation with anatomic alterations*. It is most often associated with patellofemoral dysplasia and patella alta. Treatment is surgical and for those not wishing to undergo surgery there are conservative options like orthoses and rehabilitation, however if the latter fails, treatment is surgical.
3. *Recurrent dislocation*. Surgical treatment.

35.2.1 History

The soccer player usually presents with:

1. Primary or acute dislocation of the patella
2. Several episodes of subluxation or dislocation of the patella

35.3 Acute Dislocation

35.3.1 Anatomy and Biomechanics

The articular mobility of the patella is conditioned by:

1. Bony anatomy of the patella and trochlea
2. The capsuloligamentous apparatus
3. The muscle-tendon complex
4. Lower extremity alignment [2, 3]

The patellofemoral retinacular ligaments are important stabilizers of the patella, especially the medial patellofemoral ligament (MPFL) which is the primary ligamentous stabilizer against lateral dislocation of the patella between 20° and 30° flexion [4, 5]. The lateral aspect of the femoral trochlea is also a patellar stabilizer as when the patella is within the trochlear groove, the lateral aspect provides the principal resistance for the patella not to dislocate laterally [6].

There are several studies on the influence lower limb alignment has on patellar instability. Fithian et al. [5] show that the alignment of the lower extremity and the femoropatellar joint themselves alone may not be the cause of a patellar dislocation, without the coexistence of a soft tissue insufficiency, hyperlaxity, or injury.

35.3.2 Injury Mechanism

1. Sudden change of direction. When the right foot is planted firmly on the ground, the body moves to the left with slight valgus-flexion of the knee, the femur rotating internally, and the tibia rotating externally. This is the most common mechanism.
2. Direct blow to the medial aspect of the patella which dislocates externally.
3. Blow to the lateral side of the knee with the knee in forced valgus, producing a lateral dislocation.

When the soccer player arrives at the hospital, the patella has in most cases already been reduced, and the player explains that a member of medical staff reduced it on the field and that the knee was in flexion and that a bruise appeared on the lateral side. Sometimes the players explain that the dislocation reduced spontaneously.

35.3.3 Examination

1. Moderate joint effusion is usually observed. If, on the contrary, there is a large amount of articular fluid and tension and the player describes that the knee swelled up rapidly, then an osteochondral fracture should be suspected.
2. Pain upon palpation on the medial aspect of the patella and an injury to the MPFL that can be at the adductor tubercle, at the patellar insertion, or all the way along. Injury can also be found in the most oblique fibers of the vastus medialis.

35.3.4 Complementary Examinations

To certify osteochondral fracture, radiological examination is mandatory.

For examining soft tissue, MRI and ultrasonography exams are effective (Fig. 35.2).

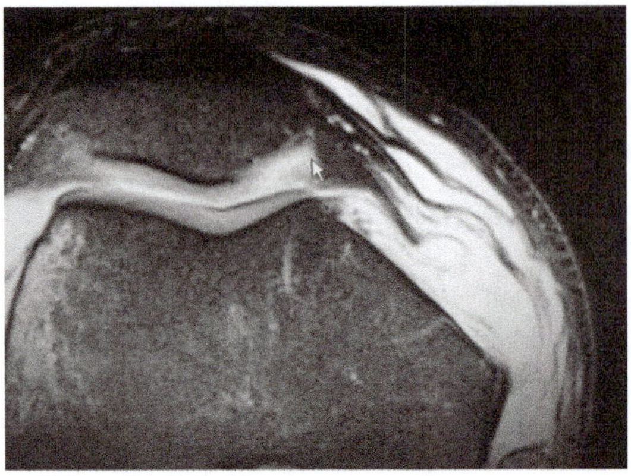

Fig. 35.2 MPFL avulsion tear

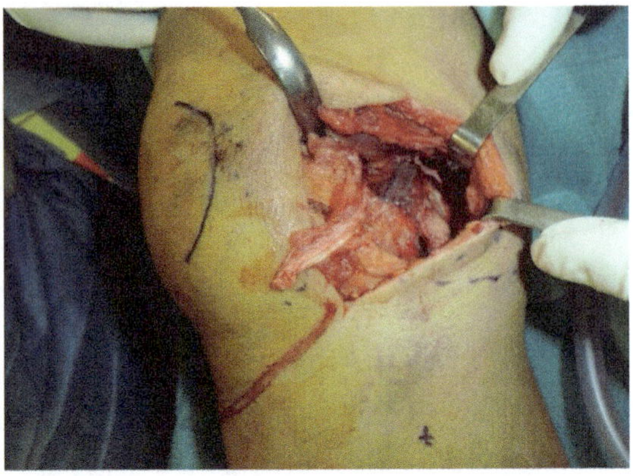

Fig. 35.3 MPFL avulsion from the femur

35.3.5 Treatment

Players with open physis:

Conservative treatment is only followed in very young soccer players with open physis, from 10 to 13 years old, and only when radiological femoropatellar parameters are near to normal for patellar location: no patella alta or trochlear dysplasia.

In young players who have followed conservative treatment – rehabilitation and retraining but suffer redislocation – or have a sense of instability that keeps them from playing soccer, they will have to undergo surgical treatment: elective, informed, and consented.

The objective of surgery is to restore the anatomy, with reconstruction of the MPFL as one of the preferred options [7] (Figs. 35.3, 35.4 and 35.5).

Due to the high incidence of redislocation in young soccer players, with open physis, it is believed that better and more extensive studies of the incidence factors are needed in these redislocations.

If an osteochondral fragment is diagnosed with patellar dislocation, the recommended treatment is surgical. If the fragment is large, it must be set, and if small, it should be removed via arthroscopy.

Post-surgery, the extremity is immobilized in extension for 3–4 weeks, a rehabilitation program is followed, and the player progressively returns to playing soccer.

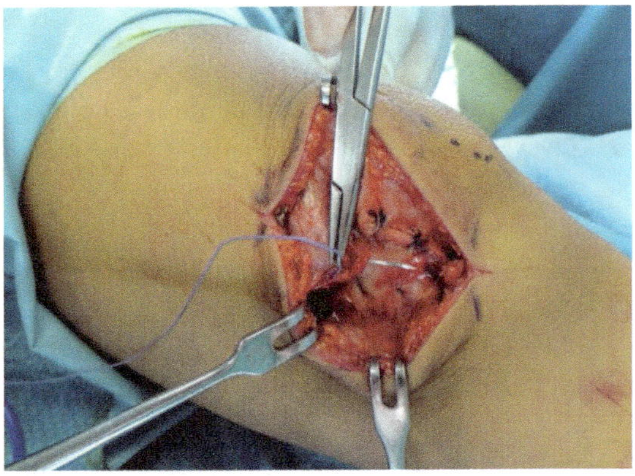

Fig. 35.4 MPFL suture

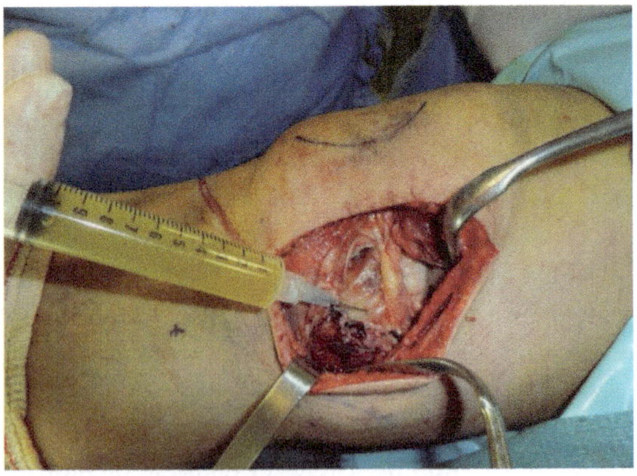

Fig. 35.5 PRGF injection

35.4 Recurrent Dislocation

35.4.1 Injury History

Did the patella come out of place? Is this the first time or have there been previous episodes? Did it go back in place alone? Quickly? Was it reduced with help? Did it occur after suffering a blow or traumatism? Did the knee swell up: immediately/quickly or only later? Did it occur during an accidental sprain or distension?

The injury mechanism for patellar dislocation and anterior cruciate ligament rupture is a sprain, strain or twist with the knee in valgus-flexion, internal femoral rotation, and external tibial rotation; the differential diagnosis should be performed.

35.4.2 Examination

With the patient supine, both knees are examined and compared. Start with inspection, palpation, and the quadriceps circumferential measurement, comparing the two thighs to see if there is a degree of atrophy.

It is followed by examination of:

(a) Patellar reflex to know if there is joint effusion
(b) Patellar mobility: proximal, distal, medial, and lateral
(c) Peripatellar soft tissue
(d) Tendons: the patellar tendon is a common source of pain in adolescent, adult, amateur, and professional soccer players whereas the *quadriceps* more commonly affects adult, amateur or professional, soccer players due to overloading
(e) Ligament stability: MCL, LCL, ACL, PCL, posteromedial corner, and posterolateral corner
(f) Hip mobility: internal-external rotation in extension and internal-external rotation at 90° flexion
(g) Mobility and stability of both ankles, paying attention to possible instability that is a common cause of chronic or recurrent sprains by injury to the anterior talofibular ligaments

The clinical exam is concluded with the patient sitting on the examination bed with legs hanging over the edge and knees at 90° flexion. This will provide the following information: audible creeking or crackling, a visual of the patellar instability at knee flexion, degree of pain, extent of powerlessness, etc.

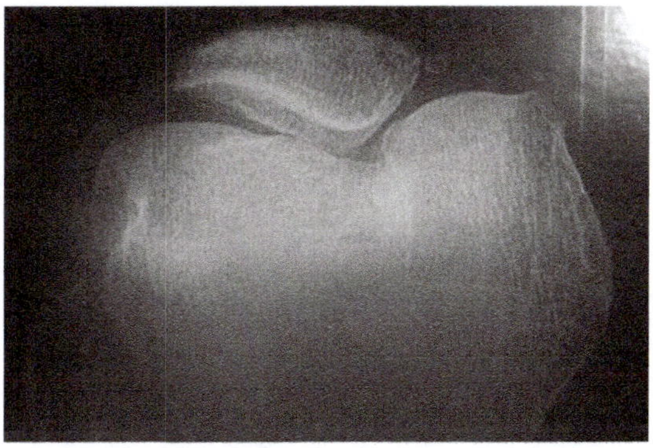

Fig. 35.6 ATT distalization

Once the injury history is taken and clinical examination has been performed, an initial diagnosis can be reached, and to consolidate it further, complementary exams should be carried out and other possible pathologies like bone tumors can be ruled out.

35.4.3 Complementary Examinations

(a) Radiological studies help in two ways: They strengthen the diagnosis and rule out certain diseases.

The radiographs requested are:

- AP of both knees
- Axial view of both patellas at 30°
- Profile views of both knees

The images taken in the axial radiographs show the types of patellar and trochlear disorders and can help diagnose trochlear dysplasia, calcification in the medial patella, and signs of chronic medial patellar retinaculum tear (Fig. 35.6).

The images taken in the profile radiographs show the positioning of the patella: normal, high, or low.

In soccer injuries it is seen that instability, subluxation, or dislocation of the patella is usually accompanied by a flattened femoral trochlea and patella alta.

(b) CT scan helps examine the bones and bone axis.
(c) MRI is useful for studying the cartilage, subchondral bone, and soft tissue.
(d) Ultrasonography is useful for the study of soft tissue, tendons, the vascular system, ligaments, muscles, etc. It allows for static and dynamic examination.

35.5 Differential Diagnosis

Anterior cruciate ligament injury
Has the same injury mechanism: knee in valgus-flexion, internal femoral rotation, external tibial rotation.
Negative ACL injury tests when it is a patellar dislocation.
Radiology: flattened trochlea and patella alta facilitate patellar dislocation.
Patellar fracture
Contusion or direct frontal trauma should lead one to think of a fracture which can be confirmed with radiology.
Medial collateral ligament injury

35.5.1 Treatment

Conservative orthopedic treatment has a 15–44 % patellar redislocation rate with persistent anterior pain, sense of instability, and activity limitation in more than 50 % of cases [8, 9].

In contrast, comparative studies between conservative and surgical treatment for patellar instability have not shown better clinical outcomes for surgical treatment [10, 11].

Studies have shown an increase in patellofemoral osteoarthritis after surgery despite reducing dislocation [12].

Surgical treatment requires:

1. Understanding of the patellofemoral joint anatomic pathology and the alignment of the lower extremity, ultimately trying to determine the instability or redislocation causal mechanism
2. Considering the risk that surgery could worsen the functional outcome of the patellofemoral joint

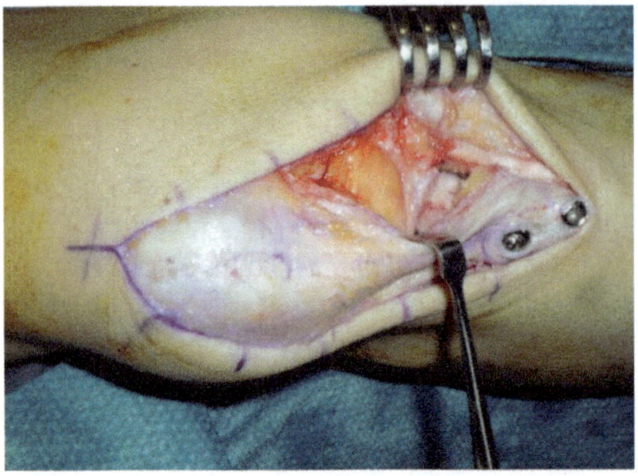

Fig. 35.7 Calcification of the medial retinaculum

3. Keeping in mind that the realignment of proximal and distal soft tissue gives good results in patellar instability [13]

In patients with poor or limited medial retinaculum tissue and hypermobility, reconstruction of the MPFL is advised with autologous tendons, like the semitendinosus tendon, for example [14–17].

It has already been said that patellar dislocation, in a large percentage of cases, is associated with patella alta. In some rare cases, distalization and centralization of the patella are performed when the physis is closed [17]. However, the authors have rarely performed osteotomy and distalization of the ATT in soccer players. The results are good and in agreement with other authors [18] (Fig. 35.7).

Post surgery, the extremity is immobilized in extension for 3–4 weeks. The rehabilitation program has the objective of good ROM, hamstring and quadriceps strengthening to achieve and maintain a well-balanced patella, and then readaptation and progressive return to playing soccer.

In conclusion, after clinical and radiological examination, considering that the soccer player morphotype is genu varus and that the injury is often at the MPFL femoral insertion between the adductor magnus insertion and the epicondyle where the MCL attaches, rather than at the patella, surgery is advised on soft tissue: capsule, medial retinaculum, insertion of the vastus medialis obliquus, and lateral retinaculum lengthening.

References

1. Teitge RA. Patellofemoral syndrome a paradigm for current surgical strategies. Orthop Clin North Am. 2008;39:287–311.
2. Grelsamer RP, Weinstein CH. Applied biomechanics of the patellar. Clin Orthop. 2001;389:9–14.
3. Andrish J. The biomechanics of patellofemoral stability. J Knee Surg. 2004;17(1):35–9.
4. Conlan T, Garth WP, Lemus JE. Evaluation of the medial soft-tissue restraints of the extensión mechanism of the knee. J Bone Joint Surg Am. 1993;75:682–93.
5. Fithian DC, Nomura E, Arendt E. Anatomy of patellar dislocation. Oper Tech Sports Med. 2001;9:102–11.
6. Ahmed AM, Duncam NA. Correlation of Patella tracking pattern with trochlear and retropatellar surface topographies. J Biomech Eng. 2000;122:652–60.
7. Lippacher S, Nelitz M. Femorale Insertion des medialen patellofemoralen Ligaments in Relation zur distalen femoralen Wachstumsfuge. Arthroskopie. 2012;25(4):275–9.
8. Hawkins RJ, Bell RH, Anisette G. Acute patellar dislocations: the natural history. Am J Sports Med. 1986; 14:117–20.
9. Harilainen A, Myllynen P. Operative treatment in acute patellar dislocation: radiological predisposing factors, diagnosis and results. Am J Knee Surg. 1988; 1:178–85.
10. Nikku R, Nietosvaara Y, Kallio PE, et al. Operative versus closed treatment of primary dislocation of the patella. Acta Orthop Scand. 1997;68:419–23.
11. Powers JA. Natural history of recurrent dislocation of the patella. Clin Orthop. 1976;68:119–81.
12. Arnbjornsson A, Eglund N, Rydling O, et al. The natural history of recurrent dislocation of the patella: long-term results of conservative and operative treatment. J Bone Joint Surg Br. 1992;74:140–2.
13. Aglietti P, Buzzi R, De Biase P, et al. Surgical treatment of recurrent dislocation of the patella. Clin Orthop. 1994;308:8–17.
14. Avikainen VJ, Nikku RK, Seppanen-Lehmonen TK. Adductor magnus tenodesis for patellar dislocation. Clin Orthop. 1993;297:12–6.
15. Gomes JLE. Medial patellofemoral ligament reconstruction for recurrent dislocation of the patella: a preliminary report. Arthroscopy. 1992;8:335–40.
16. Muneta T, Sekiya I, Tsuchiya M, et al. A technique for reconstruction of the medial patellofemoral ligament. Clin Orthop. 1999;359:151–5.
17. Deie M, Ochi M, Sumen Y, et al. Reconstruction of the medial patellofemoral ligament for the treatment of habitual or recurrent dislocation of the patella in children. J Bone Joint Surg Br. 2003;85:887–90.
18. Simmons Jr E, Cameron JC. Patella alta and recurrent dislocation of the patella. Clin Orthop. 1992;274: 265–9.

Patellofemoral Postoperative Rehabilitation

36

Lorenzo Boldrini, Furio Danelon,
Francesco Della Villa, and Stefano Della Villa

Contents

L. Boldrini (✉) • F. Danelon
Isokinetic Medical Group – FIFA Medical Centre of Excellence, Milano, Italy
e-mail: lorenzoboldrini71@gmail.com, l.boldrini@isokinetic.com

F. Della Villa • S. Della Villa
Isokinetic Medical Group – FIFA Medical Centre of Excellence, Bologna, Italy

A. Gobbi et al. (eds.), *The Patellofemoral Joint*,
DOI 10.1007/978-3-642-54965-6_36, © ISAKOS 2014

The postoperative rehabilitation of patellofemoral joint is dependent on many factors that must be known by the specialist of rehabilitation:

- Type of surgery
- Site of lesion
- Previous surgery and pathologies
- Characteristics of the patient (age, sex, functional level and expectations)
- Compliance with a rehabilitation programme (social and psychological aspects)

Following the indications of the surgeon, the rehabilitation programme can start the day after surgery and can be resumed in the progression of the five rehabilitation phases:

Phase 1: Resolution of swelling and inflammation
Phase 2: Recovery of range of motion and muscle flexibility
Phase 3: Recovery of muscle strength and resistance
Phase 4: Recovery of neuromuscular control and coordination
Phase 5: Recovery of specific gestures

The time needed to complete the recovery process can be different for each patient. Rehabilitation however progresses by a functional rather than a time criteria approach.

This means that patients move through the rehabilitation phases depending on functional goals which must be reached to allow the recovery of normal walking, running and finally sport activities.

The rehabilitation environments include the pool for hydrokinesis therapy (Fig. 36.1), particularly useful in the first rehabilitation phases

Fig. 36.1 Pool session

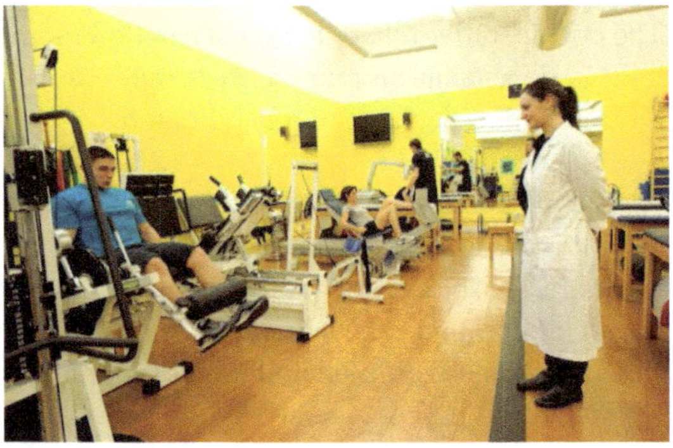

Fig. 36.2 Gym session

Fig. 36.3 On field session

(phases 1–2) when the patient must protect weight bearing, the gym (Fig. 36.2) for all treatments of phases 1–4 and the field (Fig. 36.3) to recover functional and athletic gestures (phases 4 and 5).

An example of functional rehabilitation progression following cartilage surgery in sport subjects is reported in Table 36.1.

As in nonoperative treatment the rehabilitation progresses on the responses of the knee joint according to the principle of Specific Adaptation to Imposed Demands (SAID).

We want to focus on some specifics aspects that must be highlighted during the recovery process after surgery:

- Negative effects of immobilization and passive movement
- Biomechanics considerations
- Management of pain
- Functional aspects

36.1 Negative Effects of Immobilization and Passive Movement

There is a wide literature on the negative effects of prolonged joint immobilization [1, 2]. At the same time, the beneficial use of passive movement to help joint homeostasis and the adaptations of cartilage tissue to proper exercise stimuli are well documented [2–4].

Immobilization of synovial joints decreases synthetic activities of the chondrocytes and proteoglycan content and reduces water content of the cartilage [2]. Moreover, the diminished amount of collagen fibrils and joint lubrication reduces cartilage and tendon nutrition, leading to joint stiffness and capsular contraction [5].

Thus, extended immobilization can compromise a normal recovery of strength, ROM and patellar position. A proper postoperative rehabilitation must minimize these detrimental effects while protecting the surgically repaired tissue.

An early mobilization that can be performed in a selected ROM, depending on biomechanical site and type of surgery, is therefore indicated. Continuous passive movement (CPM) and intermittent movement may protect and stimulate the repair process of the articular cartilage matrix. In the first 1–3 weeks after surgery, the use of CPM for 6–8 h per day in the ROM allowed by the surgeon is generally recommended; a gradual increase in the ROM (i.e. 5°/day) is normally suggested. At the same time rehabilitation treatments

Table 36.1 Example of functional rehabilitation progression following cartilage surgery in sport subjects

Phase	Goals	Criteria to progress
Phase 1: recovery of walk	Protect the site of surgery Decrease pain and effusion Increase range of movement Retard muscle atrophy Recovery of a correct gait pattern	Full active knee extension Knee flexion >120° No swelling No pain during weight bearing Correct walk pattern Adequate muscle recruitment (quadriceps)
Phase 2: recovery of running	Recovery of full range of motion Increase of muscular strength Increase of neuromuscular control Recovery of a correct running pattern	Running without pain/swelling at 8 km/h for 10′ Adequate recovery of coordination and neuromuscular control Recovery of strength >80 % contralateral limb Single leg hop test: >80 % contralateral limb
Phase 3: athletic recovery	Stimulate the cartilage tissue by exercise with progressive resistance Sustain high loads and impact activities Prepare the athlete for a return to competition with good recovery of the aerobic endurance Recovery of sport specific skills	Go up and down stairs and for athletes running without pain/effusion at 10 km/h for 15′ without a significant increase of blood lactic acid concentration above resting value Correct execution of sport-specific skills Recovery of strength: 100% of contralateral limb Single leg hop test: 100% of contralateral limb
Phase 4: athletic maintaining and return to sport	Maintain a good quality of life and a good physical condition Avoid excess of body fat Prevent risk of reinjury	

of phases 1 and 2 (treatment modalities, manual therapy, flexibility and ROM assisted exercises) must be gradually introduced to enhance the recovery process and delay the negative effects of prolonged immobilization.

36.1.1 Biomechanics Considerations

The knowledge of patellofemoral joint biomechanics is important to introduce proper exercises during the rehabilitation programme. In fact, there is no contact between the patella and the femur with a fully extended knee; thus straight leg raise exercises are normally safe and well tolerated early after surgery. Depending on the site and type of surgery, progressive exercises in closed kinetic chain (CKC) and open kinetic chain (OKC) can be introduced. In CKC exercises, the safest range of flexion is between 0° and 45°, especially if the site of treatment is on the proximal portion of the patella. Exercises performed in a deeper angle of flexion can overstress the patellofemoral joint and must be introduced gradually in a pain-free ROM. For OKC exercises the safest ROM is between 0 and 20° degrees and more than 90°, because the peak of stress contact for patellofemoral joint is at a mid-range of motion (30–60°) [2, 6].

36.1.2 Management of Pain

Pain is usually referred in patellofemoral disorders and may delay the process of recovery. The persistence of pain can have negative effects on the ability to perform exercises. Pain inhibits the quadriceps muscle making difficult the recovery of muscle tone and mass [7]. This can provoke a vicious cycle that maintains an alteration of normal biomechanics, loss of muscle function and pain with progressive psychological complications.

Therefore it is necessary to avoid pain during rehabilitation exercises, especially in phase 3. The control of pain can be done in phases 1 and 2 with analgesic and anti-inflammatory medicines, ice and other treatment modalities, manual therapy and progressive recovery of ROM and function, taking care of instructing properly the patient in the management of activities of daily living. The physiatrist and physiotherapist must strictly supervise the rehabilitation process, looking daily at knee responses to the imposed rehabilitation stimuli.

36.2 Functional Aspects

Functional aspects are really important in the recovery after patellofemoral pathologies. The goal of rehabilitation is to recover a correct knee function for ADL and for sport activities depending on patient expectations. Strengthening, proprioceptive and dynamic exercises (phases 3, 4 and 5) are crucial for a complete recovery of the patient. CKC exercises are normally preferred and emphasized as more functional compared to OKC exercises. The control of knee valgus and hip rotation must be trained and supervised. Step up-down exercises, proprioceptive pathways, eccentric movements and plyometrics are useful methods to regain a better control of knee functional movements as already explained in the nonoperative section for patellofemoral disorders.

To achieve a complete recovery of the patient, it is suggested to measure clinical and functional parameters by clinical examination (i.e. full range of motion, no pain, symmetric tight circumferences, good muscle flexibility, symmetric squat test), muscle knee strength tests (i.e. isokinetic knee test) and dynamic and neuromuscular tests (i.e. single and triple leg hop test, functional sport specific tests on the field).

References

1. Vanwanseeele B, et al. The effects of immobilization on the characteristics of articular cartilage: current concepts and future directions. Osteoarthritis Cartilage. 2002;10(5):408–19.
2. Hambly K, et al. Autologous chondrocyte implantation postoperative care and rehabilitation. Am J Sports Med. 2006;34:1020–37.
3. Eckstein F, et al. The effects of exercise on human articular cartilage. J Anat. 2006;208:491–512.
4. Bader DL, et al. Biomechanical influence of cartilage homeostasis in health and disease. Arthritis. 2011; 2011:979032.
5. Manske RC. Postsurgical orthopedic – sport rehabilitation – knee & shoulder. St Louis: Mosby Elsevier Publisher; 2006.
6. Andrews JR, Harrelson GL, Wilk KE. Physical Rehabilitation of the Injured Athlete. Elsevier Saunders Publisher; 1998.
7. Suter E, et al. Quadriceps inhibition following arthroscopy in patients with anterior knee pain. Clin Biomech (Bristol, Avon). 1998;13(4–5):314–9.

Complications of Patellofemoral Surgeries: Prevention and Management Strategies

37

Timothy J. Luchetti, Lauren H. Redler, and Michael R. Redler

Contents

T.J. Luchetti, BS
Columbia University College of Physicians and Surgeons, New York, NY, USA

L.H. Redler, MD
Department of Orthopaedics, Columbia University Medical Center, New York, NY, USA

M.R. Redler, MD (✉)
The OSM Center, Trumbull, CT, USA

Department of Orthopaedics, University of Virginia, Charlottesville, VA, USA

Department of Physical Therapy and Athletic Training, Sacred Heart University, Fairfield, CT, USA

Frank H Netter
School of Medicine,
Quinnipiac University, Hamden, CT, USA
e-mail: mredler@osmcenter.com

A. Gobbi et al. (eds.), *The Patellofemoral Joint*,
DOI 10.1007/978-3-642-54965-6_37, © ISAKOS 2014

37.1 Introduction

Disorders of the patellofemoral joint include a wide array of anatomic disorders, overuse injuries, instability, trauma, and pain disorders. The etiology of many of these disorders is not well understood. Patients with patellofemoral pain and instability are difficult for clinicians to manage because of difficulty with accurate diagnosis and reproducible treatments. Caring for patients with patellofemoral disorders has been a challenge for orthopedic surgeons.

Complications in patellofemoral surgery can be attributed to a variety of sources. Incorrect diagnosis, poor patient selection, insufficient preoperative physical therapy, incorrect choice of intervention, technical error, and incorrect postoperative physical therapy can all lead to failures and complications. The goal of this review is to summarize prevention and management strategies of common complications that occur in the most frequently performed procedures of the patellofemoral joint. The content of this review includes tibial tubercle osteotomy, medial patellofemoral ligament surgery, soft tissue procedures of the lateral patella, trochleoplasty, and patellofemoral arthroplasty.

37.2 Tibial Tubercle Osteotomy

The "quadriceps vector," or lateral offset of the patella tendon insertion on the tibial tubercle, is implicated in issues of the patellofemoral joint

[1]. Patellofemoral surgical procedures can involve moving the tibial tubercle anterior, medial, distal, or a combination of these directions. Many different procedures have been described involving osteotomy of the tibial tubercle [2].

There are a variety of complications that can occur after tibial tubercle osteotomy. Anteromedialization (AMZ) osteotomy aims to unload the inferior and lateral patellar facets. Overcorrection in the medial or anterior direction can lead to increased pressure on the lateral and superior facets. This complication can be avoided if the surgeon considers the location of the chondral lesion on the patella [2]. Skin closure compromise or necrosis is more common in the Maquet procedure, and this technique should therefore be avoided if possible [3]. Hardware-related symptoms can be reduced with the use of cortical screws with low profile heads. Countersinking the screw head helps, but the surgeon should avoid sinking the screw past the cortical margin because this can lead to bicortical fixation loss. Patella infera can be prevented by utilizing precise imaging preoperatively. The distal cut must be completed as well and secure fixation with a minimum of two bicortical screws is mandated. Postoperatively, look for quick return of the quadriceps muscle tone because prolonged inactivation of the quadriceps muscle can promote patella infera. The risk of nonunion can be minimized by obtaining adequate fixation, and the surgeon should perform bicortical screw fixation with screws angled 90°. Smoking cessation prior to the surgery can help optimize healing factors. Avoiding overaggressive compression of bicortical fixation can minimize the risk of tubercle fracture, and the surgeon should also be cognizant to avoid creating an osteotomy fragment that is too thin. Diaphyseal tibial fractures can be prevented if the surgeon avoids the creation of a stress riser at the time of osteotomy [2]. It is recommended to use an oscillating saw for deep cuts rather than an osteotome [4–6]. As you proceed distally on the tibia, thin out the osteotomy cut [2]. One author suggested avoiding full weight-bearing, but there is not a consensus postoperative protocol for this procedure [6].

37.3 Medial Patellofemoral Ligament (MPFL) Surgery

Recurrent patellar instability is a disabling condition that usually affects younger people [7]. The primary soft tissue restraint for lateral subluxation of the patella near full extension is the medial patellofemoral ligament. The consensus surgical technique to treat recurrent patellar dislocation is not established. However, the reconstruction of the MPFL has become a popular procedure to treat these patients [8]. This procedure is associated with a high complication rate. A systematic review by Shah et al. [9] reported a pooled complication rate of 25.7 % across all studies. Common complications include restricted knee range of motion [10], arthrofibrosis [11], recurrent lateral instability [11], medial instability [12], patellofemoral arthrosis [13], patellar fractures [14], hemarthrosis [15], wound complications [16], and implant pain [15, 17].

In order to help insure successful surgery and to reduce complication rates, a particular focus should be applied to patient selection and the use of good technique. MPFL reconstruction should be reserved for lateral instability of the patella and should be regarded as a stabilization procedure, specifically. Symptoms of excessive lateral patellar pressure, maltracking, and arthrosis would be better addressed by a distal realignment procedure to unload the lateral patellofemoral cartilage. Patellofemoral pain alone must be carefully differentiated from patellofemoral instability in the history and physical. Medial instability is a contraindication to MPFL reconstruction and is especially likely to occur after overly aggressive lateral release.

Patellar fracture is a rare but dreaded complication of MPFL surgery. Type I fractures follow a transverse fracture pattern through the patellar tunnel or drill hole. A transverse bone tunnel that traverses the entire width of the patella from the medial to lateral border is biomechanically strong [18], but it is recommended that this technique only be used with caution. The anterior cortex of the patella must also be preserved to avoid the risk of fracture [19]. These are customarily treated with tension-band wiring.

The risk of a fracture can be decreased by avoiding transverse patellar tunnels that traverse the entire patellar width, minimizing the tunnel size, using a single bone tunnel, maintaining an adequate bone bridge, avoiding devascularization of the superior pole of the patella, and using anatomic tunnel placement in the femur and the patella [21]. Because of the pathologic nature of any patellar fracture and the weakness of the patellar bone, immobilization of the knee is recommended for the first 6 weeks after fracture fixation. This leads to stiffness, especially since most fractures occur relatively early (within 3 months) in the postoperative period following MPFL reconstruction [22]. Once the patellar fracture shows signs of healing, aggressive physical therapy should begin.

37.3.1 Factors for Successful Surgery

Correct tunnel positioning is essential to restoring the correct isometry and function of the graft [23–25]. The most important point affecting isometry is the femoral attachment site of the medial patellofemoral ligament [26]. The tunnel should be placed at the anatomic insertion point of the ligament between the adductor tubercle and the medial femoral epicondyle [16, 26, 27]. It is most common to incorrectly place the femoral tunnel anteriorly or proximally [25]. Proximal malpositioning of the femoral tunnel has been shown to increase medial tilt and medial patellofemoral contact pressures, which can potentially lead to medial compartment arthritis [28]. Anterior malpositioning can also result in overload of the medial facet.

The patellar tunnel should begin at the anatomic insertion of the ligament on the proximal half of the medial border of the patella. It should extend through the center of the patella in the sagittal plane. The ideal position of the patellar tunnel has not been elucidated, but the correct placement may vary in patients with patella alta. Anterior placement of the patellar tunnel can result in violation of the anterior cortex. Iatrogenic patellar fracture has been reported 6 weeks after surgery when a patient arose from a chair without support. It was found that violation of the anterior patellar cortex had occurred intraoperatively during drilling of the patellar tunnels [17]. Posterior placement of the patellar tunnel can violate the articular surface of the patella. Digital palpation of the articular surface of the patella and use of fluoroscopy can both aid in guiding the tunnel positioning in the patella to avoid this complication [12].

Decisions about graft type and isometry are key to a successful surgery. Options for graft type include quadriceps tendon, hamstring tendon, adductor tendon, and synthetic grafts. A cadaver study showed that hamstring grafts are substantially stiffer and stronger than the native medial patellofemoral ligament. When tensioning a hamstring graft, this finding should be kept in mind to avoid the risk of overloading the medial patellofemoral cartilage [29]. Graft tensioning is paramount to performing a successful medial patellofemoral ligament reconstruction. Overtensioning of the graft can lead to increased medial patellofemoral contact pressures. A graft that is in the correct position but is 3 mm too short will lead to increased medial pressure. This results in increased medial patellar tilt from overtensioning of the graft, and it can lead to medial patellar subluxation. Undertensioning of the graft can result in a lack of adequate medial restraint and may lead to recurrent lateral patellofemoral instability. Correct tensioning should approximate the patellar motion in the contralateral extremity. Normal patellar motion should allow for two to three patellar quadrants of lateral translation. In cases of trochlear dysplasia, there is a tendency to tension the graft too tightly because of the lack of normal osseous landmarks [23]. Patella alta can change the normal relationship between the patellar and femoral attachments of the MPFL because it increases the required length of the MPFL graft.

Graft fixation is also important. There are multiple methods of femoral fixation, but a study by Mountney et al. found that the through-tunnel tendon graft provides the strongest reconstruction, with a strength equal to that of the native ligament [18]. Implant pain is a common postoperative complication with 57 % of patients with staple fixation and 23 % of patients who had fixation with an integrated double staple experiencing pain at the femoral fixation site [16]. Ten percent of patients may require removal of a prominent or painful screw after MPFL reconstruction [15]. Selection of a fixation method that has a low profile and is secure can minimize pain at the femoral fixation site [12]. A reliable method described by Schottle et al. used radiographic landmarks that can easily be found under fluoroscopy. Their point is 1 mm anterior to the posterior cortex extension line, 2.5 mm distal to the posterior origin of the medial femoral condyle, and proximal to the posterior point of the Blumensaat line on the lateral radiograph [30]. There are also many methods of patellar fixation. At least one author recommends the use of a docking anchor-based or suture fixation technique because these may be less likely to lead to patellar fracture [9], although the incidence of patellar fractures is too low to draw statistical conclusions. Implant pain is less common in the patella than the femur, but it is possible with prominent fixation methods such as the EndoButton [31].

The need for concurrent or alternate procedures should be recognized. If there is substantial malalignment, MPFL reconstruction alone is often not enough to correct patellofemoral instability. Failure to address malalignment when reconstructing this ligament can lead to early failure and recurrence of lateral instability. Appropriate identification of the exact pathology and utilization of the appropriate treatment algorithm will help define the optimum treatment for each patient. More than 25 mm between the tibial tubercle and the trochlear groove with a Q angle of >14° in males and >17° in females is indicative of malalignment. This may require tibial tubercle osteotomy. Patella alta, defined as an Insall-Salvati ratio of greater than 1.2, may be better treated with distal advancement of a tibial tubercle osteotomy because the MPFL reconstruction alone could result in increased tension on the graft. For lateral tilt and substantial lateral retinacular tightness, a concurrent lateral release may be indicated.

37.4 Soft Tissue Procedures of the Lateral Patella

Lateral patellar hypercompression syndrome (LPHS) may be caused by a tight lateral patellar retinaculum with lateral patellar tracking, lateral patellofemoral joint overload, degeneration, and the development of pain [32–34]. After the failure of nonsurgical therapy, this condition is commonly treated with lateral retinacular lengthening or release [35]. The most relevant complications of these procedures, often requiring revisions, were recurrence of LPHS [32], patellar instability with medial subluxation, quadriceps weakness, and severe atrophy caused by muscular detachment [36–41]. The most frequent complication after lateral retinacular release is subcutaneous or articular hematoma. These hematomas hinder knee motion, may require needle aspiration and surgical revision, and may also promote wound healing problems, superficial or deep infection, or pain syndromes. Some authors favor open over arthroscopic lateral retinacular release to simplify bleeding control from the superior lateral genicular artery [42]. One author recommended a 3-cm open approach over the two 1-cm incisions performed during arthroscopic lateral retinacular release [35]. Meticulous open hemostasis with a deflated tourniquet may lead to lower complication rates.

Recurrence of pain is a frequent complication [32]. Extensive release and retinacular resection has been recommended until lateral patellar turnup of 90° can be performed [32, 37, 43]. However, some authors have noted that 90° of turnup may be too aggressive because extensive proximal release may divide the vastus lateralis, causing quadriceps atrophy and even tendon rupture [41, 44]. Recurrence may be prevented at the expense of medial patellar subluxation. At least one author recommends limiting the turnup test to 70° to reduce extensive soft tissue release [35].

Lateral retinacular lengthening is a modification of the popular lateral retinacular release procedure. Lateral retinacular lengthening has been shown to be a safer surgical method to treat symptomatic LPHS when compared with lateral retinacular release, in terms of reduced quadriceps atrophy, reduced incidence of medial patellar subluxation, and improved clinical outcome scores at 2 years [35]. This may be explained by the controlled preservation of the lateral patellar muscle-capsuloligamentous continuity after retinacular lengthening.

37.5 Trochleoplasty

Trochleoplasty is still an uncommon surgical procedure, especially in the United States, despite the fact that trochlear dysplasia is a central risk factor in patellar instability and patellofemoral arthritis. Several procedures have been described to “reshape” the trochlea to help address high-grade dysplasia associated with patella instability [45–47]

The common complications of trochleoplasty surgery are arthrofibrosis, supratrochlear bony prominence with catching, worsening pain, and over- or undercorrection with patellar incongruence leading to continued instability and pain. The risk of all these complications can be reduced if the procedure is only performed on patients with high-grade trochlear dysplasia. Do not use this procedure to treat pain symptoms alone. The surgeon should be sure to carefully handle the cartilage during the corrective process. Given the technical difficulties of this procedure, surgeons should consider a referral to a patellofemoral surgery expert [2].

37.6 Patellofemoral Arthroplasty

Isolated patellofemoral arthritis is relatively uncommon [48], and various surgical options have yielded only marginal results [49]. Patellofemoral arthroplasty is gaining in popularity [50–52] despite controversy due to inconsistent results and high early failure rates in some reports [53–56].

Early postoperative complications of patellofemoral arthroplasty include persistent anterior knee pain, patellar catching or snapping, and extensor mechanism disruption. Increased peripatellar pain may be an early consequence of overstuffing of the patellofemoral joint through the placement of an implant that is thicker than the amount of bone and cartilage that is resected. There is a greater increase in patellar thickness after surgery in patients who experience poor results [57]. A larger radius of curvature may increase the risk of patellar catching or maltracking [58]. This is particularly problematic when the trochlear implant is placed in flexion or in cases in which the patellar prosthesis articulates with the femur proximal to the trochlear component in full extension. This risk is increased in prostheses with relatively short proximal extension on the anterior femoral cortex. Alternatively, narrow implants may increase the risk of patellar catching or maltracking. This risk is decreased in implants with a relatively unconstrained geometry of the trochlear component. With more recent patellofemoral arthroplasty designs, the incidence of many early complications has decreased [59].

Poor technique is a major driver of complications in patellofemoral arthroplasty. Malpositioning of the femoral prosthesis may result in the component overhanging the femoral condyle, patellar maltracking, and patellar instability [60]. Placing the femoral component in internal rotation is associated with a higher risk of reoperation [61]. Malalignment of the extensor mechanism and overstuffing of the anterior compartment are two frequent causes of anterior knee pain. Both of these scenarios can cause symptoms quite early and lead to high rates of revision surgery [61, 62]. Postoperative patellofemoral instability is commonly caused by poor soft tissue balancing [58].

Historically the most frequent late complication has been patellar maltracking, but this has been reduced with the introduction of newer prostheses [60]. Today, the most common late complication of patellofemoral arthroplasty is progression of tibiofemoral osteoarthritis. One study of 103 Avon prostheses showed a 12 % revision rate due to progression of tibiofemoral osteoarthritis at 7.1 years of follow-up [63]. This complication may not be predicted by the functional results or pain control achieved by the prosthesis in the first 2 years following surgery

[62]. Loosening is a rare complication of patellofemoral arthroplasty. Most reports of implant loosening have involved uncemented prostheses [64]. Chronic effusion is another late complication, and one author reported a chronic effusion rate of 33 % [65]. He noted the importance of avoiding placement of the femoral prosthesis in internal rotation or recurvatum.

Success of patellofemoral prostheses is highly dependent on selection criteria [66, 67]. This surgery can be offered in cases of isolated PF osteoarthritis, posttraumatic patellofemoral arthritis, or for patients with patellofemoral arthritis associated with trochlear dysplasia and patellar subluxation [68]. Isolated patellofemoral arthroplasty cannot by itself stabilize a patellofemoral joint with severe malalignment [59]. In cases of significant patellar subluxation or tilt, consider medialization of the tibial tubercle with or without lateral release and associated facetectomy. A distalization procedure should be considered in cases of severe patella alta if the patella does not engage with the trochlea at maximal knee extension.

Contraindications for patellofemoral arthroplasty are numerous and need to be carefully considered. They include chondrocalcinosis, pain, cartilage defects, evidence of significant osteoarthritis in the medial or lateral tibiofemoral compartments [69]. Obesity with BMI greater than 30 kg/m^2 is a relative contraindication due to potential for the progression of tibiofemoral disease [68]. Some authors limit this procedure to patients under 60 years [70], but there is currently no definitive evidence that this is necessary. There is insufficient data to determine if chronic anterior laxity leads to reduced function and longevity of the patellofemoral prosthesis. Inflammatory joint disease remains an absolute contraindication because of global involvement of the knee joint [71]. Some authors have proposed combining PF prosthesis with cartilage restoration procedures in cases of a defect in the load-bearing portions of the femoral condyles [58] or with unicompartmental knee arthroplasty in cases of significant degenerative disease in an additional compartment [72].

There is some evidence that suggests that implant choice is a major driver of outcomes [73]. Similarly, the etiology of arthritis may influence outcomes. Patients with arthritis secondary to trochlear dysplasia experience better clinical outcomes than other groups [51]. Trochlear dysplasia is an ideal indication for patellofemoral arthroplasty [74]. Recently, custom implants that match the radius of curvature of PF joint based on 3D CT reconstructions have been developed. These designs may help optimize coverage of bone surfaces without causing overconstraint of the patella [75]. There is also renewed interest in the combination of medial unicompartmental and patellofemoral prostheses [72, 76]. Recent publications suggest that this option yields good long-term results and would extend indications for patellofemoral prosthesis, especially in young, active patients [72].

Conclusions

The patellofemoral compartment of the knee has a unique geometry and complex static and dynamic stabilizing features. Many procedures exist to address varying pathology. The key to successful PF surgery is patient selection and understanding the goals of the selected procedure. Surgical complications can be minimized with meticulous operative technique and attention to post-operative protocols.

References

1. Insall J, Falvo KA, Wise DW. Chondromalacia patellae. A prospective study. J Bone Joint Surg Am. 1976;58(1):1–8.
2. Tompkins M, Arendt EA. Complications in patellofemoral surgery. Sports Med Arthrosc. 2012;20(3):187–93.
3. Kadambande SS, et al. A review of wound healing following Maquet osteotomy. Knee. 2004;11(6):463–7.
4. Eager MR, et al. Delayed fracture of the tibia following anteromedialization osteotomy of the tibial tubercle: a report of 5 cases. Am J Sports Med. 2004;32(4):1041–8.
5. Kohn D, Steimer O, Seil R. Anterior medial correction the tibial tuberosity. Orthopade. 2004;33(2):218–23.

6. Stetson WB, et al. Fracture of the proximal tibia with immediate weightbearing after a Fulkerson osteotomy. Am J Sports Med. 1997;25(4):570–4.
7. Fithian DC, et al. Epidemiology and natural history of acute patellar dislocation. Am J Sports Med. 2004; 32(5):1114–21.
8. Hautamaa PV, et al. Medial soft tissue restraints in lateral patellar instability and repair. Clin Orthop Relat Res. 1998;349:174–82.
9. Shah JN, et al. A systematic review of complications and failures associated with medial patellofemoral ligament reconstruction for recurrent patellar dislocation. Am J Sports Med. 2012;40(8):1916–23.
10. Thaunat M, Erasmus PJ. Management of overtight medial patellofemoral ligament reconstruction. Knee Surg Sports Traumatol Arthrosc. 2009;17(5):480–3.
11. Drez Jr D, Edwards TB, Williams CS. Results of medial patellofemoral ligament reconstruction in the treatment of patellar dislocation. Arthroscopy. 2001;17(3):298–306.
12. Tanaka MJ, et al. Complications of medial patellofemoral ligament reconstruction: common technical errors and factors for success: AAOS exhibit selection. J Bone Joint Surg Am. 2012;94(12):e87.
13. Beck P, Brown N, Greis PE, Burks RT. Patellofemoral contact pressure and lateral patellar translation after medial patellofemoral ligament reconstruction. AM J Sports Med. 2006;35:1557–63.
14. Thaunat M, Erasmus PJ. Recurrent patellar dislocation after medial patellofemoral ligament reconstruction. Knee Surg Sports Traumatol Arthrosc. 2008;16(1):40–3.
15. Steiner TM, Torga-Spak R, Teitge RA. Medial patellofemoral ligament reconstruction in patients with lateral patellar instability and trochlear dysplasia. Am J Sports Med. 2006;34(8):1254–61.
16. Nomura E, Horiuchi Y, Kihara M. A mid-term follow-up of medial patellofemoral ligament reconstruction using an artificial ligament for recurrent patellar dislocation. Knee. 2000;7(4):211–5.
17. Christiansen SE, et al. Reconstruction of the medial patellofemoral ligament with gracilis tendon autograft in transverse patellar drill holes. Arthroscopy. 2008; 24(1):82–7.
18. Mountney J, et al. Tensile strength of the medial patellofemoral ligament before and after repair or reconstruction. J Bone Joint Surg Br. 2005;87(1):36–40.
19. Ellera Gomes JL. Medial patellofemoral ligament reconstruction for recurrent dislocation of the patella: a preliminary report. Arthroscopy. 1992;8(3):335–40.
20. Ritter MA, et al. Patellofemoral complications following total knee arthroplasty. Effect of a lateral release and sacrifice of the superior lateral geniculate artery. J Arthroplasty. 1996;11(4):368–72.
21. Parikh SN, Wall EJ. Patellar fracture after medial patellofemoral ligament surgery: a report of five cases. J Bone Joint Surg Am. 2011;93(17):e97(1–8).
22. Fithian DC, Powers CM, Khan N. Rehabilitation of the knee after medial patellofemoral ligament reconstruction. Clin Sports Med. 2010;29(2):283–90, ix.
23. Bicos J, Fulkerson JP, Amis A. Current concepts review: the medial patellofemoral ligament. Am J Sports Med. 2007;35(3):484–92.
24. Philippot R, et al. Medial patellofemoral ligament anatomy: implications for its surgical reconstruction. Knee Surg Sports Traumatol Arthrosc. 2009;17(5): 475–9.
25. Servien E, et al. In vivo positioning analysis of medial patellofemoral ligament reconstruction. Am J Sports Med. 2011;39(1):134–9.
26. Steensen RN, Dopirak RM, Maurus PB. A simple technique for reconstruction of the medial patellofemoral ligament using a quadriceps tendon graft. Arthroscopy. 2005;21(3):365–70.
27. Smirk C, Morris H. The anatomy and reconstruction of the medial patellofemoral ligament. Knee. 2003; 10(3):221–7.
28. Elias JJ, Cosgarea AJ. Technical errors during medial patellofemoral ligament reconstruction could overload medial patellofemoral cartilage: a computational analysis. Am J Sports Med. 2006;34(9):1478–85.
29. Hamner DL, et al. Hamstring tendon grafts for reconstruction of the anterior cruciate ligament: biomechanical evaluation of the use of multiple strands and tensioning techniques. J Bone Joint Surg Am. 1999; 81(4):549–57.
30. Schottle PB, et al. Technical note: anatomical reconstruction of the medial patellofemoral ligament using a free gracilis autograft. Arch Orthop Trauma Surg. 2008;128(5):479–84.
31. Shocks EJ, Burks R. Medial patellofemoral ligament reconstruction using a hamstring graft. Oper Tech Sports Med. 2001;9:169–75.
32. Johnson RP. Lateral facet syndrome of the patella. Lateral restraint analysis and use of lateral resection. Clin Orthop Relat Res. 1989;238:148–58.
33. Krompinger WJ, Fulkerson JP. Lateral retinacular release for intractable lateral retinacular pain. Clin Orthop Relat Res. 1983;179:191–3.
34. Metcalf RW. An arthroscopic method for lateral release of subluxating or dislocating patella. Clin Orthop Relat Res. 1982;167:9–18.
35. Pagenstert G, et al. Open lateral patellar retinacular lengthening versus open retinacular release in lateral patellar hypercompression syndrome: a prospective double-blinded comparative study on complications and outcome. Arthroscopy. 2012;28(6): 788–97.
36. Fulkerson JP. Diagnosis and treatment of patients with patellofemoral pain. Am J Sports Med. 2002; 30(3):447–56.
37. Henry JH, Goletz TH, Williamson B. Lateral retinacular release in patellofemoral subluxation. Indications, results, and comparison to open patellofemoral reconstruction. Am J Sports Med. 1986;14(2):121–9.
38. Hughston JC, Deese M. Medial subluxation of the patella as a complication of lateral retinacular release. Am J Sports Med. 1988;16(4):383–8.
39. Laurin CA, Dussault R, Levesque HP. The tangential x-ray investigation of the patellofemoral joint: x-ray

technique, diagnostic criteria and their interpretation. Clin Orthop Relat Res. 1979;144:16–26.
40. Nonweiler DE, DeLee JC. The diagnosis and treatment of medial subluxation of the patella after lateral retinacular release. Am J Sports Med. 1994;22(5): 680–6.
41. O'Neill DB. Open lateral retinacular lengthening compared with arthroscopic release. A prospective, randomized outcome study. J Bone Joint Surg Am. 1997;79(12):1759–69.
42. Small NC. An analysis of complications in lateral retinacular release procedures. Arthroscopy. 1989;5(4): 282–6.
43. Ficat P, Ficat C, Bailleux A. External hypertension syndrome of the patella. Its significance in the recognition of arthrosis. Rev Chir Orthop Reparatrice Appar Mot. 1975;61(1):39–59.
44. Blasier RB, Ciullo JV. Rupture of the quadriceps tendon after arthroscopic lateral release. Arthroscopy. 1986;2(4):262–3.
45. Dejour D, Saggin P. The sulcus deepening trochleoplasty-the Lyon's procedure. Int Orthop. 2010;34(2):311–6.
46. Lippacher S, et al. Observer agreement on the Dejour trochlear dysplasia classification: a comparison of true lateral radiographs and axial magnetic resonance images. Am J Sports Med. 2012;40(4):837–43.
47. Saggin PR, Saggin JI, Dejour D. Imaging in patellofemoral instability: an abnormality-based approach. Sports Med Arthrosc. 2012;20(3):145–51.
48. Davies AP, et al. The radiologic prevalence of patellofemoral osteoarthritis. Clin Orthop Relat Res. 2002;402:206–12.
49. Becker R, et al. Surgical treatment of isolated patellofemoral osteoarthritis. Clin Orthop Relat Res. 2008;466(2):443–9.
50. Lonner JH. Patellofemoral arthroplasty. Orthopedics. 2010;33(9):653.
51. Leadbetter WB, et al. Patellofemoral arthroplasty: a multi-centre study with minimum 2-year follow-up. Int Orthop. 2009;33(6):1597–601.
52. Newman JH. Patellofemoral arthritis and its management with isolated patellofemoral replacement: a personal experience. Orthopedics. 2007;30(8 Suppl):58–61.
53. Hassaballa MA, Porteous AJ, Newman JH. Observed kneeling ability after total, unicompartmental and patellofemoral knee arthroplasty: perception versus reality. Knee Surg Sports Traumatol Arthrosc. 2004;12(2):136–9.
54. Hendrix MR, Ackroyd CE, Lonner JH. Revision patellofemoral arthroplasty: three- to seven-year follow-up. J Arthroplasty. 2008;23(7):977–83.
55. Leadbetter WB, Ragland PS, Mont MA. The appropriate use of patellofemoral arthroplasty: an analysis of reported indications, contraindications, and failures. Clin Orthop Relat Res. 2005;436:91–9.
56. van Jonbergen HP, Werkman DM, van Kampen A. Conversion of patellofemoral arthroplasty to total knee arthroplasty: a matched case–control study of 13 patients. Acta Orthop. 2009;80(1):62–6.
57. Mofidi A, et al. Functional relevance of patellofemoral thickness before and after unicompartmental patellofemoral replacement. Knee. 2012;19(3):180–4.
58. Lonner JH. Patellofemoral arthroplasty. J Am Acad Orthop Surg. 2007;15(8):495–506.
59. Farr 2nd J, Barrett D. Optimizing patellofemoral arthroplasty. Knee. 2008;15(5):339–47.
60. Ackroyd CE, et al. The Avon patellofemoral arthroplasty: five-year survivorship and functional results. J Bone Joint Surg Br. 2007;89(3):310–5.
61. Gadeyne S, et al. Results of self-centering patellofemoral prosthesis: a retrospective study of 57 implants. Rev Chir Orthop Reparatrice Appar Mot. 2008;94(3):228–40.
62. Ackroyd CE, Chir B. The Avon patellofemoral arthroplasty: two to five year results. J Bone Joint Surg Br. 2003;85:162–3.
63. Nicol SG, et al. Arthritis progression after patellofemoral joint replacement. Knee. 2006;13(4):290–5.
64. Argenson JN, Guillaume JM, Aubaniac JM. Is there a place for patellofemoral arthroplasty? Clin Orthop Relat Res. 1995;321:162–7.
65. Leadbetter WB, et al. Indications, contraindications, and pitfalls of patellofemoral arthroplasty. J Bone Joint Surg Am. 2006;88 Suppl 4:122–37.
66. Arciero RA, Toomey HE. Patellofemoral arthroplasty. A three- to nine-year follow-up study. Clin Orthop Relat Res. 1988;236:60–71.
67. Blazina ME, et al. Patellofemoral replacement. Clin Orthop Relat Res. 1979;144:98–102.
68. van Jonbergen HP, et al. Long-term outcomes of patellofemoral arthroplasty. J Arthroplasty. 2010; 25(7):1066–71.
69. Lonner JH. Patellofemoral arthroplasty. Instr Course Lect. 2010;59:67–84.
70. Lonner JH, Mehta S, Booth Jr RE. Ipsilateral patellofemoral arthroplasty and autogenous osteochondral femoral condylar transplantation. J Arthroplasty. 2007;22(8):1130–6.
71. Lustig S, et al. Patellofemoral arthroplasty, where are we today? Knee Surg Sports Traumatol Arthrosc. 2012;20(7):1216–26.
72. Parratte S, et al. Survival of bicompartmental knee arthroplasty at 5 to 23 years. Clin Orthop Relat Res. 2010;468(1):64–72.
73. Lonner JH. Patellofemoral arthroplasty: pros, cons, and design considerations. Clin Orthop Relat Res. 2004;428:158–65.
74. Odumenya M, et al. The Avon patellofemoral joint replacement: five-year results from an independent centre. J Bone Joint Surg Br. 2010;92(1):56–60.
75. Sisto DJ, Sarin VK. Custom patellofemoral arthroplasty of the knee. J Bone Joint Surg Am. 2006;88(7): 1475–80.
76. Lonner JH. Modular bicompartmental knee arthroplasty with robotic arm assistance. Am J Orthop (Belle Mead NJ). 2009;38(2 Suppl):28–31.

38 Conclusions

Alberto Gobbi, João Espregueira-Mendes, and Norimasa Nakamura

Medicine does not stand still and is in a constant state of evolution in order to provide safe, effective, and beneficial treatments and techniques to improve quality of life. Patellofemoral pathologies are among the most common causes of knee pain and disability, and current literature offers a plethora of options for assessment and treatments. This book is an effort to help with decision making and selection of the appropriate treatment, putting to rest some of the existing controversies.

The contributors of *The Patellofemoral Joint, State of the Art in Evaluation and Management* are internationally recognized orthopedic surgeons, basic scientists, sport doctors, and physiatrists that devoted many years studying the patellofemoral joint. ISAKOS as international scientific society offered the opportunity to gather the group of experts that contributed to this book, initiated by Alberto Gobbi, Norimasa Nakamura, and João Espregueira-Mendes.

The book offers a comprehensive outlook ranging from basic morphogenesis and anatomy to an in-depth discussion about current concepts which define the treatment approach, keeping in mind the importance of an evidence-based outlook.

We would like to thank all the authors from across the globe for their valuable contribution to this book which we hope will be a guide to many surgeons in their practice.

A. Gobbi (✉)
Orthopaedic Arthroscopic Surgery International (O.A.S.I.) Bioresearch Foundation, Gobbi N.P.O., Milan, Italy
e-mail: gobbi@cartilagedoctor.it

J. Espregueira-Mendes
Clínica do Dragão – Espregueira-Mendes Sports Centre – FIFA Medical Centre of Excellence, F.C. Porto Stadium, 4350-415 Porto, Portugal

ICVS/3B's Research Group – Biomaterials, Biodegradables and Biomimetics, University of Minho, Headquarters of the European Institute of Excellence on Tissue Engineering and Regenerative Medicine, Caldas das Taipas, Guimarães, Portugal

Department of Orthopaedics, Minho University, Braga, Portugal
e-mail: jem@espregueira.com

N. Nakamura
Department of Orthopaedic Surgery, Osaka University Graduate School of Medicine, 2-2 Yamadaoka, Suita, Osaka 565-0871, Japan
e-mail: norimasa.nakamura@ohsu.ac.jp

A. Gobbi et al. (eds.), *The Patellofemoral Joint*,
DOI 10.1007/978-3-642-54965-6_38, © ISAKOS 2014

GPSR Compliance

The European Union's (EU) General Product Safety Regulation (GPSR) is a set of rules that requires consumer products to be safe and our obligations to ensure this.

If you have any concerns about our products, you can contact us on ProductSafety@springernature.com

In case Publisher is established outside the EU, the EU authorized representative is:

Springer Nature Customer Service Center GmbH
Europaplatz 3
69115 Heidelberg, Germany

Batch number: 10732675

Printed by Printforce, the Netherlands